15th Edition

CISM

Review Manual

ISACA®

ISACA (isaca.org) helps global professionals lead, adapt and assure trust in an evolving digital world by offering innovative and world-class knowledge, standards, networking, credentialing and career development. Established in 1969, ISACA is a global nonprofit association of 140,000 professionals in 180 countries. ISACA also offers the Cybersecurity Nexus™ (CSX), a holistic cybersecurity resource, and COBIT®, a business framework to govern enterprise technology.

In addition, ISACA advances and validates business-critical skills and knowledge through the globally respected Certified Information Systems Auditor® (CISA®), Certified Information Security Manager® (CISM®), Certified in the Governance of Enterprise IT® (CGEIT®) and Certified in Risk and Information Systems Control™ (CRISC™) credentials.

Disclaimer

ISACA has designed and created *CISM® Review Manual 15th Edition* primarily as an educational resource to assist individuals preparing to take the CISM certification exam. It was produced independently from the CISM exam and the CISM Certification Working Group, which has had no responsibility for its content. Copies of past exams are not released to the public and were not made available to ISACA for preparation of this publication. ISACA makes no representations or warranties whatsoever with regard to these or other ISACA publications assuring candidates' passage of the CISM exam.

ISACA

3701 Algonquin Road, Suite 1010
Rolling Meadows, Illinois 60008 USA
Phone: +1.847.253.1545
Fax: +1.847.253.1443
Email: *info@isaca.org*
Web site: *www.isaca.org*

Participate in the ISACA Knowledge Center: *www.isaca.org/knowledge-center*
Follow ISACA on Twitter: *https://twitter.com/ISACANews*
Join ISACA on LinkedIn: ISACA (Official), *http://linkd.in/ISACAOfficial*
Like ISACA on Facebook: *www.facebook.com/ISACAHQ*

ISBN 978-1-60420-508-4
CISM® Review Manual 15th Edition
Printed in the United States of America

CISM REVIEW MANUAL 15TH EDITION

ISACA is pleased to offer the 15th edition of the *CISM® Review Manual*. The purpose of this manual is to provide CISM candidates with updated technical information and references to assist in the preparation and study for the Certified Information Security Manager exam.

The *CISM® Review Manual* is updated to keep pace with rapid changes in the management, design, oversight and assessment of information security. As with previous manuals, the 15th edition is the result of contributions from many qualified authorities who have generously volunteered their time and expertise. We respect and appreciate their contributions and feel certain their efforts will provide extensive educational value to CISM manual readers.

Your comments and suggestions regarding this manual are welcome. After taking the exam, please take a moment to complete the online questionnaire *(www.isaca.org/studyaidsevaluation)*. Your observations will be invaluable for the preparation of the 16th edition of the *CISM® Review Manual*.

The self-assessment questions contained in this manual are designed to depict the type of questions typically found on the CISM exam and to provide further clarity to the content presented in this manual. The CISM exam is a practice-based exam. Simply reading the reference material in this manual will not properly prepare candidates for the exam. The self-assessment questions are included for guidance only. Scoring results do not indicate future individual exam success.

Certification has resulted in a positive impact on many careers. CISM is designed to provide executive management with assurance that those earning the designation have the required knowledge and ability to provide effective information security management and consulting. While the central focus of the CISM certification is information security management, all those in the IT profession with security experience will certainly find value in the CISM designation. ISACA wishes you success with the CISM exam.

ACKNOWLEDGMENTS

The 15th edition of the *CISM® Review Manual* is the result of the collective efforts of many volunteers. ISACA members from throughout the global information security management profession participated, generously offering their talent and expertise. This international team exhibited a spirit and selflessness that has become the hallmark of contributors to this manual. Their participation and insight are truly appreciated.

All of the ISACA members who participated in the review of the *CISM® Review Manual* deserve our thanks and gratitude.

Special thanks go to W. Krag Brotby, CISM, CGEIT, a senior security consultant from the Sacramento (California) Chapter, USA, who served as technical content project leader and editor.

Expert Reviewers
Michael Broady, CISM, CRISC, ACE, US Southern Command/Exeter Cooperation, USA
Mouhamed Diop, CISA, CISM, CGEIT, CRISC, Senegal
Sandeep Godbole, CISA, CISM, CGEIT, CEH, CISSP, Syntel, India
Mohamed Gohar, CISA, CISM, COBIT Foundation, CPDE, ISO 27001, 27034, 38500 and 24762 certified, ITIL Expert, PECB-CLPTP, PMP, Resilia Practitioner, TOGAF Practitioner, Itpreneurs, Global Knowledge, AUC, El-Khalij Institute or New Horizons and Egybyte, Egypt
Robert T. Hanson, CISA, CISM, CRISC, CRMA, Australian Government, Australia
Foster Henderson, CISM, CRISC, CISSP, USA
Kevin Henry, CISA, CISM, CRISC, CISSP, Canada
Abdus Sami Khan, CISA, CISM, CIA, SALE Advanced Co. Ltd., Saudi Arabia
Israel Rosales M., CISA, CISM, CRISC, CEH, CHFI, CISSP, COBIT 5, ISO 27001LA, ITIL, COSIM TI, Bolivia
Cory Missimore, CISM, USA
Juan Carlos Morales, CISA, CISM, CGEIT, CRISC, Guatemala
Balakrishnan Natarajan, CISM, Pivotal Software Inc., USA
S. Peter Nota, CISA, CISM, APMP, CISSP, MBCS, PCI-ISA, Premier Farnell plc, UK
Opeyemi Onifade, CISA, CISM, CGEIT, CISSP, COBIT Accredited Trainer, COBIT Certified Assessor, Afenoid Enterprise Limited, Nigeria
Vaibhav Patkar, CISA, CISM, CRISC, CGEIT, CISSP, India
Abdul Jaleel Puthenpurayil, CISM, United Arab Emirates
Ravikumar Ramachandran, CISA, CISM, CGEIT, CRISC, CAP, CEH, CFE, CHFI, CIA, CISSP-ISSAP, CIA, CIMA-Adv. Dip. MA, CRMA, ECSA, FCMA , PMP, SSCP, Hewlett-Packard India Sales Pvt. Ltd, India
James C. Samans, CISA, CISM, CRISC, CISSP-ISSEP, CPP, CIPT, CEH, PMP, XENSHA LLC, USA
Pavel Strongin, CISA, CISM, CPA, Charter Communications, USA
Darlene M. Tester, CISM, CISSP, Mystic Lake Hotel & Casino, USA
Larry G. Wlosinski, CISA, CISM, CRISC, CAP, CCSP, CBCP, CDP, CISSP, ITIL v3, Veris Group, LLC, USA

ISACA has begun planning the 16th edition of the *CISM® Review Manual*. Volunteer participation drives the success of the manual. If you are interested in becoming a member of the select group of professionals involved in this global project, we want to hear from you. Please email us at *studymaterials@isaca.org.*

NEW—CISM JOB PRACTICE

BEGINNING IN 2017, THE CISM EXAM WILL TEST THE NEW CISM JOB PRACTICE.

An international job practice analysis is conducted at least every five years or sooner to maintain the validity of the CISM certification program. A new job practice forms the basis of the CISM exam beginning in 2017.

The primary focus of the job practice is the current tasks performed and the knowledge used by CISMs. By gathering evidence of the current work practice of CISMs, ISACA is able to ensure that the CISM program continues to meet the high standards for the certification of professionals throughout the world.

The findings of the CISM job practice analysis are carefully considered and directly influence the development of new test specifications to ensure that the CISM exam reflects the most current best practices.

The new 2017 job practice reflects the areas of study to be tested and is compared below to the previous job practice. The complete CISM job practice can be found at *www.isaca.org/cismjobpractice*.

Previous CISM Job Practice	New 2017 CISM Job Practice
Domain 1: Information Security Governance (24%) Domain 2: Information Risk Management and Compliance (33%) Domain 3: Information Security Program Development and Management (25%) Domain 4: Information Security Incident Management (18%)	**Domain 1: Information Security Governance (24%)** **Domain 2: Information Risk Management (30%)** **Domain 3: Information Security Program Development and Management (27%)** **Domain 4: Information Security Incident Management (19%)**

Page intentionally left blank

Table of Contents

CISM
Certified Information Security Manager
An ISACA Certification

Chapter 2:

CISM
Certified Information Security Manager
An ISACA Certification

CISM
Certified Information Security Manager
An ISACA Certification

About This Manual

Overview

The *CISM® Review Manual 15th Edition* is a reference guide designed to assist candidates in preparing for the CISM examination. **The manual is one source of preparation for the exam and should not be thought of as the only source nor viewed as a comprehensive collection of all the information and experience that are required to pass the exam.** No single publication offers such coverage and detail.

As candidates read through the manual and encounter topics that are new to them or ones in which they feel their knowledge and experience are limited, additional references should be sought. The examination will be composed of questions testing the candidate's technical and practical knowledge and his/her ability to apply the knowledge (based on experience) in given situations.

Organization of This Manual

The *CISM® Review Manual 15th Edition* is divided into four chapters covering the CISM domains tested on the exam in the percentages listed below:

Domain 1	Information Security Governance	24 percent
Domain 2	Information Risk Management	30 percent
Domain 3	Information Security Program Development and Management	27 percent
Domain 4	Information Security Incident Management	19 percent

Note: Each chapter defines the tasks that CISM candidates are expected to know how to do and includes a series of knowledge statements required to perform those tasks. These constitute the current practices for the information security manager. **The detailed CISM job practice can be viewed at *www.isaca.org/cismjobpractice*. The exam is based on these task and knowledge statements.**

The manual has been developed and organized to assist in the study of these areas. Exam candidates should evaluate their strengths, based on knowledge and experience, in each of these areas.

Format of This Manual

Each of the four chapters of the *CISM® Review Manual 15th Edition* is divided into two sections for focused study.

Section one includes:
- A definition of the domain
- Learning objectives for the domain as a practice area
- A listing of the task and knowledge statements for the domain
- A map of the relationship of each task to the knowledge statements for the domain
- A reference guide for the knowledge statements for the domain, including the relevant concepts and explanations
- References to specific content in section two for each knowledge statement
- Sample self-assessment questions and answers with explanations
- Suggested resources for further study of the domain

Section two includes:
- Reference material and content that supports the task and knowledge statements
- Definitions of terms most commonly found on the exam
- Learning activities to reinforce concepts and knowledge

Material included is pertinent for the CISM candidate's knowledge and/or understanding when preparing for the CISM certification exam.

The structure of the content includes numbering to identify the chapter where a topic is located and headings of the subsequent levels of topics addressed in the chapter (i.e., 2.1.1 The Importance of Risk Management is a subtopic of Risk Management Overview in chapter 2). Relevant content in a subtopic is bolded for specific attention.

Understanding the material is a barometer of the candidate's knowledge, strengths and weaknesses, and is an indication of areas in which the candidate needs to seek additional external reference sources. However, written material is not a substitute for experience. **CISM exam questions will test the candidate's practical application of this knowledge.**

Although every effort is made to address the majority of information that candidates are expected to know, not all examination questions are necessarily covered in the manual, and candidates will need to rely on professional experience to provide the best answer.

Throughout the manual, "association" refers to ISACA, formerly known as Information Systems Audit and Control Association, and "institute" or "ITGI®" refers to the IT Governance Institute®. Also, please note that the manual has been written using standard American English.

Note: The *CISM® Review Manual 15th Edition* is a living document. As technology advances and information security management practices evolve, the manual will be updated to reflect such changes. Further updates to this document before the date of the exam may be viewed at *www.isaca.org/studyaidupdates.*

Evaluation of This Manual

ISACA continuously monitors the swift and profound professional, technological and environmental advances affecting the information security management profession. Recognizing these rapid advances, the *CISM® Review Manual* is updated annually.

To assist ISACA in keeping abreast of these advances, please take a moment to evaluate the *CISM® Review Manual 15th Edition.* Such feedback is valuable to fully serve the profession and future CISM exam registrants.

To complete the evaluation on the web site, please go to *www.isaca.org/studyaidsevaluation.*

Thank you for your support and assistance.

Preparing for the CISM Exam

The CISM exam evaluates a candidate's practical knowledge, including experience and application, of the job practice domains as described in this Review Manual. We recommend that the exam candidate look to multiple resources to prepare for the exam, including this Review Manual and the Questions, Answers & Explanation Manual or database, along with external publications. This section will cover some tips for studying for the exam and how best to use this Review Manual in conjunction with other resources.

GETTING STARTED

Having adequate time to prepare for the CISM exam is critical. Most candidates spend between three and six months studying prior to taking the exam. Make sure you set aside a designated time each week to study, which you may wish to increase as your exam date approaches.

Developing a plan for your study efforts can also help you make the most effective use of your time prior to taking the exam.

CISM Self-assessment

In order to effectively study for the CISM exam, you should first identify the job practice areas in which you are weak. A good starting point is the CISM self-assessment, available at http://*www.isaca.org/Certification/CISM-Certified-Information-Security-Manager/Prepare-for-the-Exam/Pages/CISM-Self-Assessment.aspx*

This 50-question sample exam is based upon the question distribution of the CISM exam and can provide you with a high-level evaluation of your areas of needs. When you complete the self-assessment, you will receive a summary of how you performed in each of the four job practice domains. You can use this summary to review the task and knowledge statements in the job practice and get an idea of where you should primarily focus your study efforts.

USING THE CISM REVIEW MANUAL

The *CISM Review Manual* is divided into four chapters, each corresponding with a domain in the CISM job practice. The content of the chapters is organized around the task statements for each chapter. While the Review Manual does not include every concept that could be tested on the CISM exam, it does cover a breadth of knowledge that provides a solid base for the exam candidate. **The manual is one source of preparation for the exam and should not be thought of as the only source nor viewed as a comprehensive collection of all the information and experience that are required to pass the exam.**

Manual Features

The *CISM Review Manual* includes several features to help you navigate the CISM job practice and enhance your learning and retention of the material.

Task Statement Reference Guide

The Task Statement Reference Guide maps the task statement in each domain to relevant sections in the Review Manual. This guide can be used in conjunction with other study materials, such as the *CISM Review Questions, Answers & Explanations Manual 9th Edition*, to help you easily find content related to the task statements you want to review.

Self-assessment Questions and Answers

The self-assessment questions at the end of section one of each chapter assist in understanding how a CISM question could be presented on the CISM exam and should not be used independently as a source of knowledge. Self-assessment questions should not be considered a measurement of the candidate's ability to answer questions correctly on the CISM exam for that area. The questions are intended to familiarize the candidate with question structure, and may or may not be similar to questions that will appear on the actual examination.

Suggested Resources for Further Study

As many of the concepts presented within the Review Manual are complex, you may find it useful to refer to external sources to supplement your understanding of these concepts. The suggested resources are references you can use to help to enhance your study efforts as they relate to each chapter.

In Practice

The In Practice questions are designed for you to further explore concepts from the Review Manual in your own practice. These questions are prompts that may require you to look into your organization's practices to reinforce the material presented in that specific session. For further exploration, consider interacting with colleagues on the ISACA forums or social media platforms.

Knowledge Checks

Knowledge Checks are activities designed to put the material from the Review Manual into practice. These include matching questions, scenarios, recall questions and other activities to further enhance your learning. Answers are provided at the end of each chapter, but it is suggested that you attempt to complete the Knowledge Check prior to referring to the answer key.

Case Studies

Case studies provide scenario-based learning that focuses on the concepts presented within each chapter. Each case study includes an information security management scenario related to each domain and questions related to the scenario. The purpose of these cases studies is to provide a real-world perspective on the content of each domain and how it relates to the CISM's practice.

Glossary

A glossary is included at the end of the manual and contains terms that apply to the material included in the chapters. Also included are terms that apply to related areas not specifically discussed. The glossary is an extension of the text in the manual and can, therefore, be another indication of areas in which the candidate may need to seek additional references.

USING THE CISM REVIEW MANUAL WITH OTHER ISACA RESOURCES

The *CISM Review Manual* can be used in conjunction with other CISM exam preparation. These products are based on the CISM job practice, and referenced task and knowledge statements can be used to find related content within the *CISM Review Manual.* These resources include:

- *CISM Review Questions, Answers and Explanations Manual 9th Edition*
- CISM Review Questions, Answers and Explanations Database – 12 Month Subscription
- Chapter CISM Review Courses

ABOUT THE CISM REVIEW QUESTIONS, ANSWERS AND EXPLANATIONS PRODUCTS

The *CISM® Review Questions, Answers & Explanations Manual 9th Edition* consists of 1,000 multiple-choice study questions, answers and explanations arranged in the domains of the current CISM job practice.

Another study aid that is available is the **CISM® Review Questions, Answers & Explanations Database – 12 Month Subscription.** The database consists of the 1,000 questions, answers and explanations included in the *CISM® Review Questions, Answers & Explanations Manual 9th Edition.* With this product, CISM candidates can quickly identify their strengths and weaknesses by taking random sample exams of varying length and breaking the results down by domain. Sample exams also can be chosen by domain, allowing for concentrated study, one domain at a time, and other sorting features such as the omission of previous correctly answered questions are available.

Questions in these products are representative of the types of questions that could appear on the exam and include explanations of the correct and incorrect answers. Questions are sorted by the CISM domains and as a sample test. These products are ideal for use in conjunction with the *CISM® Review Manual 15th Edition.* These products can be used as study sources throughout the study process or as part of a final review to determine where candidates may need additional study. It should be noted that these questions and suggested answers are provided as examples; they are not actual questions from the examination and may differ in content from those that actually appear on the exam.

TYPES OF QUESTIONS ON THE CISM EXAM

CISM exam questions are developed with the intent of measuring and testing practical knowledge and the application of information security managerial principles and standards. As previously mentioned, all questions are presented in a multiple-choice format and are designed for one best answer.

The candidate is cautioned to read each question carefully. Many times a CISM exam question will require the candidate to choose the appropriate answer that is **MOST** likely or **BEST**, or the candidate may be asked to choose a practice or procedure that would be performed **FIRST** related to the other answers. In every case, the candidate is required to read the question carefully, eliminate known wrong answers and then make the best choice possible. Knowing that these types of questions are asked and how to study to answer them will go a long way toward answering them correctly. The best answer is of the choices provided. There can be many potential solutions to the scenarios posed in the questions, depending on industry, geographical location, etc. It is advisable to consider the information provided in the question and to determine the best answer of the options provided.

Each CISM question has a stem (question) and four options (answer choices). The candidate is asked to choose the correct or best answer from the options. The stem may be in the form of a question or incomplete statement. In some instances, a scenario or description also may be included. These questions normally include a description of a situation and require the candidate to answer two or more questions based on the information provided.

A helpful approach to these questions includes the following:

- Read the entire stem and determine what the question is asking. Look for key words such as "BEST," "MOST," "FIRST," etc. and key terms that may indicate what domain or concept that is being tested.
- Read all of the options, and then read the stem again to see if you can eliminate any of the options based on your immediate understanding of the question.
- Re-read the remaining options and bring in any personal experience to determine which is the best answer to the question.

Another condition the candidate should consider when preparing for the exam is to recognize that information security is a global profession, and individual perceptions and experiences may not reflect the more global position or circumstance. Because the exam and CISM manuals are written for the international information security community, the candidate will be required to be somewhat flexible when reading a condition that may be contrary to the candidate's experience. It should be noted that CISM exam questions are written by experienced information security managers from around the world. Each question on the exam is reviewed by ISACA's CISM Exam Item Development Working Group, which consists of international members. This geographic representation ensures that all exam questions are understood equally in every country and language.

Note: When using the CISM review materials to prepare for the exam, it should be noted that they cover a broad spectrum of information security management issues. **Again, candidates should not assume that reading these manuals and answering review questions will fully prepare them for the examination.** Since actual exam questions often relate to practical experiences, candidates should refer to their own experiences and other reference sources, and draw upon the experiences of colleagues and others who have earned the CISM designation.

Page intentionally left blank

Chapter 1:

Information Security Governance

Section One: Overview

Section Two: Content

rview

knowledge and associated tasks on security governance structure organizational objectives.

DOMAIN DEFINITION

Establish and/or maintain an information security governance framework and supporting processes to ensure that the information security strategy is aligned with organizational goals and objectives.

LEARNING OBJECTIVES

The objective of this domain is to ensure that the CISM candidate has the knowledge necessary to:

- Understand the purpose of information security governance, what it consists of and how to accomplish it
- Understand the purpose of an information security strategy, its objectives, and the reasons and steps required to develop one
- Understand the meaning, content, creation and use of policies, standards, procedures and guidelines and how they relate to each other
- Develop business cases and gain commitment from senior leadership
- Define governance metrics requirements, selection and creation

CISM EXAM REFERENCE

This domain represents 24 percent of the CISM examination (approximately 36 questions).

TASK AND KNOWLEDGE STATEMENTS

TASK STATEMENTS

There are nine tasks within this domain that a CISM candidate must know how to perform:

T1.1 Establish and/or maintain an information security strategy in alignment with organizational goals and objectives to guide the establishment and/or ongoing management of the information security program.

T1.2 Establish and/or maintain an information security governance framework to guide activities that support the information security strategy.

T1.3 Integrate information security governance into corporate governance to ensure that organizational goals and objectives are supported by the information security program.

T1.4 Establish and maintain information security policies to guide the development of standards, procedures and guidelines in alignment with enterprise goals and objectives.

T1.5 Develop business cases to support investments in information security.

T1.6 Identify internal and external influences to the organization (e.g., emerging technologies, social media, business environment, risk tolerance, regulatory requirements, third-party considerations, threat landscape) to ensure that these factors are continually addressed by the information security strategy.

T1.7 Gain ongoing commitment from senior leadership and other stakeholders to support the successful implementation of the information security strategy.

T1.8 Define, communicate and monitor information security responsibilities throughout the organization (e.g., data owners, data custodians, end users, privileged or high-risk users) and lines of authority.

T1.9 Establish, monitor, evaluate and report key information security metrics to provide management with accurate and meaningful information regarding the effectiveness of the information security strategy.

KNOWLEDGE STATEMENTS

The CISM candidate must have a good understanding of each of the areas delineated by the knowledge statements. These statements are the basis for the exam.

There are 19 knowledge statements within the information security governance domain:

K1.1 Knowledge of techniques used to develop an information security strategy (e.g., SWOT [strengths, weaknesses, opportunities, threats] analysis, gap analysis, threat research)

K1.2 Knowledge of the relationship of information security to business goals, objectives, functions, processes and practices

K1.3 Knowledge of available information security governance frameworks

K1.4 Knowledge of globally recognized standards, frameworks and industry best practices related to information security governance and strategy development

K1.5 Knowledge of the fundamental concepts of governance and how they relate to information security

K1.6 Knowledge of methods to assess, plan, design and implement an information security governance framework

K1.7 Knowledge of methods to integrate information security governance into corporate governance

K1.8 Knowledge of contributing factors and parameters (e.g., organizational structure and culture, tone at the top, regulations) for information security policy development

K1.9 Knowledge of content in, and techniques to develop, business cases

K1.10 Knowledge of strategic budgetary planning and reporting methods

K1.11 Knowledge of the internal and external influences to the organization (e.g., emerging technologies, social media, business environment, risk tolerance, regulatory requirements, third-party considerations, threat landscape) and how they impact the information security strategy

K1.12 Knowledge of key information needed to obtain commitment from senior leadership and support from other stakeholders (e.g., how information security supports organizational goals and objectives, criteria for determining successful implementation, business impact)

K1.13 Knowledge of methods and considerations for communicating with senior leadership and other stakeholders (e.g., organizational culture, channels of communication, highlighting essential aspects of information security)

K1.14 Knowledge of roles and responsibilities of the information security manager

K1.15 Knowledge of organizational structures, lines of authority and escalation points

K1.16 Knowledge of information security responsibilities of staff across the organization (e.g., data owners, end users, privileged or high-risk users)

K1.17 Knowledge of processes to monitor performance of information security responsibilities

K1.18 Knowledge of methods to establish new, or utilize existing, reporting and communication channels throughout an organization

K1.19 Knowledge of methods to select, implement and interpret key information security metrics (e.g., key goal indicators [KGIs], key performance indicators [KPIs], key risk indicators [KRIs])

RELATIONSHIP OF TASK TO KNOWLEDGE STATEMENTS

The task statements are what the CISM candidate is expected to know how to perform. The knowledge statements delineate each of the areas in which the CISM candidate must have a good understanding to perform the tasks. The task and knowledge statements are mapped, insofar as it is possible to do so. Note that although there is often overlap, each task statement will generally map to several knowledge statements.

Task and Knowledge Statements Mapping

Task Statement	Knowledge Statements
T1.1 Establish and/or maintain an information security strategy in alignment with organizational goals and objectives to guide the establishment and/or ongoing management of the information security program.	K1.1 Knowledge of techniques used to develop an information security strategy (e.g., SWOT [strengths, weaknesses, opportunities, threats] analysis, gap analysis, threat research) K1.2 Knowledge of the relationship of information security to business goals, objectives, functions, processes and practices K1.4 Knowledge of globally recognized standards, frameworks and industry best practices related to information security governance and strategy development K1.10 Knowledge of strategic budgetary planning and reporting methods K1.11 Knowledge of the internal and external influences to the organization (e.g., emerging technologies, social media, business environment, risk tolerance, regulatory requirements, third-party considerations, threat landscape) and how they impact the information security strategy K1.12 Knowledge of key information needed to obtain commitment from senior leadership and support from other stakeholders (e.g., how information security supports organizational goals and objectives, criteria for determining successful implementation, business impact)
T1.2 Establish and/or maintain an information security governance framework to guide activities that support the information security strategy.	K1.3 Knowledge of available information security governance frameworks K1.4 Knowledge of globally recognized standards, frameworks and industry best practices related to information security governance and strategy development K1.5 Knowledge of the fundamental concepts of governance and how they relate to information security K1.6 Knowledge of methods to assess, plan, design and implement an information security governance framework
T1.3 Integrate information security governance into corporate governance to ensure that organizational goals and objectives are supported by the information security program.	K1.2 Knowledge of the relationship of information security to business goals, objectives, functions, processes and practices K1.3 Knowledge of available information security governance frameworks K1.4 Knowledge of globally recognized standards, frameworks and industry best practices related to information security governance and strategy development K1.5 Knowledge of the fundamental concepts of governance and how they relate to information security K1.6 Knowledge of methods to assess, plan, design and implement an information security governance framework K1.7 Knowledge of methods to integrate information security governance into corporate governance
T1.4 Establish and maintain information security policies to guide the development of standards, procedures and guidelines in alignment with enterprise goals and objectives.	K1.2 Knowledge of the relationship of information security to business goals, objectives, functions, processes and practices K1.8 Knowledge of contributing factors and parameters (e.g., organizational structure and culture, tone at the top, regulations) for information security policy development
T1.5 Develop business cases to support investments in information security.	K1.9 Knowledge of content in, and techniques to develop, business cases K1.10 Knowledge of strategic budgetary planning and reporting methods K1.12 Knowledge of key information needed to obtain commitment from senior leadership and support from other stakeholders (e.g., how information security supports organizational goals and objectives, criteria for determining successful implementation, business impact)

Task and Knowledge Statements Mapping *(cont.)*	
Task Statement	**Knowledge Statements**
T1.6 Identify internal and external influences to the organization (e.g., emerging technologies, social media, business environment, risk tolerance, regulatory requirements, third-party considerations, threat landscape) to ensure that these factors are continually addressed by the information security strategy.	K1.2 Knowledge of the relationship of information security to business goals, objectives, functions, processes and practices K1.7 Knowledge of methods to integrate information security governance into corporate governance K1.11 Knowledge of the internal and external influences to the organization (e.g., emerging technologies, social media, business environment, risk tolerance, regulatory requirements, third-party considerations, threat landscape) and how they impact the information security strategy
T1.7 Gain ongoing commitment from senior leadership and other stakeholders to support the successful implementation of the information security strategy.	K1.12 Knowledge of key information needed to obtain commitment from senior leadership and support from other stakeholders (e.g., how information security supports organizational goals and objectives, criteria for determining successful implementation, business impact) K1.13 Knowledge of methods and considerations for communicating with senior leadership and other stakeholders (e.g., organizational culture, channels of communication, highlighting essential aspects of information security)
T1.8 Define, communicate and monitor information security responsibilities throughout the organization (e.g., data owners, data custodians, end users, privileged or high-risk users) and lines of authority.	K1.15 Knowledge of organizational structures, lines of authority and escalation points K1.16 Knowledge of information security responsibilities of staff across the organization (e.g., data owners, end users, privileged or high-risk users) K1.17 Knowledge of processes to monitor performance of information security responsibilities K1.18 Knowledge of methods to establish new, or utilize existing, reporting and communication channels throughout an organization
T1.9 Establish, monitor, evaluate and report key information security metrics to provide management with accurate and meaningful information regarding the effectiveness of the information security strategy.	K1.10 Knowledge of strategic budgetary planning and reporting methods K1.19 Knowledge of methods to select, implement and interpret key information security metrics (e.g., key goal indicators [KGIs], key performance indicators [KPIs], key risk indicators [KRIs])

TASK STATEMENT REFERENCE GUIDE

The following section contains the task statements a CISM candidate is expected to know how to accomplish mapped to and the areas in the review manual with information that supports the execution of the task. The references in the manual focus on the knowledge the information security manager must know to accomplish the tasks and successfully negotiate the exam.

Task Statement Reference Guide	
Task Statement	**Reference in Manual**
T1.1 Establish and/or maintain an information security strategy in alignment with organizational goals and objectives to guide the establishment and/or ongoing management of the information security program.	1.2 Effective Information Security Governance 1.7 Information Security Strategy Overview 1.7.1 Developing an Information Security Strategy 1.8 Information Security Strategy Objectives 1.13 Action Plan to Implement Strategy
T1.2 Establish and/or maintain an information security governance framework to guide activities that support the information security strategy.	1.2.5 Business Model for Information Security 1.8.3 The Desired State
T1.3 Integrate information security governance into corporate governance to ensure that organizational goals and objectives are supported by the information security program.	1.1 Information Security Governance Overview 1.2.1 Business Goals and Objectives 1.8.3 The Desired State 1.9 Determining the Current State of Security 1.10 Information Security Strategy Development
T1.4 Establish and maintain information security policies to guide the development of standards, procedures and guidelines in alignment with enterprise goals and objectives.	1.3 Roles and Responsibilities 1.11.1 Policies and Standards
T1.5 Develop business cases to support investments in information security.	1.4.2 Developing and Presenting the Business Case 3.10.8 Business Case Development

Task Statement Reference Guide *(cont.)*			
Task Statement		**Reference in Manual**	
T1.6	Identify internal and external influences to the organization (e.g., emerging technologies, social media, business environment, risk tolerance, regulatory requirements, third-party considerations, threat landscape) to ensure that these factors are continually addressed by the information security strategy.	1.5 1.12	Governance of Third-party Relationships Strategy Constraints
T1.7	Gain ongoing commitment from senior leadership and other stakeholders to support the successful implementation of the information security strategy.	1.4.2	Obtaining Senior Management Commitment
T1.8	Define, communicate and monitor information security responsibilities throughout the organization (e.g., data owners, data custodians, end users, privileged or high-risk users) and lines of authority.	1.3	Roles and Responsibilities
T1.9	Establish, monitor, evaluate and report key information security metrics to provide management with accurate and meaningful information regarding the effectiveness of thc information security strategy.	1.6	Information Security Governance Metrics

SUGGESTED RESOURCES FOR FURTHER STUDY

Brotby, W. Krag, and IT Governance Institute; *Information Security Governance: Guidance for Boards of Directors and Executive Management, 2nd Edition*, ISACA, USA, 2006

Brotby, W. Krag, and IT Governance Institute; *Information Security Governance: Guidance for Information Security Managers*, ISACA, USA, 2008

Brotby, W. Krag; *Information Security Governance: A Practical Development and Implementation Approach*, Wiley & Sons, 2009

Committee of Sponsoring Organizations of the Treadway Commission (COSO), *Internal Control—Integrated Framework*, USA, 2013

International Organization for Standardization (ISO), *ISO/IEC 27001:2013 Information technology—Security techniques—Information security management systems—Requirements,* Switzerland, 2013

ISO, *ISO/IEC 27002:2013 Information technology—Security techniques—Code of practice for information security controls*, Switzerland, 2013

ISO, *ISO/IEC 27014:2013 Information technology—Security techniques—Governance of information security*, Switzerland, 2013

ISACA, *The Business Model for Information Security,* USA, 2010

ISACA, COBIT 5, USA, 2012, *www.isaca.org/cobit*

ISACA, *COBIT® 5: Enabling Processes*, USA, 2012, *www.isaca.org/cobit*

ISACA, *COBIT® 5 for Information Security,* USA, 2012, *www.isaca.org/cobit*

National Institute of Standards and Technology (NIST), *NIST Special Publication 800-53, Revision 4 Security and Privacy Controls for Federal Information Systems and Organizations*, USA, 2013

PricewaterhouseCoopers, *The Global State of Information Security Survey 2016, www.pwc.com/gx/en/consulting-services/information-security-survey*

Note: Publications in bold are stocked in the ISACA Bookstore.

SELF-ASSESSMENT QUESTIONS

CISM exam questions are developed with the intent of measuring and testing practical knowledge in information security management. All questions are multiple choice and are designed for one best answer. Every CISM question has a stem (question) and four options (answer choices). The candidate is asked to choose the correct or best answer from the options. The stem may be in the form of a question or incomplete statement. In some instances, a scenario or a description problem may also be included. These questions normally include a description of a situation and require the candidate to answer two or more questions based on the information provided. Many times a CISM examination question will require the candidate to choose the most likely or best answer.

In every case, the candidate is required to read the question carefully, eliminate known incorrect answers and then make the best choice possible. Knowing the format in which questions are asked, and how to study to gain knowledge of what will be tested, will go a long way toward answering them correctly.

1-1 A security strategy is important for an organization **PRIMARILY** because it:

A. provides a basis for determining the best logical security architecture for the organization.
B. provides the approach to achieving the outcomes management wants.
C. provides users guidance on how to operate securely in everyday tasks.
D. helps IS auditors ensure compliance.

1-2 Which of the following is the **MOST** important reason to provide effective communication about information security?

A. It makes information security more palatable to resistant employees.
B. It mitigates the weakest link in the information security landscape.
C. It informs business units about the information security strategy.
D. It helps the organization conform to regulatory information security requirements.

1-3 Which of the following approaches **BEST** helps the information security manager achieve compliance with various regulatory requirements?

A. Rely on corporate counsel to advise which regulations are the most relevant.
B. Stay current with all relevant regulations and request legal interpretation.
C. Involve all impacted departments and treat regulations as just another risk.
D. Ignore many of the regulations that have no penalties.

1-4 The **MOST** important consideration in developing security policies is that:

A. they are based on a threat profile.
B. they are complete and no detail is left out.
C. management signs off on them.
D. all employees read and understand them.

1-5 The **PRIMARY** security objective in creating good procedures is:

A. to make sure they work as intended.
B. that they are unambiguous and meet the standards.
C. that they are written in plain language and widely distributed.
D. that compliance can be monitored.

1-6 Which of the following **MOST** helps ensure that assignment of roles and responsibilities is effective?

A. Senior management is in support of the assignments.
B. The assignments are consistent with existing proficiencies.
C. The assignments are mapped to required skills.
D. The assignments are given on a voluntary basis.

1-7 Which of the following benefits is the **MOST** important to an organization with effective information security governance?

A. Maintaining appropriate regulatory compliance
B. Ensuring disruptions are within acceptable levels
C. Prioritizing allocation of remedial resources
D. Maximizing return on security investments

1-8 From an information security manager's perspective, the **MOST** important factors regarding data retention are:

A. business and regulatory requirements.
B. document integrity and destruction.
C. media availability and storage.
D. data confidentiality and encryption.

1-9 Which role is in the **BEST** position to review and confirm the appropriateness of a user access list?

A. Data owner
B. Information security manager
C. Domain administrator
D. Business manager

1-10 In implementing information security governance, the information security manager is **PRIMARILY** responsible for:

A. developing the security strategy.
B. reviewing the security strategy.
C. communicating the security strategy.
D. approving the security strategy.

ANSWERS TO SELF-ASSESSMENT QUESTIONS

1-1 A. Policies have to be developed to support the security strategy, and an architecture can only be developed after policies are completed (i.e., the security strategy is not the basis for architecture; policies are).

B. A security strategy will define the approach to achieving the security program outcomes management wants. It should also be a statement of how security aligns with and supports business objectives, and it provides the basis for good security governance.

C. A security strategy may include requirements for users to operate securely, but it does not address how that is to be accomplished.

D. IS auditors do not determine compliance based on strategy, but rather on elements such as standards and control objectives.

1-2 A. Effective communication may assist in making information security more palatable, but that is not the most important aspect.

B. Security failures are, in the majority of instances, directly attributable to lack of awareness or failure of employees to follow policies or procedures. Communication is important to ensure continued awareness of security policies and procedures among staff and business partners.

C. Effective communication will allow business units to be informed about various aspects of information security, including the strategy, but it is not the most important aspect.

D. Effective communication will assist in achieving compliance because it is unlikely that employees will be compliant with regulations unless they are informed about them. However, it is not the most important consideration.

1-3 A. Corporate counsel is generally involved primarily with stock issues and the associated filings required by regulators and with contract matters. It is unlikely that legal staff will be current on information security regulations and legal requirements.

B. While it can be useful to stay abreast of all current and emerging regulations, it is, as a practical matter, nearly impossible, especially for a multinational company.

C. Departments such as human resources, finance and legal are most often subject to new regulations and, therefore, must be involved in determining how best to meet the existing and emerging requirements and, typically, would be most aware of these regulations. Treating regulations as another risk puts them in the proper perspective, and the mechanisms to deal with them should already exist. The fact that there are so many regulations makes it unlikely that they can all be specifically addressed efficiently. Many do not currently have significant consequences and, in fact, may be addressed by compliance with other regulations. The most relevant response to regulatory requirements is to determine potential impact to the organization just as must be done with any other risk.

D. Even if certain regulations have few or no penalties, ignoring them without consideration for other potential impacts (e.g., reputational damage) and whether they might be relevant to the organization is not generally a prudent approach.

1-4 **A. The basis for developing relevant security policies is addressing viable threats to the organization, prioritized by the likelihood of occurrence and their potential impact on the business. The strictest policies apply to the areas of greatest business value. This ensures that protection proportionality is maintained.**

B. Policies are a statement of management's intent and direction at a high level and provide little, if any, detail.

C. While the policies are being developed, management would not be asked to sign them until they have been completed.

D. Employees would not be reading and understanding the policies while they are being developed.

1-5 A. While it is important to make sure that procedures work as intended, the fact that they do not may not be a security issue.

B. All of the answers are important, but the first criterion must be to ensure that there is no ambiguity in the procedures and that, from a security perspective, they meet the applicable standards and, therefore, comply with policy.

C. Of importance, but not as critical, is that procedures are clearly written and that they are provided to all staff as needed.

D. Compliance is important, but it is essential that it is compliant with a correct procedure.

1-6 A. Senior management support is always important, but it is not of as significant importance to the effectiveness of employee activities.

B. The level of effectiveness of employees will be determined by their existing knowledge and capabilities—in other words, their proficiencies.

C. Mapping roles to the tasks that are required can be useful but it is no guarantee that people can perform the required tasks.

D. While employees are more likely to be enthusiastic about a job they have volunteered for, it is not a requirement for them to be effective.

1-7 A. Maintaining appropriate regulatory compliance is a useful, but subordinate, outcome.

B. The bottom line of security efforts is to ensure that business can continue to operate with an acceptable level of disruption that does not unduly constrain revenue-producing activities.

C. Prioritizing allocation of remedial resources is a useful, but subordinate, outcome.

D. Maximizing return on security investments is a useful, but subordinate, outcome.

1-8 **A. Business and regulatory requirements are the driving factors for data retention.**
B. Integrity is a key factor for information security; however, business and regulatory requirements are the driving factors for data retention.
C. Availability is a key factor for information security; however, business and regulatory requirements are the driving factors for data retention.
D. Confidentiality is a key factor for information security; however, business and regulatory requirements are the driving factors for data retention.

1-9 **A. The data owner is responsible for periodic reconfirmation of the access lists for systems he/she owns.**
B. The information security manager is in charge of the coordination of the user access list reviews but he/she does not have any responsibility for data access.
C. The domain administrator may technically provide the access, but he/she does not approve it.
D. The business manager is incorrect because the business manager may not be the data owner.

1-10 **A. The information security manager is responsible for developing a security strategy based on business objectives with the help of business process owners and senior management.**
B. The information security strategy is the responsibility of a steering committee and/or senior management.
C. The information security manager is not necessarily responsible for communicating the security strategy.
D. Final approval of the information security strategy must be made by senior management.

Section Two: Content

1.0 INTRODUCTION

For information security to effectively address the ever-growing challenges of providing adequate protection for information assets, an information security strategy is essential. This strategy documents the direction and goals for the information security program, as determined by senior management. Subsequently, the strategy provides the basis to implement effective information security governance. **Governance** is broadly defined as the rules that run the organization, including policies, standards and procedures that are used to set the direction and control the organization's activities.

The first step in establishing information security governance is senior management determining the outcomes it wants from the information security program. Often, those outcomes are stated in terms of risk management and the levels of acceptable risk. This can be accomplished through a set of facilitated meetings with senior management and business unit leaders. The information security manager then has the information needed to develop a set of requirements for a security program. This is followed by setting a series of specific objectives that, when achieved, will satisfy the requirements.

An element of developing the strategy is to develop objectives, or the desired state of the enterprise's information security. The desired state is based on the outcomes set by senior management, and a variety of frameworks are available to assist in defining this desired state. The outcomes and levels of acceptable risk should be determined and used to set control objectives. Next, the information security manager can determine what needs to be done to move from the current to the desired state by using a gap analysis. This, then, becomes the basis of the strategy.

Next, a road map is created to identify the specifics needed to achieve the objectives. This is followed by identifying the resources needed to navigate the road map and implement the strategy. At the same time, the constraints need to be identified and considered. Constraints include time limits, skills available, funding, laws and regulations.

Many resources must be considered for achieving the strategy's objectives, including controls such as technologies, standards and processes.

Information security governance needs to be integrated into the overall enterprise governance structure to ensure that the organizational goals are supported by the information security program. The governance framework is an outline or skeleton of interlinked items that supports a particular approach to a specific objective as stated in the strategy. Several governance frameworks may be suitable for an organization to implement, including COBIT 5 and ISO/IEC 27000. High-level architecture can serve as a framework as well. The framework will serve to integrate and guide activities needed to implement the information security strategy.

Information security governance is a subset of corporate governance and must be consistent with the enterprise's governance. If enterprise governance is structured using a particular framework, it would make sense to use the same framework for information security governance to facilitate integration.

Security policies are designed to mitigate risk and are usually developed in response to an actual or perceived threat. Policies state management intent and direction at a high level. With the development of an information security strategy, the policies have to be developed or modified to support the strategic objectives.

Standards are developed or modified to set boundaries for people, processes, procedures and technologies to maintain compliance with policies and support the achievement of the organization's goals and objectives. There are usually several standards for each policy, depending on the classification levels of the asset related to the standard. Collectively, standards are combined with other controls (i.e., technical, physical, administrative) to create the security baselines.

Baseline security, control objectives, acceptable risk, risk appetite and residual risk are related to one another. Control objectives are set to fall within the boundaries of acceptable risk. Risk mitigation should keep the organization within the level of acceptable risk and achieving the control objectives. Residual risk levels set the security baselines. These topics are discussed in greater detail in chapters 2 and 3.

New information security projects should support the information security strategy. A **business case** is used to capture the business reasoning for initiating a project or task. In the context of information security, it should concisely identify the needs and the business purpose for implementing various governance processes, including technologies and the cost benefit (i.e., the value proposition) of doing so.

The business case should include all the factors that can materially affect the project's success or failure. The method of presentation should be consistent with the organization's usual approach (i.e., slide show, electronic or written document). It must persuasively encompass benefits, costs and risk. The benefits must be tangible, supportable and relevant to the organization. Particular attention must be given to the financial aspects of the proposal. The total cost of ownership (TCO) and risk must be realistically represented for the full life cycle of the project. It is important to avoid overconfidence, overly optimistic projections and excessive precision for what are likely to be somewhat speculative results.

The strategy should keep in mind that information security is never static. Threats, vulnerabilities and exposures are always changing because of internal and external factors. Internal factors include changes in technology, personnel, business activities and financial capacity to absorb loss. External factors include changing markets, emerging threats, new technologies, increasing sophistication of attackers and new regulations. Risk and impact assessments are often a good indicator of changes that may need to be made to strategy elements. Another indicator is number of security incidents.

To continue to be effective, the strategy must be a living document with objectives, approaches and methods changing to meet new conditions. In many cases, the high-level objectives

may not change significantly while the approaches, methods and technologies may evolve as conditions warrant.

Internal metrics and ongoing monitoring will aid the information security manager in keeping aware of changing conditions. Keeping current on external events that may affect the enterprise and its risk is also key.

For an effective information security program and strategy, it is necessary to have ongoing support from senior management, business owners and department heads. Information security cannot be driven up from the middle of an organization. Without support of senior management, the issues of inadequate funding, insufficient staffing and poor compliance will be ongoing. An organization's culture is largely a reflection of senior management, and senior management's lack of support will be reflected in the entire organization. This is evidenced by the organization's perception of and relationship to information security specifically and risk management generally.

A lack of commitment and support from senior management cascades down and result in little support from business owners and department heads. Because this group is largely responsible for policy administration and compliance enforcement, it is likely that information security policy compliance will be poor.

Support can be gained by educating senior management or developing persuasive business cases. Unfortunately, more often, it is a serious incident or major compromise that is needed to generate management commitment. In organizations that are subject to strong governmental regulation, frequent audits and costly sanctions, support and commitment may be high as a result.

It is often said that security is only as strong as the weakest link, and this is certainly true for information security. For information security to be effective across the entire organization, everyone should have responsibilities related to information security or risk management. It is up to the information security manager to define those responsibilities for everyone in the organization. The next step is to inform employees of those responsibilities on an ongoing or frequent basis. Most employees will do most of what is required most of the time if they are clearly aware of their responsibilities and the consequences of failing to do so. To reinforce responsible information security behavior, it is necessary to monitor compliance on a regular basis because information security is often an afterthought for busy employees focusing on other tasks.

What cannot be measured cannot be managed. Information security has been managed for some time without good metrics in an *ad hoc*, haphazard and often ineffective way. The resistance to developing good metrics appears to be related to management's desire to avoid knowing too much that might require taking action; this enables management to sidestep being held accountable and supports plausible deniability. For some information security managers, a suite of effective information security metrics may increase oversight, performance accountability and workload. Whatever the underlying causes, information security suffers from a lack of or poor metrics especially at the management and strategic levels—precisely where informed decision making must take place.

No profession or activity has achieved reliable and effective maturity prior to the development of a suite of good metrics. Metrics must be considered at three levels—operational, management and strategic.

One of the most important requirements for good metrics is to find out what information is needed at the different organizational levels to make informed decisions. This information becomes the basis for effective information security metrics. Of course, the information needed by the chief executive officer (CEO) or managing director is different from what is needed by a system administrator to make informed information security decisions. The first step to establish a system of metrics is to determine what is meaningful to the recipients. Then those metrics must be monitored, evaluated and communicated to the appropriate people on a timely basis.

1.1 INFORMATION SECURITY GOVERNANCE OVERVIEW

Information can be defined as "data endowed with meaning and purpose." It plays a critical role in all aspects of our lives. Information has become an indispensable component of conducting business for virtually all organizations. In a growing number of companies, information *is* the business.

Approximately 80 percent of national critical infrastructures in the developed world are controlled by the private sector. Coupled with often ineffective bureaucracies, countless conflicting jurisdictions and aging institutions unable to adapt to dealing with burgeoning global information crime, a preponderance of the task of protecting information resources critical to survival is falling squarely on corporate shoulders. However, studies continue to show that a large percentage of organizations do not adequately address either existing or emerging security issues. The findings of the *2015 Ernst and Young Global Information Security Survey* highlight the problem. According to the survey:

- Only 12 percent of organizations believe that information security meets the needs of the organization, and 67 percent are still making improvements.
- Sixty-nine percent noted that the information security budget should increase by as much as 50 percent to be able to protect the organization in line with the risk tolerance set by management.

The conclusion that can be drawn from this report is that information security failures are predominantly a failure of governance and cannot be solved by technology alone. The need for adequate protection of information resources needs to be raised at the board level, like other critical governance functions, to be successful. The complexity, relevance and criticality of information security and its governance mandate that it be addressed and supported at the highest organizational levels.

Ultimately, to achieve significant improvement in information security, senior management and the board of directors must be held accountable for information security governance and must provide the necessary leadership, organizational structures, oversight, resources and processes to ensure that information security governance is an integral and transparent part of enterprise governance.

Increasingly, those who understand the scope and depth of risk to information take the position that, as a critical resource, information must be treated with the same care, caution and prudence that any other asset essential to the survival of the organization and, perhaps society, would receive.

Often, the focus of protection has been on the IT systems that process and store the vast majority of information rather than on the information itself. But this approach is too narrow to accomplish the level of integration, process assurance and overall security that are required. Information security takes the larger view that the content, information and the knowledge based on it must be adequately protected, regardless of how it is handled, processed, transported or stored.

Senior management is increasingly confronted by the need to stay competitive in the global economy and heed the promise of ever greater gains from the deployment of more information resources. Even as organizations reap those gains, the increasing dependence on information and the systems that support it, and advancing risk from an increasing host of threats, are forcing management to make difficult and often costly decisions about how to effectively address information security. In addition, scores of new and existing laws and regulations are increasingly demanding compliance and higher levels of accountability in an effort of governments to address increasingly sophisticated attacks and growing losses that pose mounting threats to their national critical infrastructures.

1.1.1 IMPORTANCE OF INFORMATION SECURITY GOVERNANCE

From an organization's perspective, information security governance is increasingly critical as dependence on information has grown essential to survival.

For most organizations, information, and the knowledge based on it, is one of their most important assets because conducting business without it would not be possible. The systems and processes that handle this information have become pervasive throughout organizations across the globe. This reliance on information and their related systems have made information security governance a critical facet of overall enterprise governance. In addition to addressing legal and regulatory requirements, effective information security governance is good business. Prudent management understands that it provides a series of significant benefits, including:

- Addressing the increasing potential for civil or legal liability inuring to the organization and senior management as a result of information inaccuracy or the absence of due care in its protection or inadequate regulatory compliance
- Providing assurance of policy compliance
- Increasing predictability and reducing uncertainty of business operations by lowering risk to definable and acceptable levels
- Providing the structure and framework to optimize allocations of limited security resources
- Providing a level of assurance that critical decisions are not based on faulty information
- Providing a firm foundation for efficient and effective risk management, process improvement, rapid incident response and continuity management
- Providing greater confidence in interactions with trading partners
- Improving trust in customer relationships
- Protecting the organization's reputation
- Enabling new and better ways to process electronic transactions
- Providing accountability for safeguarding information during critical business activities, such as mergers and acquisitions, business process recovery, and regulatory response
- Effective management of information security resources

In summary, because new information technology provides the potential for dramatically enhanced business performance, effective information security can add significant value to the organization by reducing losses from security-related events and providing assurance that security incidents and breaches are not catastrophic. In addition, evidence suggests that improved perception in the market has resulted in increased share value.

1.1.2 OUTCOMES OF INFORMATION SECURITY GOVERNANCE

Information security governance includes the elements required to provide senior management assurance that its direction and intent are reflected in the security posture of the organization by using a structured approach to implementing a security program. Once those elements are in place, senior management can be confident that adequate and effective information security will protect the organization's vital information assets.

The objective of information security governance is to develop, implement and manage a security program that achieves the following six basic outcomes:

1. **Strategic alignment**—Aligning information security with business strategy to support organizational objectives such as:
   - Security requirements driven by enterprise requirements that are thoroughly developed to provide guidance on what must be done and a measure of when it has been achieved
   - Security solutions fit for enterprise processes that take into account the culture, governance style, technology and structure of the organization
   - Investment in information security that is aligned with the enterprise strategy; enterprise operations; and a well-defined threat, vulnerability and risk profile
2. **Risk management**—Executing appropriate measures to mitigate risk and reduce potential impacts on information resources to an acceptable level such as:
   - Collective understanding of the organization's threat, vulnerability and risk profile
   - Understanding of risk exposure and potential consequences of compromise
   - Awareness of risk management priorities based on potential consequences
   - Risk mitigation sufficient to achieve acceptable consequences from residual risk
   - Risk acceptance/deference based on an understanding of the potential consequences of residual risk
3. **Value delivery**—Optimizing security investments in support of business objectives such as:
   - A standard set of security practices (i.e., baseline security requirements following adequate and sufficient practices proportionate to risk and potential impact)

- Information security overheads that are maintained at a minimum level while maintaining a security program that enables the organization to achieve its objectives
- A properly prioritized and distributed effort to areas with the greatest probability and highest impact and business benefit
- Institutionalized and commoditized standards-based solutions with the greatest cost-effectiveness
- Complete solutions, covering organization and process as well as technology based on an understanding of the end-to-end business of the organization
- A continuous improvement culture based on the understanding that security is an ongoing process, not an event

4. **Resource optimization**—Using information security knowledge and infrastructure efficiently and effectively to:
   - Ensure that knowledge is captured and available
   - Document security processes and practices
   - Develop security architecture(s) to define and utilize infrastructure resources efficiently
5. **Performance measurement**—Monitoring and reporting on information security processes to ensure that objectives are achieved, including:
   - A defined, agreed-upon and meaningful set of metrics that are properly aligned with strategic objectives and provide the information needed for effective decisions at the strategic, management and operational levels
   - Measurement process that helps identify shortcomings and provides feedback on progress made resolving issues
   - Independent assurance provided by external assessments and audits
   - Criteria for separating the most useful metrics from the variety of things that can be measured
6. **Assurance process integration**—Integrating all relevant assurance factors to ensure that processes operate as intended from end to end by:
   - Determining all organizational assurance functions
   - Developing formal relationships with other assurance functions
   - Coordinating all assurance functions for more cost-effective security
   - Ensuring that roles and responsibilities between assurance functions overlap and leave no gaps in protection
   - Employing a systems approach to information security planning, deployment, metrics and management

1.2 EFFECTIVE INFORMATION SECURITY GOVERNANCE

Information security governance is the responsibility of the board of directors and senior management. It must be an integral and transparent part of enterprise governance and complement or encompass the IT governance framework. While senior management has the responsibility to consider and respond to increasingly complex and potentially destructive information security issues, boards of directors will be required to make information security an intrinsic part of governance, integrated with the processes they have in place to govern other critical organizational resources. This includes monitoring and reporting processes to ensure that governance processes are effective and compliance enforcement is sufficient to reduce risk to acceptable levels.

According to the developers of the Business Model for Information Security™ (BMIS™) (see section 1.2.5), "It is no longer enough to communicate to the world of stakeholders why we [the organization] exist and what constitutes success, we must also communicate how we are going to protect our existence." This suggests that a clear organizational strategy for preservation is equally important to, and must accompany, a strategy for progress.

In addition to protection of information assets, effective information security governance is required to address legal and regulatory requirements and is becoming mandatory in the exercise of due care.

1.2.1 BUSINESS GOALS AND OBJECTIVES

Corporate governance is the set of responsibilities and practices exercised by the board and senior management with the goals of providing strategic direction, ensuring that objectives are achieved, ascertaining that risk is managed appropriately and verifying that the enterprise's resources are used responsibly. Strategy is the plan to achieve an objective. The strategic direction of the business is defined by the objectives set forth in the strategy. To be of value to the organization, information security must support the business strategy and the activities that take place to achieve the objectives.

Information security governance is a subset of corporate governance. It provides strategic direction for security activities and ensures that objectives are achieved. It ensures that information security risk is appropriately managed and enterprise information resources are used effectively and efficiently.

To achieve effective information security governance, management must establish and ensure maintenance of a framework to guide the development and management of a comprehensive information security program that supports the business objectives.

The governance framework will generally consist of the following:

1. A comprehensive security strategy intrinsically linked with business objectives
2. Governing security policies that clearly express management intent and address each aspect of strategy, controls and regulation
3. A complete set of standards for each policy to ensure that people, procedures, practices and technologies comply with policy requirements and set appropriate security baselines for the enterprise
4. An effective security organizational structure with sufficient authority and adequate resources, void of conflicts of interest
5. Defined workflows and structures that assist in defining responsibilities and accountability for information security governance
6. Institutionalized metrics and monitoring processes to ensure compliance, provide feedback on control effectiveness and provide the basis for appropriate management decisions

This framework provides the basis for developing a cost-effective information security program that supports the organization's business goals. Implementing and managing a security program is covered in chapter 3. The objective of the program is a set of activities that provide assurance that information assets are given a level of protection commensurate with their business value or with the risk their compromise poses to the organization. The relationships among IT, information security, controls, architecture and the other components of a governance framework are represented in **figure 1.1**. While linkage between IT and information security may occur at various higher levels, strategy is the point where information security must integrate IT to achieve its objectives.

Note that the main requirement for IT is an adequate level of performance; for information security, it is managing risk to an acceptable level. Once outcomes, requirements and the objectives that will meet the requirements are established, high-level conceptual security architecture can be considered. The security architecture should simply be an overlay on the enterprise architecture, although ideally these would be created simultaneously to "bake in" the security elements.

When the basic information security/risk management strategy is formulated, a logical architecture showing the various functions and data flows can be developed. A road map can then be laid out with the specific steps needed to attain the goals defined in the strategy. The specifics include the control objectives necessary to achieve acceptable risk levels. Some of these objectives will best be achieved with technologies defined in the physical architecture. The physical architecture comprises the specific devices such as servers, databases, firewalls, switches and routers, along with their interconnections. Controls must be developed that are consistent with the standards and control choices will affect the specifics of both the physical and the operational architectures. Finally, both the strategy and information security procedures are input into the operational architecture.

Figure 1.1 shows the relationship between IT and information security governance and the sequence of the development process. Once the information security strategy has been developed, it must inform the IT strategy to ensure adequate protection for information assets and align subsequent IT development and management processes with information security requirements.

In Practice: Identify an IT-related goal in your organization. Next, consider the IT-related goal in the context of the enterprise's goals. How does the IT-related goal serve to support the enterprise's overall goals? If it does not, what could be done to change the goal so it better aligns with the organization? Use the information provided in this chapter to guide your recommendations.

Figure 1.1—Relationship of Governance Elements

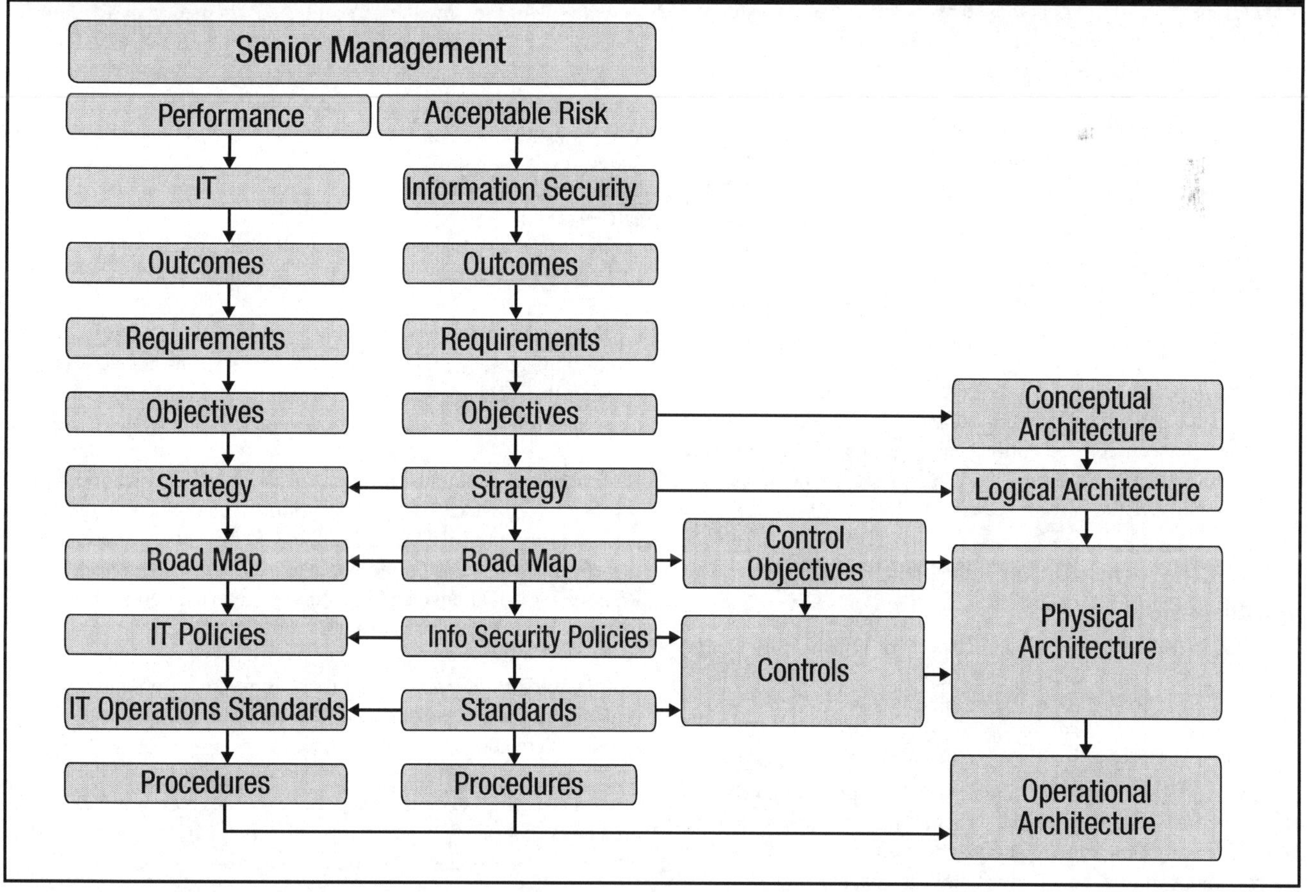

Senior Management
Performance
Acceptable Risk
IT
Information Security
Outcomes
Outcomes
Requirements
Requirements
Objectives
Objectives
Conceptual Architecture
Strategy
Strategy
Logical Architecture
Road Map
Road Map
Control Objectives
IT Policies
Info Security Policies
Physical Architecture
Controls
IT Operations Standards
Standards
Procedures
Procedures
Operational Architecture

1.2.2 DETERMINING RISK CAPACITY AND ACCEPTABLE RISK (RISK APPETITE)

Every organization has a particular risk capacity, defined as the objective amount of loss an enterprise can tolerate without its continued existence being called into question. Subject to the absolute maximum imposed by this risk capacity, the owners or board of directors of an organization set the risk appetite for the organization. Risk appetite is defined as the amount of risk, on a broad basis, that an entity is willing to accept in pursuit of its mission. In some cases, setting the risk appetite may be delegated by the board of directors to senior management as part of strategic planning.

Acceptable risk determination or risk appetite and the criteria by which it can be assessed is an essential element for virtually all aspects of information security as well as most other aspects of organizational activities. It will determine many aspects of strategy including control objectives, control implementation, baseline security, cost-benefit calculations, risk management options, severity criteria determinations, required incident response capabilities, insurance requirements and feasibility assessments, among others.

Risk appetite is translated into a number of standards and policies to contain the risk level within the boundaries set by the risk appetite. These boundaries need to be regularly adjusted or confirmed. Within these boundaries, risk may be accepted, a formal and explicit process that affirms that the risk requires and warrants no additional response by the organization as long as it and the risk environment stay substantially the same and accountability for the risk is assigned to a specific owner.

Risk acceptance generally should not exceed the risk appetite of the organization, but it must not exceed the risk capacity (which would threaten the continued existence of the organization). Risk tolerance levels are deviations from risk appetite, which are not desirable but are known to be sufficiently below the risk capacity that acceptance of risk is still possible when there is compelling business need and other options are too costly. Risk tolerance may be defined using IT process metrics or adherence to defined IT procedures and policies, which are a translation of the IT goals that need to be achieved. Like risk appetite, risk tolerance is defined at the enterprise level and reflected in the policies created by senior management. Exceptions can be tolerated at lower levels of the enterprise as long as the overall exposure does not exceed the risk appetite at the enterprise level.

1.2.3 SCOPE AND CHARTER OF INFORMATION SECURITY GOVERNANCE

Information security deals with all aspects of information, in any medium (e.g., written, spoken, electronic), regardless of whether it is being created, viewed, transported, stored or destroyed. This is contrasted with IT security, which is concerned with security of information within the boundaries of the technology domain, usually in a custodial capacity. It is important to note this distinction. IT usually is not the owner of most of the information in its systems; rather, it owns the machinery that processes it. The information is in IT's care, control and custody, and, therefore, IT functions as a custodian for the data owners.

Many organizations have a corporate security function as well as an IT security operation dealing specifically with security of the IT systems. While this is an essential function that must be governed by the information security policies and standards, confidential information disclosed in an elevator conversation or sent via the postal mail would be outside the scope of IT security but squarely in the purview of information security.

Cybersecurity is an area of concern and importance to information security governance and management. It includes both information and IT security, and although definitions vary widely, a common position holds that cybersecurity is a subdiscipline of information security. Specific areas of concern for cybersecurity include advanced persistent threats (APTs), malware, ransomware, phishing in all its forms, and the host of other threats related to, and enabled by, cyberspace.

In the context of information security governance, it is important that the scope and responsibilities of information security are clearly set forth in the information security strategy and reflected in the policies. It is also essential that information security is fully supported by senior management and the various organizational units. Without clearly defined information security responsibilities, it is impossible to determine accountability.

1.2.4 GOVERNANCE, RISK MANAGEMENT AND COMPLIANCE

Governance, risk management and compliance (GRC) is an example of the growing recognition of the necessity for convergence, or assurance process integration, as discussed in section 1.2.6 Assurance Process Integration—Convergence.

GRC is a term that reflects an approach that organizations can adopt to integrate these three areas. Often stated as a single business activity, GRC includes multiple overlapping and related activities within an organization, which may include internal audit, compliance programs such as the US Sarbanes-Oxley Act, ERM, operational risk, incident management and others.

Governance, as discussed previously, is the responsibility of senior management and the board of directors and focuses on creating the mechanisms an organization uses to ensure that personnel follow established processes and policies.

Risk management is the process by which an organization manages risk to acceptable levels within acceptable tolerances, identifies potential risk and its associated impacts, and prioritizes their mitigation based on the organization's business objectives. Risk management develops and deploys internal controls to manage and mitigate risk throughout the organization.

Compliance is the process that records and monitors the policies, procedures and controls needed to ensure that policies and standards are adequately adhered to.

It is important to recognize that effective integration of GRC processes requires that governance is in place before risk can be effectively managed and compliance enforced.

According to Michael Rasmussen, an industry GRC analyst, the challenge in defining GRC is that, individually, each term has "many different meanings within organizations." Development of GRC was initially a response to the US Sarbanes-Oxley Act, but has evolved as an approach to enterprise risk management (ERM).

While a GRC program can be used in any area of an organization, it is usually focused on financial, IT and legal areas. Financial GRC is used to ensure proper operation of financial processes and compliance with regulatory requirements. In a similar fashion, IT GRC seeks to ensure proper operation and policy compliance of IT processes. Legal GRC may focus on overall regulatory compliance.

1.2.5 BUSINESS MODEL FOR INFORMATION SECURITY

The BMIS model uses systems thinking to clarify complex relationships within the enterprise to more effectively manage security. The elements and dynamic interconnections that form the basis of the model establish the boundaries of an information security program and model how the program functions and reacts to internal and external change. BMIS provides the context for frameworks such as COBIT.

A system needs to be viewed holistically—not merely as a sum of its parts—to be accurately understood. This is the essence of systems theory. A holistic approach examines the system as a complete functioning unit. Another tenet of systems theory is that one part of the system enables understanding of other parts of the system. "Systems thinking" is a widely recognized term that refers to the examination of how systems interact, how complex systems work and why "the whole is more than the sum of its parts."

Systems theory is most accurately described as a complex network of events, relationships, reactions, consequences, technologies, processes and people that interact in often unseen and unexpected ways. Studying the behaviors and results of the interactions can assist the manager to better understand the organizational system and the way it functions. While management of any discipline within the enterprise can be enhanced by approaching it from a systems thinking perspective, its implementation will certainly help with managing risk.

The success that the systems approach has achieved in other fields bodes well for the benefits it can bring to security. The often dramatic failures of enterprises to adequately address security issues in recent years are due, to a significant extent, to their inability to define security and present it in a way that is comprehensible and relevant to all stakeholders. Using a systems approach to information security management will help information security managers address complex and dynamic environments and will generate a beneficial effect on collaboration within the enterprise, adaptation to operational change, navigation of strategic uncertainty and tolerance of the impact of external factors.

As illustrated in **figure 1.2**, BMIS is best viewed as a flexible, three-dimensional, pyramid-shaped structure made up of four elements linked together by six dynamic interconnections. All aspects of the model interact with each other. If any one part of the model is changed, not addressed or managed inappropriately, the equilibrium of the model is potentially at risk. The dynamic interconnections act as tensions, exerting a push/pull force in reaction to changes in the enterprise, allowing the model to adapt as needed.

Figure 1.2—Business Model for Information Security

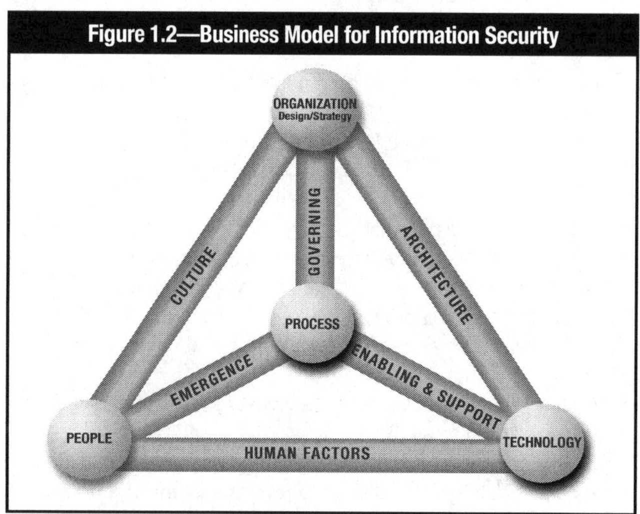

The four elements of the model are:

1. **Organization design and strategy**—An organization is a network of people, assets and processes interacting with each other in defined roles and working toward a common goal.
   - An enterprise's strategy specifies its business goals and the objectives to be achieved as well as the values and missions to be pursued. It sets the basic direction of the enterprise. The strategy should adapt to external and internal factors. Resources are the primary material to design the strategy and can be of different types (e.g., people, equipment, know-how).
   - Design defines how the organization implements its strategy. Processes, culture and architecture are important in determining the design.
2. **People**—The human resources and the security issues that surround them. It defines who implements (through design) each part of the strategy. It represents a human collective and must take into account values, behaviors and biases.
   - Internally, it is critical for the information security manager to work with the human resources and legal departments to address issues such as:
     - Recruitment strategies (access, background checks, interviews, roles and responsibilities)
     - Employment issues (location of office, access to tools and data, training and awareness, movement within the enterprise)
     - Termination (reasons for leaving, timing of exit, roles and responsibilities, access to systems, access to other employees)
   - Externally, customers, suppliers, media, stakeholders and others can have a strong influence on the enterprise and need to be considered within the security posture.

3. **Process**—Includes formal and informal mechanisms (large and small, simple and complex) to get things done and provides a vital link to all of the dynamic interconnections. Processes identify, measure, manage and control risk, availability, integrity and confidentiality, and they also ensure accountability. They derive from the strategy and implement the operational part of the organization element. To be advantageous to the enterprise, processes must:
   - Meet business requirements and align with policy
   - Consider emergence and be adaptable to changing requirements
   - Be well documented and communicated to appropriate human resources
   - Be reviewed periodically, once they are in place, to ensure efficiency and effectiveness
4. **Technology**—Composed of all of the tools, applications and infrastructure that make processes more efficient. As an evolving element that experiences frequent changes, it has its own dynamic risk. Given the typical enterprise's dependence on technology, technology constitutes a core part of the enterprise's infrastructure and a critical component in accomplishing its mission.
   - Technology is often seen by the enterprise's management team as a way to resolve security threats and risk. While technical controls are helpful in mitigating some types of risk, technology should not be viewed as an information security solution.
   - Technology is greatly impacted by users and by organizational culture. Some individuals still mistrust technology; some have not learned to use it; and others feel it slows them down. Regardless of the reason, information security managers must be aware that, just as with physical and administrative controls, some people will try to sidestep technical controls.

Dynamic Interconnections

The dynamic interconnections link the elements together and exert a multidirectional force that pushes and pulls as things change. Actions and behaviors that occur in the dynamic interconnections can force the model out of balance or bring it back to equilibrium. The six dynamic interconnections are:

1. **Governance**—Steering the enterprise and demanding strategic leadership. Governing sets limits within which an enterprise operates and is implemented within processes to monitor performance, describe activities and achieve compliance while also providing adaptability to emergent conditions. Governing incorporates ensuring that objectives are determined and defined, ascertaining that risk is managed appropriately, and verifying that the enterprise's resources are used responsibly.
2. **Culture**—A pattern of behaviors, beliefs, assumptions, attitudes and ways of doing things. It is emergent and learned, and it creates a sense of comfort.
   - It is important to understand the culture of the enterprise because it profoundly influences what information is considered, how it is interpreted and what will be done with it.
   - Culture may exist on many levels, such as national (legislation/regulation, political and traditional), organizational (policies, hierarchical style and expectations) and social (family, etiquette). It is created from both external and internal factors, and is influenced by and influences organizational patterns.
3. **Enablement and support**—Connects the technology element to the process element.
   - One way to help ensure that people comply with technical security measures, policies and procedures is to make processes usable and easy. Transparency can help generate acceptance for security controls by assuring users that security will not inhibit their ability to work effectively.
   - Many of the actions that affect both technology and processes occur in the enablement and support dynamic interconnection. Policies, standards and guidelines must be designed to support the needs of the business by reducing or eliminating conflicts of interest, remaining flexible to support changing business objectives, and being acceptable and easy for people to follow.
4. **Emergence**—Connotes surfacing, developing, growing and evolving, and refers to patterns that arise in the life of the enterprise that appear to have no obvious cause and whose outcomes seem impossible to predict and control. The emergence dynamic interconnection (between people and processes) is a place to introduce possible solutions such as feedback loops; alignment with process improvement; and consideration of emergent issues in system design life cycle, change control and risk management.
5. **Human factors**—Represents the interaction and gap between technology and people and, as such, is critical to an information security program. If people do not understand how to use technology, do not embrace technology or will not follow pertinent policies, serious security problems can evolve. Internal threats such as data leakage, data theft and misuse of data can occur within this dynamic interconnection. Human factors may arise because of experience level, cultural experiences and differing generational perspectives. Because human factors are critical components in maintaining balance within the model, it is important to train all of the enterprise's human resources on pertinent skills.
6. **Architecture**—A comprehensive and formal encapsulation of the people, processes, policies and technology that comprise an enterprise's security practices.
   - A robust business information architecture is essential to understanding the need for security and designing the security architecture.
   - It is within the architecture dynamic interconnection that the enterprise can ensure defense in depth. The design describes how the security controls are positioned and how they relate to the overall enterprise architecture. An enterprise security architecture facilitates security capabilities across lines of businesses in a consistent and cost-effective manner and enables enterprises to be proactive with their security investment decisions.

1.2.6 ASSURANCE PROCESS INTEGRATION—CONVERGENCE

The tendency to segment security into separate but related functions has created the need to integrate the variety of assurance processes found in a typical organization. These activities are often fragmented and segmented in silos with different reporting structures. They tend to use different terminology and generally reflect different understandings of their processes and outcomes with, at times, little in common. These assurance silos can include risk management, change management, internal and external audit, privacy offices, insurance offices, human resources (HR), legal, and others. Evaluating business processes from start to finish (along with their controls), regardless of which particular assurance process is involved, can mitigate the tendency for security gaps to exist among various assurance functions.

GRC, discussed in section 1.2.4, is a recent effort to address integration issues among the major functions of governance, risk management and compliance.

BMIS, covered in section 1.2.5, is an effort by ISACA to develop a fully integrated approach to information security using a systems approach.

ISO/IEC 27001:2013, covered in section 3.5.2, specifies the requirements for establishing, implementing, operating, monitoring, reviewing, maintaining and improving a documented information security management system (ISMS) within the context of the organization's overall business risk.

Convergence

For several decades, security-related activities have been divided into separate divisions, such as physical security, IT security, risk management, privacy, compliance and information security. This has not been conducive to achieving optimal results, which has led professionals and managers to reconsider this separation. It is becoming more common to see CISOs elevated to CSOs or the functions combined to better integrate the main security elements. Physical security has become fairly routine and is more easily integrated into information security than the other way around.

Because it is not possible to effectively deliver information security without a number of physical considerations, the evolution is natural. It is reasonable to expect that integration of many security activities will center around information security in the coming decade. As a result, integration of physical security and other assurance functions will be increasingly relevant for management to consider.

In other words, this is a holistic and encompassing approach that looks beyond assets and considers such factors as culture, organizational structure and processes such as the systems approach discussed in section 1.2.5 Business Model for Information Security. Survey findings clearly show convergence as a business trend with a great deal of momentum. Delivering on convergence is not just about organizational integration; rather, it is about integrating the security disciplines with the business's mission to deliver shareholder value through consistent, predictable operations and optimizing the allocation of security resources.

As new technologies emerge and threats become increasingly complex and unpredictable, senior security executives recognize the need to merge security functions throughout the entire enterprise. An incident at the Sumitomo Mitsui Bank in London, England, in which hackers attempted to steal £220 million from the bank, underlines this principle. Even though the bank had strong IT security measures in place, a physical security lapse occurred. Adversaries posing as janitors installed devices on computer keyboards (i.e., key loggers) that allowed them to obtain valuable login information. This situation highlights and reinforces the need to bring together—in fact, converge—all components of an organization's security through an integrated and deliberate approach. To be effective, this converged approach should reach across people, processes and technology, and enable enterprises to prevent, detect, respond to and recover from any type of security incident.

In addition to the costs that companies face to deal with the immediate effects of an incident, security incidents can cause more costly, long-term harm such as damage to reputation and brand. Beyond the impact to market capitalization, if the issue threatens the public good, regulators may intervene, enacting stricter requirements to govern future business practices.

1.3 ROLES AND RESPONSIBILITIES

Part of effective information security governance is having clearly defined roles and responsibilities. Because everyone in the organization has some information security responsibilities, it is essential these responsibilities are clearly defined and communicated throughout the enterprise on a recurring basis.

A role is a designation assigned to an individual by virtue of a job function or other label. A responsibility is a description of some procedure or function related to the role that someone is accountable to perform. Roles are important to information security because they allow responsibilities and/or access rights to be assigned based on the fact that an individual performs a function rather than having to assign them to individual people. There are typically many job functions in the organization that support security functions and the ability to assign access authorizations based on roles simplifies administration.

RACI (responsible, accountable, consulted, informed) charts can be used effectively in defining the various roles associated with aspects of developing an information security program. Clear designation of roles and responsibilities is necessary to ensure effective implementation. **Figure 1.3** shows a number of typical organizational roles and related governance activities. The specific requirement is shown by the letter: R—responsible, A—accountable, C—consulted, I—informed.

Skills

Skills must necessarily be considered when developing RACI charts. Skills are the training, expertise and experience held by the personnel in a given job function. It is important to understand the proficiencies of available personnel to ensure that they map to competencies required for program implementation. Specific skills needed for program implementation can be acquired through training or utilizing external resources. External resources, such as consultants, are often a more cost-effective choice for skills required for only a short time for specific projects.

Once it has been agreed that certain personnel will have specific information security responsibilities, formal employment agreements should be established that reference those responsibilities, and these must be considered when screening applicants for employment.

Figure 1.3—Sample RACI Chart

EDM01 RACI Chart

Governance Practice	Board	Chief Executive Officer	Chief Financial Officer	Chief Operating Officer	Business Executives	Business Process Owners	Strategy Executive Committee	Steering (Programmes/Projects) Committee	Project Management Office	Value Management Office	Chief Risk Officer	Chief Information Security Officer	Architecture Board	Enterprise Risk Committee	Head Human Resources	Compliance	Audit	Chief Information Officer	Head Architect	Head Development	Head IT Operations	Head IT Administration	Service Manager	Information Security Manager	Business Continuity Manager	Privacy Officer
EDM01.01 Evaluate the governance system.	A	R	C	C	R		R				C		C	C	C	C	C	R	C	C	C					
EDM01.02 Direct the governance system.	A	R	C	C	R	I	R	I	I	I	C	I	I	I	I	C	C	R	C	I	I	I	I	I	I	I
EDM01.03 Monitor the governance system.	A	R	C	C	R	I	R	I	I	I	C	I	I	I	I	C	C	R	C	I	I	I	I	I	I	I

Source: ISACA, *COBIT 5: Enabling Processes*, USA, 2012

Culture

Culture is another issue that warrants consideration when determining responsibilities and accountability as it represents organizational behavior, methods for navigating and influencing the organization's formal and informal structures to get the work done, attitudes, norms, level of teamwork, existence or lack of turf issues, and geographic dispersion. Culture is impacted by the individual backgrounds, work ethics, values, past experiences, individual filters/blind spots and perceptions of life that individuals bring to the workplace. Every organization has a culture, whether it has been purposely designed or simply emerged over time as a reflection of the leadership, and it must be considered in determining roles and responsibilities.

Information security primarily involves logical and analytical activities; however, building relationships, fostering teamwork and influencing organizational attitudes toward a positive security culture rely more on good interpersonal skills. The astute information security program manager will recognize the importance of developing both types of skills as essential to being an effective manager.

Building a security-aware culture depends on individuals in their respective roles performing their jobs in a way that protects information assets. Each person, no matter what level or role within the organization, should be able to articulate how information security relates to their role. For this to happen, the security manager must plan communications, participate in committees and projects, and provide individual attention to the end users' or managers' needs. The security department should be able to answer "What is in it for me?" or "Why should I care?" for every person in the organization. Once these questions have been answered, effective communications can be tailored to these messages.

Some indicators of a successful security culture are the information security department being brought into projects at the appropriate times, end users knowing how to identify and report incidents, the organization's ability to identify the security manager, and people knowing their role in protecting the information assets of the organization and integrating information security into their daily practices.

1.3.1 BOARD OF DIRECTORS

Board members need to be aware of the organization's information assets, risk to those assets and their criticality to ongoing business operations. According to the PricewaterhouseCoopers' *Global State of Information Security Survey 2016*, 45 percent of respondents say the board participates in the overall security strategy of an organization. The board should periodically be provided with the high-level results of comprehensive risk assessments and business impact analysis (BIA). A result of these activities should include board members validating/ratifying the key assets they want protected and that protection levels and priorities are appropriate to a standard of due care.

Security expectations should be met at all levels of the enterprise. Penalties for noncompliance must be defined, communicated and enforced from the board level down.

Beyond these requirements, the board has an ongoing obligation to provide a level of oversight of the activities of information security. Given the legal and ethical responsibility of directors to exercise due care in protecting the organization's key assets, which include its confidential and other critical information, an ongoing level of involvement and oversight of information security are required.

More specifically, there are a number of reasons why it is becoming essential for the board of directors to be involved with, and provide oversight of, information security activities. A common concern is liability. Most organizations, to protect themselves from shareholder lawsuits, provide specific insurance to create a level of protection for the board in exercising its governance responsibilities. However, a typical requirement of this insurance requires directors to exercise a good faith effort at exercising due care in the discharge of their duties. Neglecting to address information security risk may be found to be a failure to exercise due care and may void the protection provided by insurance.

The US Sarbanes-Oxley Act mandates that every company listed on a US stock exchange maintain an audit committee with a required level of experience and demonstrable competence. This committee is often comprised of members of the company's board of directors. One of the committee's key responsibilities is the ongoing monitoring of the organization's internal controls that directly affect the reliability of financial statements. Most financial controls are technical as well as procedural, with the technical portion generally under the scope of the information security manager. It is important for the security manager to maintain an open channel of communication with this committee.

Finally, institutional investors and others have come to understand that the long-term prospects for an organization are heavily impacted by the overall state of governance. A number of corporate rating organizations now provide a governance rating or metric based on a number of factors. Failure to provide an adequate level of governance and support of activities that protect the major assets of the organization is likely to be reflected in those scores. The weight and relevance of the scores will be driven by the impacts and consequences suffered by organizations that suffer the loss of significant critical and sensitive information. The bottom line is that a board of directors that does not provide a level of direction, oversight and the requirements for appropriate metrics will be subject to an increasing degree of liability and regulatory intervention. This is reflected in a key finding in PricewaterhouseCoopers' *Global State of Information Security® Survey 2015*: "Cyber security is now a persistent business risk. It is no longer an issue that concerns only information technology and security professionals, the impact has extended to the C-suites and boardroom."

1.3.2 SENIOR MANAGEMENT

As with any other major initiative, information security must have leadership and ongoing support from senior management to succeed. Benign neglect, indifference or outright hostility is not likely to result in satisfactory outcomes.

An organization's senior management team is responsible for ensuring that needed organizational functions, resources and supporting infrastructure are available and properly utilized to fulfill the information-security-related directives of the board, regulatory compliance and other demands.

In addition, IT is often faced with performance pressures, while security must deal with risk and regulatory issues. These imperatives all too often fall at opposite ends of the spectrum. The result can be tension between IT and security, and it is important that senior management promotes cooperation, arbitrates differences in perspective and is clear about priorities so that a suitable balance among performance, cost and security can be maintained.

1.3.3 BUSINESS PROCESS OWNERS

Developing an effective information security strategy requires integration with, and cooperation of, business process owners. A successful outcome is the alignment of information security activities in support of business objectives. The extent to which this is achieved determines the cost-effectiveness of the information security program in achieving the desired objective of providing a predictable, defined level of assurance for business processes and an acceptable level of impact from adverse events.

1.3.4 STEERING COMMITTEE

Security affects all aspects of an organization to some extent and must be pervasive throughout the enterprise to be effective. To ensure that all stakeholders impacted by security considerations are involved, many organizations use a steering committee comprised of senior representatives of affected groups. This composition helps to achieve consensus on priorities and trade-offs. It also serves as an effective communications channel and provides an ongoing basis for ensuring the alignment of the security program with business objectives. It can also be instrumental in achieving modification of behavior toward a culture more conducive to good security.

Common topics, agendas and decisions for a security steering committee include:

- Security strategy and integration efforts, especially efforts to integrate security with business unit activities
- Specific actions and progress relative to business unit support of information security program functions and vice versa
- Emerging risk, business unit security practices and compliance issues

The information security manager should make sure that the roles, responsibilities, scope and activities of the information security steering committee are clearly defined. This should include clear objectives and topics to prevent poor productivity or distractions from the priorities of the committee.

It is important that materials be distributed, reviews encouraged and solution discussions held in advance of full committee meetings so that the meeting can be focused on resolving issues and making decisions. The use of subcommittees and/or individual action assignments is appropriate to this type of management strategy.

1.3.5 CHIEF INFORMATION SECURITY OFFICER

All organizations have a chief information security officer (CISO), whether anyone officially holds that title. It may be the chief information officer (CIO), chief security officer (CSO), chief financial officer (CFO) or the chief executive officer (CEO).

The scope and breadth of information security is such that the authority required to make decisions and the responsibility to take action will inevitably be held by a C-level officer or senior manager. Legal responsibility will, by default, extend up the command structure and ultimately reside with senior management and the board of directors. Failure to recognize this and implement appropriate governance structures can result in senior management being unaware of this responsibility and the attendant liability. It also usually results in a lack of effective alignment of security activities with organizational objectives.

PricewaterhouseCoopers' *Global State of Information Security® Survey 2016* notes that 54 percent of organizations surveyed have a CISO in charge of the security program. Most frequently, this role reported to the CEO, followed by the CIO, board and chief technology officer (CTO). This finding highlights that the role of the CISO, which is fundamentally a regulatory role, is different from that of a typical CIO, which is generally focused on performance.

While the elevation of the information security manager to an executive role is a global trend, it is not consistent across sectors and the responsibilities and authority of information security managers vary dramatically between organizations. These responsibilities currently range from the CISO or vice president for security reporting to the CEO or board of directors, to system administrators who have part-time responsibility for security management who might report to the IT manager or CIO. A main concern, according to PricewaterhouseCoopers, is a CISO having an inadequate budget to successfully execute the security program and provide adequate attention to end users' or managers' needs.

1.4 RISK MANAGEMENT ROLES AND RESPONSIBILITIES

Managing risk effectively at an acceptable cost is the basis for the objectives and drivers for information security governance and the program to implement, develop and manage it. While risk management is ultimately a responsibility of senior management and the board of directors, everyone in an organization has a role to play in managing risk. In addition, it is important that all risk management activities are integrated and operating under a consistent set of rules and clear objectives. This will help prevent gaps in protection, provide a more consistent level of protection, reduce needless redundant efforts and help prevent different parts of the organization from operating at cross purposes.

1.4.1 KEY ROLES

The US National Institute of Standards and Technology (NIST) *Special Publication 800-30 Revision 1: Guide for Conducting Risk Assessments* describes the key roles of personnel who must support and participate in the risk management process. While the specifics in different organizations and different countries may vary, this high-level view generally maps to most organizations. In many organizations, the information security manager is the CISO with executive-level status, reporting directly to senior management.

- **Governing boards and senior management**—Senior management, under the standard of due care and ultimate responsibility for mission accomplishment, must ensure that the necessary resources are effectively applied to develop the capabilities needed to accomplish the mission. They must also assess and incorporate results of the risk assessment activity into the decision-making process. An effective risk management program that assesses and mitigates IT-related mission risk requires the support and involvement of senior management.
- **Chief risk officer**—The chief risk officer (CRO) is generally charged with overall ERM, which may include information security in some cases. Generally, this position includes all noninformation risk such as operational risk, environmental risk and credit risk.
- **Chief information officer**—The CIO is responsible for IT planning, budgeting and performance, often including its information security components consistent with the policies and standards under the purview of the CISO or information security manager.
- **Chief information security officer**—The CISO generally performs most of the same functions as an information security manager, but holds officer status in an organization usually accompanied with greater authority, and now typically reports to the CEO, chief operating officer (COO) or the board of directors. The position usually includes more strategic and management elements than is typical of an information security manager.
- **Information security manager**—Information security managers are responsible for their organizations' security programs, usually including information risk management. Therefore, they play a leading role in introducing an appropriate, structured methodology to help identify, evaluate and minimize risk to information resources, including the IT systems that support their organizations' missions. Information security managers also generally act as major consultants in support of senior management to ensure that these activities take place on an ongoing basis.
- **System and information owners**—The system and information owners are responsible for ensuring that proper controls are in place to address confidentiality, integrity and availability of the IT systems and data they own. Typically, the system and information owners are responsible for changes to their IT systems and are responsible for ensuring policy compliance and enforcement. Thus, they usually have to approve and sign off on changes to their IT systems and are involved in the policies and standards that govern them (e.g., system enhancement, major changes to the software and hardware, and compliance requirements). The system and information owners must, therefore, understand their role in the risk management process, the risk management objectives and the requirements to provide support for this process.
- **Business and functional managers**—The managers responsible for business operations and the IT procurement process must take an active role in the risk management process. These managers have the authority and responsibility for making the trade-off decisions essential to mission accomplishment. Their involvement in the risk management process enables the achievement of proper security for the IT systems, which, if managed properly, will provide mission effectiveness with acceptable expenditure of resources.
- **IT security practitioners**—IT security practitioners (e.g., network, system, application and database administrators;

computer specialists; security analysts; security consultants) are responsible for proper implementation of security requirements in their IT systems. As changes occur in the existing IT system environment (e.g., expansion in network connectivity, changes to the existing infrastructure and organizational policies, standards or procedures, introduction of new technologies), the IT security practitioners must support or use the risk management process to identify and assess new potential risk and ensure implementation of new security controls as needed to safeguard their IT systems.

- **Security awareness trainers (security/subject matter professionals)**—The organization's personnel are the users of the IT systems. Use of the IT systems and data according to an organization's policies, guidelines and rules of behavior is critical to mitigating risk and protecting the organization's IT resources. To manage risk to acceptable levels, it is essential that system, data and application users and custodians be provided with security awareness training and the processes and requirements for secure operations. Therefore, the IT security trainers or security/subject matter professionals must understand the organization's risk management objectives and processes so they can develop appropriate training materials and incorporate risk management requirements into training programs for end users.

Knowledge Check: Roles and Responsibilities

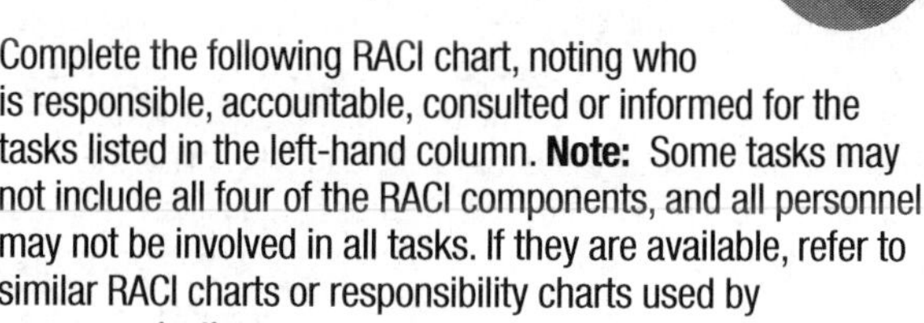
Complete the following RACI chart, noting who is responsible, accountable, consulted or informed for the tasks listed in the left-hand column. **Note:** Some tasks may not include all four of the RACI components, and all personnel may not be involved in all tasks. If they are available, refer to similar RACI charts or responsibility charts used by your organization.

	Information Security Manager	Board of Directors	Chief Information Security Officer	Chief Executive Officer	Business Process Owner
Define the target IT capabilities.					
Conduct a gap analysis.					
Define the strategic plan and road map.					
Communicate the IT strategy and direction.					

Answers on page 71.

1.4.2 INFORMATION SECURITY ROLES AND RESPONSIBILITIES

As noted previously, information security spans the enterprise and requires ongoing support from senior management to succeed. Without this support and guidance from senior management, it is difficult for the information security manager to know what goals to steer the program toward, determine optimal governance processes or develop meaningful program metrics.

Obtaining Senior Management Commitment

After an information security manager has developed an appropriate security strategy with input from business and data owners, senior management approves the strategy to ensure a required level of support and resources. Because this is typically a complicated subject, the information security manager may need to first educate senior managers on the high-level aspects of information security and submit the overall strategy for review. A presentation to senior management describing the various aspects of the security strategy usually will take place to support and explain the documentation that the information security manager submitted. Unfortunately, in many organizations, the true value of securing information systems does not become apparent until they fail.

In some organizations, there may be an inadequate level of management commitment, and information security managers may be restricted in their effectiveness. Under these circumstances, it may be useful to make an effort to educate senior management in the areas of regulatory compliance and the organization's dependence on its information assets. It may also be useful to document risk and potential impacts faced by the organization, making sure that senior management is informed of the potential for compromise and resulting consequences and finds them acceptable.

Senior management is in a better position to support security initiatives if its members are educated on how critical IT systems and information are to the continued operation of the enterprise as well as other aspects of information security. In addition, it will be helpful and will minimize confusion if senior management is provided an overview of pertinent regulations, compliance requirements and possible sanctions if the organization is out of compliance.

To address management's possible questions on what actions would constitute an appropriate level of support, the following recommendations are suggested:

- Clear approval and support for formal security strategies and policies
- Monitoring and measuring organizational performance in implementing security policies
- Supporting security awareness and training for all staff throughout the organization
- Adequate resources and sufficient authority to implement and maintain security activities
- Treating information security as a critical business issue and creating a security-positive environment
- Demonstrating to third parties that the organization deals with information security in a professional manner
- Providing high-level oversight and control
- Periodically reviewing information security effectiveness
- Setting an example by adhering to the organization's security policies and practices
- Addressing information security issues at board/senior management meetings

In many cases, insufficient management support of information security is not an issue of apathy, but a lack of understanding. Information security is rarely a part of general management

expertise or education. As such, executive managers may not fully appreciate what is expected of them, the structure of an information security management program, or how such a program should be integrated and operated. It is often productive to coordinate workshops to assist high-level managers in gaining a better understanding of these issues and establish expectations of required support and resources.

In other cases, support for information security programs may be limited for financial or other reasons. The information security manager must recognize these constraints, prioritizing and maximizing the effects of available resources in addition to working with management to develop additional resources.

Developing and Presenting the Business Case

An important consideration in any information security or IT project is the business case. The achievement of business benefits should drive projects. A business case provides the information required for an organization to decide whether a project should proceed. The essential consideration is the value proposition, or the cost-benefit analysis of moving forward with the project. Depending on the organization and often on the size of the investment, the development of a business case is either the first step in a project or a precursor to the commencement of a project.

The initial business case would normally derive from a feasibility study undertaken as part of project initiation/planning. This is an early study of a problem to assess if a solution is practical and meets requirements within established budgets and schedule requirements. The feasibility study will normally include the following six elements:

1. The **project scope** defines the business problem and/or opportunity to be addressed. It should be clear, concise and to the point.
2. The **current analysis** defines and establishes an understanding of a system, a software product, an information security control, etc. Based on this analysis, it may be determined that the current system or software product is working correctly, some minor modifications are needed, or a complete upgrade or replacement is required. At this point in the process, the strengths and weaknesses of the current system or software product are identified.
3. **Requirements** are defined based on stakeholder needs and constraints. Defining requirements for software differs from defining requirements for systems. The following are examples of needs and constraints used to define requirements:
   - Business, contractual and regulatory processes
   - End-user functional needs
   - Technical and physical attributes defining operational and engineering parameters
4. The **approach** is the recommended system and/or software solution to satisfy the requirements. This step clearly identifies the alternatives that were considered and the rationale as to why the preferred solution was selected. This is the process wherein the use of existing structures and commercial alternatives are considered (e.g., "build versus buy" decisions).
5. **Evaluation** is based on the previously completed elements within the feasibility study. The final report addresses the cost-effectiveness of the approach selected or the value proposition. Elements of the final report include:
   - The estimated total cost of the project if the preferred solution is selected, along with the alternates to provide a cost comparison, including:
     - Estimate of employee hours required to complete
     - Material and facility costs
     - Vendors and third-party contractors costs
     - Project schedule start and end dates
     - A cost and evaluation summary encompassing cost-benefit analysis, return on investment (ROI), etc.
6. A formal **review** of the feasibility study report is conducted with all stakeholders. This review both validates the completeness and accuracy of the feasibility study and renders a decision to approve or reject the project or ask for corrections before making a final decision. If the feasibility study is approved, all key stakeholders sign the document. Rationale for rejection of the feasibility study should be explained and attached to the document as part of a lessons learned list for use in future project studies.

Part of the work in developing options is to calculate and outline a business case for each solution to allow a comparison of costs and business benefits.

The business case should have sufficient detail to describe the justification for setting up and continuing a project and provide the reasons for the project by answering the question, "Why should this project be undertaken?"

The business case should also be a key element of the decision process throughout the life cycle of any project. If at any stage the business case is thought to be no longer be valid, through increased costs or reduction in the anticipated benefits, the project sponsor or IT steering committee should consider whether the project should proceed. In a well-planned project, decision points, sometimes called "stage gates" or "kill points," are points at which the business case is formally reviewed to ensure it is still valid. If the business case changes during the course of an IT project, the project should be reapproved through the departmental planning and approval process.

A formal presentation is the most widely used technique the information security manager can use to secure senior management commitment and support of information security management policies, standards and strategy.

The formal presentation to senior management is used as a means to educate and communicate key aspects of the overall security program. Acceptance is facilitated by the information security manager applying common business case aspects throughout the acceptance process. These can include:

- Aligning security objectives with business objectives, enabling senior management to understand and apply the security policies and procedures
- Identifying potential consequences of failing to achieve certain security-related objectives and regulatory compliance
- Identifying budget items so that senior management can quantify the costs of the security program
- Using commonly accepted project risk/benefit or financial models, such as total cost of ownership (TCO) or ROI, to quantify the benefits and costs of the security program

- Defining the monitoring and auditing measures that will be included in the security program

It also should be noted that, while senior management may support the security program, it is also imperative that all employees understand and abide by the security policies, standards and procedures. Without employee acceptance, it is unlikely that the security program will be successful and meet its objectives. An important aspect of gaining this acceptance is management clearly complying with the organization's security policies. For example, if a physical access control system is implemented in the organization, senior management should comply with the same access rules and restrictions that are required of other employees. If the policies require the display of badges at all times, managers should set the example and comply with this requirement.

Establishing Reporting and Communication Channels

After obtaining commitment from senior management, effective and reliable reporting and communication channels must be established throughout the organization to ensure the effectiveness and efficiency of the entire information security program. Consistent, reliable reporting from various parts of the organization is an essential monitoring tool necessary to track the status of the security program and issues like compliance and emerging risk. This can also, along with other sources of information such as metrics, serve as a part of an early warning system for potential or emerging security issues. Communication channels are needed for the dissemination of security-related material such as changes in policies, standards or procedures or new or emerging threats or vulnerabilities. Some channels may be formal and a requirement of policies, contracts or regulations. Others may be informal and primarily for keeping a finger on the security-related pulse of the organization.

Periodic formal reporting to the board of directors/senior management is important to make senior management aware of the state of the information security program and governance issues. A well-organized face-to-face presentation to senior management can be conducted periodically and include business process owners as key users of the system. The presentation should be well mapped with prior presentations used to obtain support and security program commitment and can contain:

- Status of the implementation of the system based on the approved strategy
- Overall BIA result comparison (prior to and after implementation)
- Statistics of detected and prevented threats as a means of demonstrating value
- Identifying the weakest security links in the organization and potential consequences of compromise
- Performance measurement data analysis supported with independent, external assessment or audit reports, if available
- Addressing ongoing alignment for critical business objectives, operation processes or corporate environments
- Requiring the approval for renewed plans, as well as related budget items

In addition to formal presentations, routine communication channels also are crucial to the success of the information security program. There are four groups requiring different communication listed below, followed by suggestions of actions the information security manager should undertake to be effective:

- Senior management
  - Attend business strategy meetings to become more aware and understand the updated business strategies and objectives.
  - Hold periodic one-to-one meetings with senior management to understand the business objectives from its perspective.
- Business process owners
  - Join operation review meetings to realize the challenges and requirements of daily operations and their dependencies.
  - Initiate monthly one-to-one meetings with different process owners to gain continued support in the implementation of information security governance and address current individual security-related issues.
- Other management
  - Inform line managers, supervisors and department heads charged with various security and risk management-related functions, including ensuring adequate security requirement awareness and policy compliance, of their responsibilities.
- Employees
  - Offer timely training and education programs.
  - Initiate a centralized on-board training program for new hires.
  - Distribute organizational education material on updated strategies and policies.
  - Instruct personnel to access the intranet or email-based notifications for periodic reminders or *ad hoc* adaptations.
  - Support senior management and business process owners by assigning an information security governance coordinator within each functional unit to obtain accurate feedback of daily practices in a timely manner.

The relationship between the outcomes of effective security governance and management responsibilities is shown in **figure 1.4.** These are not meant to be comprehensive, but merely indicate some levels of management and the primary tasks for which they are responsible.

1.5 GOVERNANCE OF THIRD-PARTY RELATIONSHIPS

An important aspect of information security governance is the rules and processes employed when dealing with third-party relationships. These parties include:

- Service providers
- Outsourced operations
- Trading partners
- Merged or acquired organizations

The ability to effectively manage security in these relationships poses a significant challenge for the information security manager. This is particularly true in the case of organizations being merged. These challenges include cultural differences that may result in approaches to security and behavior that are not acceptable to the security manager's organization. There may be technology incompatibilities between the organizations, process differences that do not integrate well or inadequate levels of baseline security. Other areas of concern can involve incident response, business continuity and disaster recovery capabilities.

Figure 1.4—Relationship of Information Security Governance Outcomes to Management Responsibilities

Management Level	Strategic Alignment	Risk Management	Value Delivery	Performance Measurement	Resource Management	Process Assurance
Board of directors	Require demonstrable alignment of security and business objectives.	• Establish acceptable risk and tolerance. • Oversee a policy on risk management. • Ensure adequate regulatory compliance.	Require reporting of security activity costs and benefits.	Require reporting of security effectiveness.	Oversee a policy of knowledge management and resource utilization.	Oversee a policy of assurance process integration.
Executive management	Institute processes to integrate security with business objectives.	• Ensure that roles and responsibilities include risk management in all activities. • Monitor regulatory compliance.	Require business case development for security initiatives.	Require monitoring and metrics for security activities.	Ensure processes for knowledge capture and efficiency metrics.	Provide oversight of all assurance functions and plans for integration.
Steering committee	Review and provide input to security strategy and requirements for effective business support.	Identify emerging risk, promote business unit security practices and identify compliance issues.	Review and advise on cost effectiveness of security activities needed to serve business functions.	Review and advise whether security initiatives meet business objectives.	Review processes for knowledge capture and dissemination and utilization of resources.	• Identify critical business processes and assurance providers. • Direct assurance integration efforts.
CISO/information security management	Develop the security strategy in alignment with business objectives, oversee the security program and liaise with business process owners for ongoing alignment.	• Ensure that risk and business impact assessments are conducted. • Develop risk mitigation strategies. • Enforce policy and regulatory compliance.	Monitor and optimize utilization, efficiency and effectiveness of security resources.	Develop, implement and report monitoring and metrics needed to support decisions at the strategic, management and operational levels.	• Develop methods for knowledge capture and dissemination. • Monitor and measure resource utilization and cost effectiveness.	• Liaise with other assurance providers. • Ensure that gaps and overlaps are identified and addressed. • Promote integration of assurance activities.
Audit executives	Evaluate and report on degree of alignment.	Evaluate and report on corporate risk management practices and results.	Evaluate and report on cost effectiveness of security program.	Evaluate and report on comprehensiveness and effectiveness of program monitoring and metrics.	Evaluate and report efficiency and utilization of resources.	Evaluate and report on integration and effectiveness of assurance processes.

Source: ISACA, *Information Security Governance: Guidance for Information Security Managers*, 2008. All rights reserved. Used by permission.

To ensure that the organization is adequately protected, the information security manager must assess the impacts of any of the reasonably possible security failures of any third party that may become involved with the organization. It is important to have an understanding of and a plan to manage any such potential failures sufficiently so the potential impacts are within a range acceptable to management.

The responsibilities of the information security manager to address the potential risk and possible impacts of third-party relationships should be clear and documented. Policies, standards and procedures establishing the involvement of information security should be developed prior to the creation of any third-party relationship so risk can be determined and management can decide whether it is acceptable or must be mitigated. Finally, there should be a formalized engagement model between the information security organization and those groups that establish and manage third-party relationships for the organization.

1.6 INFORMATION SECURITY GOVERNANCE METRICS

A **metric** is defined as a quantifiable entity that allows the measurement of the achievement of a process goal. Security is the protection from or absence of danger. Therefore, security metrics should tell us about the state or degree of safety relative to a reference point. It may be useful to clarify the distinction between managing the technical IT security infrastructure at the operational level and the overall management of an information security program.

Technical metrics are useful for the tactical operational management of technical security systems (i.e., intrusion detection systems [IDSs], proxy servers, firewalls, etc.). They can indicate that the infrastructure is operating soundly and technical vulnerabilities are identified and addressed. However, these metrics are of little value from a strategic management or governance standpoint. Technical metrics say nothing about strategic alignment with organizational objectives or how well risk is being managed. They provide few measures of policy compliance or whether objectives for acceptable levels of potential impact are being reached, and they provide no information on whether the information security program is headed in the right direction and achieving the desired outcomes.

From a management perspective, technical metrics cannot provide answers to questions such as:
- How secure is the organization?
- How much security is enough?
- How do we know when we have achieved an adequate level of security?
- What are the most cost-effective security solutions?
- How do we determine the degree of risk?
- How well can risk be predicted?
- Is the security program achieving its objectives?
- What impact is lack of security having on productivity?
- What impact would a catastrophic security breach have?
- What impact will security solutions have on productivity?

Attempts to provide meaningful answers to these questions can ultimately be addressed only by developing relevant measures—metrics that specifically address the requirements of management to make appropriate decisions about the organization's safety.

Full audits and comprehensive risk assessments are typically the only activities organizations undertake that provide this breadth of perspective. While important and necessary from a security management point of view, these provide only history or a snapshot—not what is ideally needed to guide day-to-day security management and provide the information needed to make prudent decisions.

1.6.1 EFFECTIVE SECURITY METRICS

It is difficult or impossible to manage any activity that cannot be measured. The fundamental purpose of metrics, measures and monitoring is decision support. For metrics to be useful, the information they provide must be relevant to the roles and responsibilities of the recipient so that informed decisions can be made. Anything that results in a change can be measured. The key to effective metrics is to use a set of criteria to determine which of the nearly infinite number of metrics candidates are the most suitable. Good metrics are SMART:
- **Specific**—Based on a clearly understood goal; clear and concise
- **Measureable**—Able to be measured; quantifiable (objective), not subjective
- **Attainable**—Realistic; based on important goals and values
- **Relevant**—Directly related to a specific activity or goal
- **Timely**—Grounded in a specific time frame

Additional considerations include (Brotby, Krag; *Information Security Management Metrics*, Auerbach, USA, 2009):
- **Accurate**—A reasonable degree of accuracy is generally adequate.
- **Cost-effective**—The measurements cannot be too expensive to acquire or maintain.
- **Repeatable**—The measure must be able to be acquired reliably over time.
- **Predictive**—Measurements should be indicative of outcomes.
- **Actionable**—It should be clear to the recipient what action must be taken.

Standard security metrics include things such as downtime due to viruses or Trojans, number of penetrations of systems, impacts and losses, recovery times, number of vulnerabilities uncovered with network scans, and percentage of servers patched. While these measures can be indicative of aspects of security and perhaps are useful to IT operations, none provides any actual information about how secure the organization is and probably will not meet most of the aforementioned criteria.

There is an increasing understanding that the lack of useful metrics is an area of information security that has hindered effective management. This has led to a variety of approaches to address this issue being developed during the past few years. Some of the major efforts include:
- ***ISO/IEC 27004:2009 Information technology—Security techniques—Information security management—Measurement***—This standard is a part of the growing ISO/IEC 27XXX series of security standards that includes *ISO/IEC 27001:2013 Information security management* and *ISO/IEC 27002:2013 Information technology—Security techniques—Code of practice for information security controls*. ISO/IEC 27004 "provides guidance on the development and use of measures and measurement in order to assess the effectiveness of an implemented information security management system and controls or groups of controls as specified in ISO/IEC 27001." This standard has not seen wide acceptance and suffers a number of deficiencies. As a result, it is currently being rewritten to address many of the issues raised by the information security community.
- **COBIT 5**—COBIT 5 offers several sample IT metrics for each of 17 suggested enterprise goals, totaling approximately 150 metrics, based on the balanced scorecard approach. *COBIT 5: Enabling Processes* states, "These metrics are samples and every enterprise should carefully review the list, decide on relevant and achievable metrics of its own environment, and design its own scorecard system." However, there currently is a lack of guidance on actual development and implementation in terms of what might be relevant and achievable metrics or the development of the scorecard.
- **The Center for Internet Security (CIS)**—In 2010, CIS released a document titled *The CIS Security Metrics* that offers a comprehensive approach to developing and implementing information security metrics. It is based on the consensus of 150 industry professionals regarding relevant metrics and the approach to implementation and includes 28 metrics definitions with a number of possible metrics for seven major business functions.
- ***NIST Special Publication 800-55 Revision 1: Performance Measurement Guide for Information Security***—This document is aligned with the security controls provided in NIST SP 800-53 to assist with the assessment of information security program implementation. It offers a comprehensive approach to selection, development and implementation of a metrics approach primarily aimed at supporting the Federal Information Security Modernization Act (FISMA), which is a high-level policy document for US federal agencies and has value for the commercial environment.

Operational risk and security are not readily measured in any absolute sense; rather, probabilities, exposures, attributes, effects and consequences are normally the gauge. Various approaches that may be useful include value at risk (VAR), return on security investment (ROSI) and annual loss expectancy (ALE). VAR is used to compute maximum probable loss in a defined period (day, week, year) with a confidence level of typically 95 percent or 99 percent. ROSI is used to calculate the return on investment based on the reduction in losses resulting from a security control compared to the cost of the control. ALE provides the likely annualized loss based on probable frequency and magnitude of security compromise. These often speculative numbers can then be used as a basis for allocating or justifying resources for security activities.

Some organizations attempt to determine the maximum impacts of potential adverse events as a measure of security. Measuring security by consequences and impacts is similar to gauging how tall a tree is by how loud a noise it makes when it falls. In other words, adverse events would have to occur to determine if security is working. An absence of adverse events provides no information on the state of security. It may mean that defenses worked, that no one attacked or that a vulnerability was not discovered. Simulated attacks with penetration testing can provide some measure of the effectiveness of defenses against those specific attacks performed. However, unless a statistically relevant percentage of all possible attacks is attempted, no prediction can be made about the state of security and the organization's ability to resist attack.

It may be that all that can be stated with certainty about security is that:

1. Some organizations are attacked more frequently and/or suffer greater losses than others.
2. There is a strong correlation between good security management and practices and relatively fewer incidents and losses.

Good management and good governance are inextricably linked. Measuring effective information security governance and management with any precision may be more difficult than measuring security. Metrics will, in most respects, be based on attributes, costs and subsequent outcomes of the security program. **Figure 1.5** presents components of an information security governance program and demonstrates that the actual course of governance, and the ability to measure and report performance, is required.

Figure 1.5—Components of Security Metrics

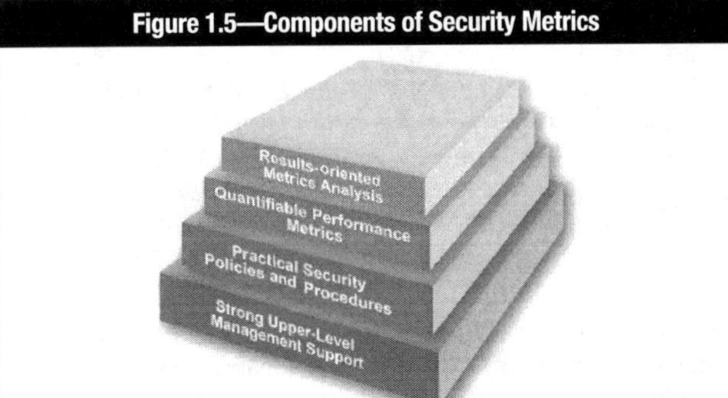

Source: National Institute of Standards and Technology (NIST); *NIST Special Publication 800-55 Revision 1: Performance Measurement Guide for Information Security*, USA, 2008. Reprinted courtesy of the NIST, US Department of Commerce. Not copyrightable in the United States of America.

Because security governance is difficult to measure by a set of objective metrics, there is a tendency to use metrics that are available, regardless of demonstrated relevancy. A typical example is the use of vulnerability scans as an indication of overall security. Arguably, if it were possible to eliminate all or most vulnerabilities (which is not possible), most risk could be avoided. The fallacy is the assumption that something can be determined about risk, threat or impact by measuring just technical vulnerabilities without considering viable threats, exposure and possible impacts.

While there is no universal objective scale for security or security governance, effective metrics can be designed to guide program development and management if clear information security objectives have been developed. Essentially, metrics can be reduced to any measure of the results of the information security program progressing toward the defined objectives, which serve as the reference point. It must also be understood that different metrics are required to provide information at the strategic, tactical and operational levels. Strategic metrics will be oriented toward high-level outcomes and business objectives for the information security program.

1.6.2 GOVERNANCE IMPLEMENTATION METRICS

Implementing an information security governance strategy and framework can require a significant effort. It is important that relevant metrics are in place during the implementation of a governance program. Performance of the overall security program will be too far downstream to provide timely information on implementation and another approach must be used. Key goal indicators (KGIs) and key performance indicators (KPIs) can be useful to provide information about the achievement of process or service goals, and can determine whether organizational milestones and objectives are being met. Because implementation of various aspects of governance will typically involve projects or initiatives, standard project measurement approaches can serve metrics requirements (e.g., achieving specific milestones or objectives, budget and time line conformance).

1.6.3 STRATEGIC ALIGNMENT METRICS

Strategic alignment of information security in support of organizational objectives is essential to the ultimate success of the information security program in providing value to the organization. It should be clear that the cost-effectiveness of the security program is inevitably tied to how well it supports the objectives of the organization and at what cost. Without organizational objectives as a reference point, any other gauge, including good practices, may be overkill, inadequate or misdirected. From a business perspective, adequate and sufficient practices proportionate to the requirements are likely to be more cost-effective than best practices. They are also likely to be received better by cost-conscious management.

The best overall indicator that security activities are in alignment with business (or organizational) objectives is the development of a security strategy that defines security objectives in business terms and ensures that the objectives are directly articulated from planning to implementation of policies, standards, procedures, processes and technology. The litmus test is the reverse order evaluation of a specific control being able to be tracked to a specific business requirement. Any control that cannot be tracked directly back to a specific business requirement is suspect and should be analyzed for relevancy and possible elimination.

Indicators of alignment include:
- The extent to which the security program demonstrably enables specific business activities
- Business activities that have not been undertaken or have been delayed because of inadequate capability to manage risk
- A security organization that is responsive to defined business requirements based on business owner input
- Organizational and security objectives that are defined and clearly understood by all involved in security and related assurance activities measured by awareness testing
- The percentage of security program activities mapped to organizational objectives and validated by senior management
- A security steering committee consisting of key executives with a charter to ensure ongoing alignment of security activities and business strategy

1.6.4 RISK MANAGEMENT METRICS

Risk management is the ultimate objective of all information security activities and organizational assurance efforts. While risk management effectiveness is not subject to direct measurement, some indicators correlate to a successful approach. A successful risk management program can be defined as one that efficiently, effectively and consistently meets expectations and attains defined objectives in maintaining risk at levels acceptable to management.

Once again, it is a requirement that expectations and objectives of risk management be defined. Otherwise, there is no basis for determining whether the program is succeeding and/or heading in the right direction and whether resource allocations are appropriate.

Indicators of appropriate risk management can include:
- A defined organizational risk appetite and tolerance in terms relevant to the organization
- The completeness of an overall security strategy and program for achieving acceptable levels of risk
- The number of defined mitigation objectives for identified significant risk
- Processes for management or reduction of adverse impacts
- Coverage of all business-critical systems by a systematic, continuous risk management processes
- Trends of periodic risk assessment indicating progress toward defined goals
- Trends in impacts
- Results from tested incident response and business continuity/disaster recovery plans
- The completeness of the asset inventory, valuation and assignment of ownership
- The percentage of BIAs of all critical or sensitive systems
- The extent of a complete and functioning asset classification process
- The ratio of security incidents from known risk compared to those from unidentified risk

The key goal of information security is to reduce adverse impacts on the organization to an acceptable level. Therefore, a key metric is the adverse impacts of information security incidents experienced by the organization and the extent to which they do or do not exceed criteria for acceptable risk and impact. An effective security program will show a trend in reduced incident frequency and impact. Quantitative measures include trend analyses of impacts over time in financial terms.

1.6.5 VALUE DELIVERY METRICS

Value delivery occurs when security investments are optimized in support of organizational objectives. Value delivery is a function of strategic alignment of security strategy and business objectives—in other words, when a business case can be convincingly made for all security activities. Optimal investment levels occur when strategic goals for security are achieved and an acceptable risk posture is attained at the lowest cost.

Key indicators (KGIs and KPIs) include:
- Security activities that are designed to achieve specific strategic objectives in a cost-effective manner
- The cost of security being proportional to the value of assets
- Security resources that are allocated by degree of assessed risk and potential impact
- Protection costs that are aggregated as a function of revenues or asset valuation
- Controls that are well designed based on defined control objectives and that achieve those control objectives and are fully utilized
- An adequate and appropriate number of controls to achieve acceptable risk and impact levels
- Control cost-effectiveness that is determined by periodic testing
- Policies in place that require all controls to be periodically reevaluated for cost, compliance and effectiveness
- The use of controls. Controls that are rarely used are not likely to be cost-effective.
- The number of controls to achieve acceptable risk and impact levels. Fewer effective controls can be expected to be more cost-effective than a greater number of less effective controls.
- The effectiveness of controls as determined by testing. Marginal controls are not likely to be cost-effective.

1.6.6 RESOURCE MANAGEMENT METRICS

Information security resource management describes the processes to plan, allocate and control information security resources, including people, processes and technologies, for improving the efficiency and effectiveness of business solutions.

As with other organizational assets and resources, these information security resources must be managed properly. Knowledge must be captured, disseminated and available when needed. Providing multiple solutions to the same problem is, obviously, inefficient and indicates a lack of resource management. Controls and processes must be standardized, when possible, to reduce administrative and training costs. Problems and solutions must be well documented, referenced and available.

Indicators of effective resource management include:
- Infrequent problem solution rediscovery
- Effective knowledge capture and dissemination
- The extent to which security-related processes are standardized
- Clearly defined roles and responsibilities for information security functions

- Information security incorporated into every project plan
- The percentage of information assets and related threats adequately addressed by security activities
- The proper organizational location, level of authority and number of personnel for the information security function
- Resource utilization levels
- Staff productivity
- Per-seat cost of security services

1.6.7 PERFORMANCE MEASUREMENT

Measuring, monitoring and reporting on information security processes is required to ensure that organizational objectives are achieved. Methods to monitor security-related events across the organization must be developed; it is critical to design metrics that provide an indication of the performance of the security machinery and, from a management perspective, information needed to make decisions to guide the security activities of the organization.

Indicators of effective performance measurement include:
- The time it takes to detect and report security-related incidents
- The number and frequency of subsequently discovered unreported incidents
- Benchmarking comparable organizations for costs and effectiveness
- The ability to determine the effectiveness/efficiency of controls
- Clear indications that security objectives are being met
- The absence of unexpected or undetected security events
- Knowledge of evolving and impending threats
- Effective means of determining organizational vulnerabilities
- Methods of tracking evolving risk
- Consistency of log review practices
- Results of business continuity planning (BCP)/disaster recovery (DR) tests
- The extent to which key controls are monitored
- The percentage of metrics achieving defined criteria (metametrics)

1.6.8 ASSURANCE PROCESS INTEGRATION (CONVERGENCE)

Organizations should consider an approach to information security governance that includes an effort to integrate assurance functions. This will serve to increase security effectiveness and efficiency by reducing duplicated efforts and gaps in protection. It will help ensure that processes operate as intended from end to end, minimizing hidden risk. KGIs include:
- No gaps in information asset protection
- The elimination of unnecessary security overlaps
- The seamless integration of assurance activities
- Well-defined roles and responsibilities
- Assurance providers understanding the relationship to other assurance functions
- All assurance functions being identified and considered in the strategy
- Effective communication and cooperation between assurance functions

The approach demonstrated by BMIS, discussed in section 1.2.5, brings together the various elements and stakeholders in the problem set to work closely together. A major objective of this activity is to understand how organizations can bring together diverse elements and get them to orient on common objectives.

Knowledge Check: Metrics

Consider the following information security metrics and answer the questions below:

A. Number of information security policy violations reported quarterly
B. Number of information security incidents that result in a disruption to key business activities
C. Percentage of systems that are patched within the required time period
D. Percentage of incidents that are responded to within the required time period
E. Maturity level of organization's information security activities

Questions:

1. Which of the above metrics are SMART?
2. Which of the above metrics could be considered strategic?
3. Which of the above metrics are related to performance management?
4. How could the metrics that are not SMART be rewritten to be SMART?

Answers on page 71.

1.7 INFORMATION SECURITY STRATEGY OVERVIEW

Strategy can be defined many different ways. While the definitions all generally point in the same direction, they vary widely in scope, emphasis and detail. Strategy derives from the military definition: the plan to achieve an objective. As is typical, the basic concept has been expanded and grown more complex, although this does not always serve to improve understanding of the concept. Two initial elements are needed to develop a plan to achieve an objective: a well-defined objective or objectives and an understanding of the current conditions. Essentially, a strategy requires knowing where you are and where you need to go, then analyzing the gap between those two locations to determine what needs to be accomplished to bridge the gap. While a straightforward notion, numerous considerations and a host of activities must be examined in developing both the objectives and what is required to achieve them.

Figure 1.6 shows the participants involved in developing a security strategy and their relationships and aligns them with business objectives. The arrow marked "Business Strategy" provides the basis for a road map to achieving the "Business Objectives." In addition, it should provide one of the primary inputs into "Risk Management" plans and the "Information Security Strategy." This flow serves to promote alignment of information security with business goals. The balance of inputs

Figure 1.6—Information Security Strategy Development Participants

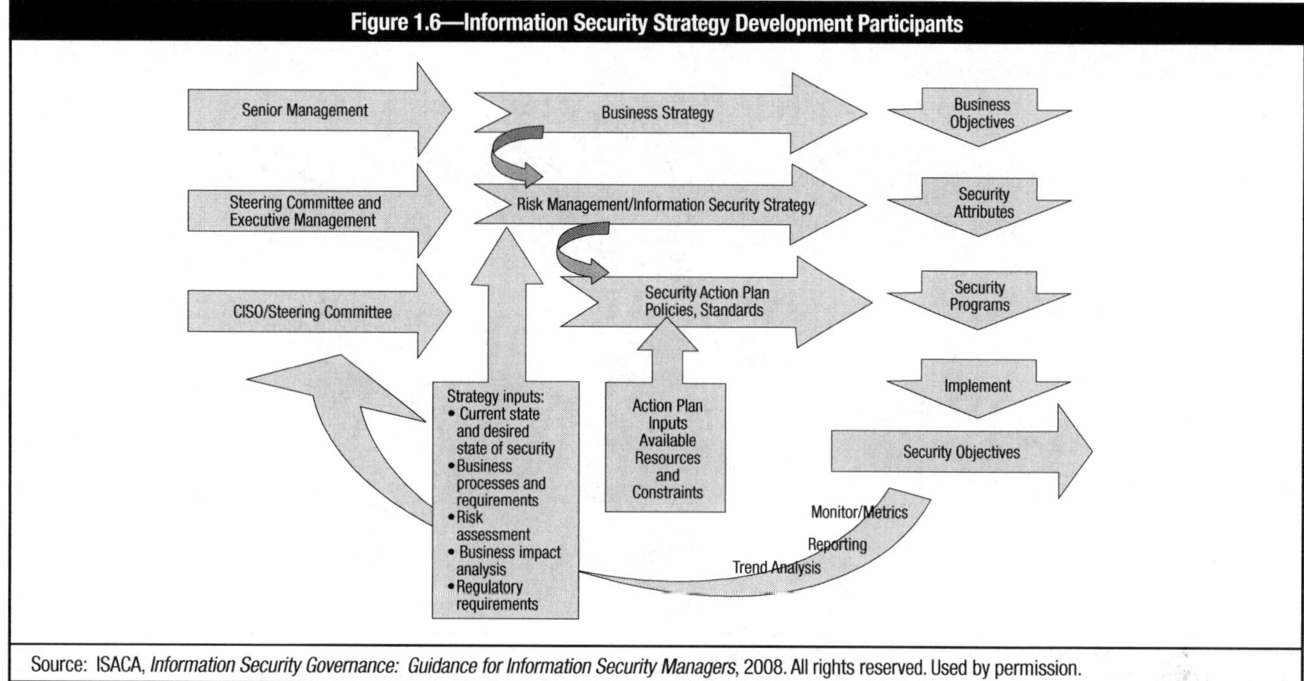

Source: ISACA, *Information Security Governance: Guidance for Information Security Managers*, 2008. All rights reserved. Used by permission.

comes from determining the desired state of security compared to the existing, or current, state. Business processes must also be considered as well as key organizational risks, including regulatory requirements, risk analysis and the associated impact analysis to determine protection levels and priorities. A constant constraint during the entire process is the organization's risk appetite and capacity discussed in section 1.2.2 Determining Risk Capacity and Acceptable Risk (Risk Appetite).

The objective of the security strategy is the desired state defined by business and security attributes. The strategy provides the basis for an action plan composed of one or more security programs that, as implemented, achieve the security objectives. The action plan(s) must be formulated based on available resources and constraints, including consideration of relevant legal and regulatory requirements.

The strategy and action plans must contain provisions for monitoring as well as defined metrics to determine the level of success. This provides feedback to the CISO and steering committee to allow for midcourse correction and ensure that security initiatives are on track to meet defined objectives.

1.7.1 DEVELOPING AN INFORMATION SECURITY STRATEGY

The process of developing an effective information security strategy requires a thorough understanding and consideration of a number of elements, as shown in **figures 1.7** and **1.8**. In addition, it is also important for the information security manager to be aware of the common failures of strategic plans to avoid the pitfalls and achieve the desired outcomes.

Figure 1.7—Prevalent Standards and Frameworks

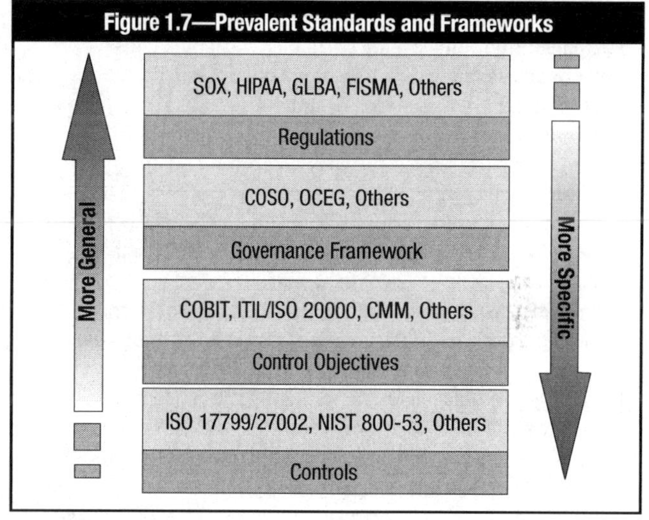

1.7.2 COMMON PITFALLS

While some strategic plans may fail for obvious reasons such as greed, poor planning, faulty execution, unanticipated events and corporate misconduct, other causes of strategy failures are not as well understood. These failures have root causes explained by behavioral economics, a branch of psychology that studies decision-making processes and departure from rational choice.

Experiments and studies have shown a variety of underlying causes for flawed decision making. Being aware of them may allow compensation to reduce adverse effects. Some of the main reasons include:

- **Overconfidence**—Research shows a tendency for people to have excessive confidence in their ability to make accurate estimates. Most people are reluctant to estimate wide ranges of possible outcomes and prefer being precisely wrong, rather than

Figure 1.8—SABSA Security Architecture

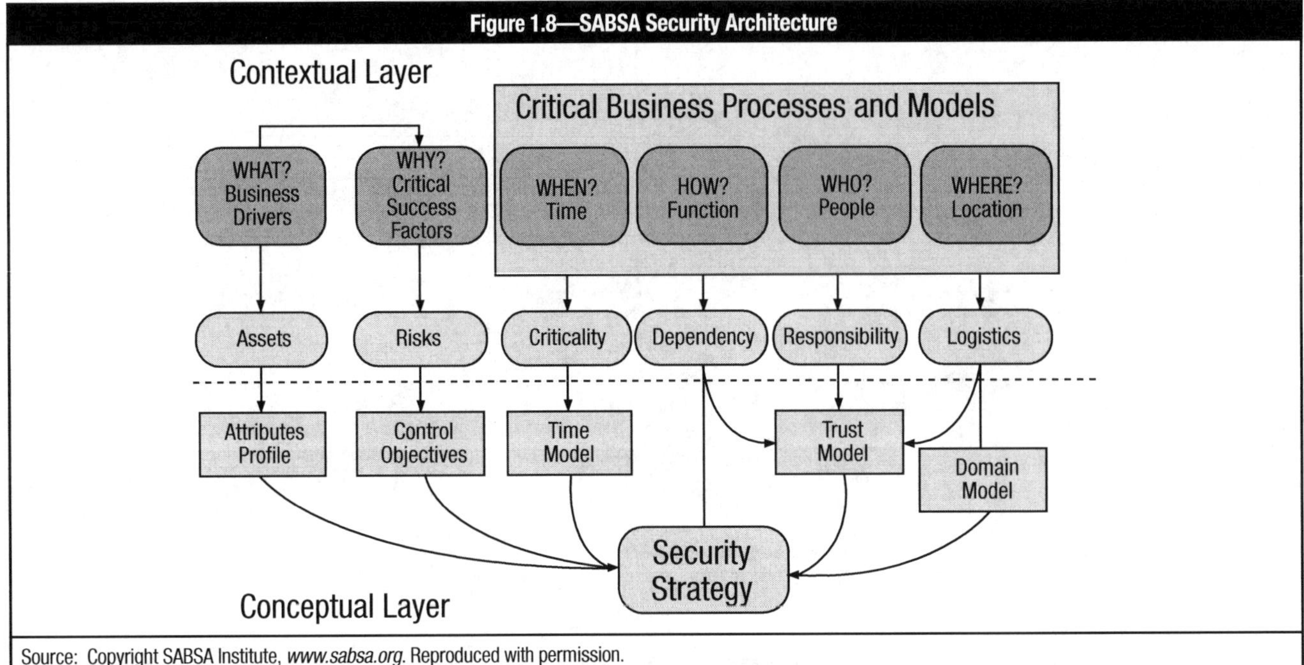

Source: Copyright SABSA Institute, *www.sabsa.org*. Reproduced with permission.

vaguely right. Most also tend to be overconfident of their own abilities. For organizational strategies based on assessments of core capabilities, this can be particularly troublesome.

- **Optimism**—People tend to be optimistic in their forecasts. A combination of overconfidence and overoptimism can have a disastrous impact on strategies based on estimates of what may happen. Typically, these estimates will be unrealistically precise and overly optimistic.
- **Anchoring**—Research shows that once a number has been presented to someone, a subsequent estimate of even a totally unrelated subject involving numbers will "anchor" on the first number. While potentially useful for marketing purposes, anchoring can have serious consequences in developing strategies when future outcomes are anchored in past experiences.
- **The *status quo* bias**—Most people show a strong tendency toward sticking with familiar and known approaches even when they are demonstrably inadequate or ineffective. Research also indicates that concern over loss is generally stronger than excitement over possible gain. The "endowment effect" is a similar bias where people prefer to keep what they own or know, and simply owning something makes it more valuable to the owner.
- **Mental accounting**—This is defined as "the inclination to categorize and treat money differently depending on where it comes from, where it is kept, and how it is spent." Mental accounting is common even in the boardrooms of conservative, and otherwise rational, corporations. Some examples of this include:
  - Being less concerned with expenses against a restructuring charge than those against the profit and loss statement
  - Imposing cost caps on a core business while spending freely on a start-up
  - Creating new categories of spending such as "revenue investment" or "strategic investment"
- **The herding instinct**—It is a fundamental human trait to conform and seek validation of others. This can be observed by the "faddism" in security (as well as all other aspects of human activity), such as everyone suddenly being involved in ID management or intrusion detection. Sometimes explained as "an idea whose time has come," it is more accurately described as the herding instinct behind thought leaders. The implications for strategy development should be clear. It is aptly demonstrated by the statement, "For senior managers, the only thing worse than making a huge strategic mistake is being the only person in the industry to make it."
- **False consensus**—There is a well-documented tendency for people to overestimate the extent that others share their views, beliefs and experiences. When developing strategies, false consensus can lead to ignoring or minimizing important threats or weaknesses in the plans and to persisting with doomed strategies.

A number of the more common causes of flawed decision making have been uncovered by research from C. F. Camerer and G. Loewenstein, including:

- **Confirmation bias**—Seeking opinions and facts that support one's own beliefs
- **Selective recall**—Remembering only facts and experiences that reinforce current assumptions
- **Biased assimilation**—Accepting only facts that support an individual's current position or perspective
- **Biased evaluation**—Easy acceptance of evidence that supports their one's hypotheses while contradictory evidence is challenged and, almost invariably, rejected. Critics are often charged with hostile motives or their competence is impugned.
- **Groupthink**—Pressure for agreement in team-based cultures

There have been numerous studies on the topic of departures from rational choice during the past several decades that may be worthy of review to reduce the risk of faulty decision making.

1.8 INFORMATION SECURITY STRATEGY OBJECTIVES

The objectives of developing an information security strategy must be defined and metrics developed to determine if those objectives are being achieved. Typically, the six defined outcomes of security governance will provide high-level guidance. As previously stated, the six outcomes are:

- Strategic alignment
- Effective risk management
- Value delivery
- Resource optimization
- Performance measurement
- Assurance process integration

The strategy needs to consider what each of the selected areas means to the organization, how they might be achieved and what will constitute success.

1.8.1 THE GOAL

The first, and often surprisingly difficult, question that must be answered by an organization seeking to develop an information security strategy is—what is the goal?

While this seems a trivial question, most organizations fail to define the objectives of information security with any specificity. This may be because it seems obvious that the goal of information security is to protect the organization's information assets. However, that answer assumes knowledge of two things. One is that information assets are known with any degree of precision, which for most organizations is not the case. The other is that there is an assumed understanding of what it means "to protect." Everyone understands the notion in general. It is considerably more difficult to state which assets need how much protection against what.

In part, this is because organizations typically have little knowledge of what information exists within the enterprise. There is generally no process to purge useless, outdated or potentially dangerous information, data or unused applications. It is extremely rare to find a comprehensive catalog or index of information/processes to define what is important, what is not important or who owns it. As a result, everything typically gets saved under the assumption that storage is cheaper than data classification, ownership assignment and the identification of users. For large organizations, this can amount to terabytes of useless data and literally thousands of outdated and unused applications accumulated over decades.

This situation makes it difficult to devise a rational data protection plan because it makes little sense to expend resources protecting useless or dangerous data and information or unused applications. In this context, dangerous data refer to information that might be used to the detriment of the organization, such as damaging evidence obtained in litigation, that could have been destroyed subject to a legal and appropriate retention policy.

Assuming relevant information is located and identified, it must be classified by criticality and sensitivity or business value. Classification provides the basis for applying protective measures in proportion to the business value, resulting in more cost-effective controls. Because a great deal of a typical organization's data and information are neither critical nor sensitive, it is wasteful to expend substantial resources to protect them. For many organizations, locating and classifying information assets may be a significant undertaking, and there is often a reluctance to allocate the resources necessary. However, this is a crucial step in developing a practical and useful information security strategy and a cost-effective security program.

Just as values are assigned to an organization's physical resources, information must be assigned a value to prioritize budget-constrained protection efforts and determine required levels of protection. Valuation of information is difficult to do with any precision. For some information, it can be the cost of creating or replacing it. In other cases, information in the form of knowledge or trade secrets is difficult or impossible to replace and may be priceless. It is obviously prudent to provide strong protection to priceless information.

One approach commonly used is to create a few rough levels of value—for example, from zero to five, with zero being of no value and five being critical. The zero-value information, including applications, would be assigned where no owner can be determined and no use has been evidenced for a period of time. Information of zero value can then be archived for a specified period, notices can be sent to business owners and, if there are no objections, it can be destroyed. Information deemed a five (critical) then becomes the priority for protection efforts on a basis proportional to the other valuations.

Another approach that may be useful and substantially easier to perform is a business dependency evaluation used as an indication of value. This process starts by defining critical business processes and then determines what information and physical assets are used in the process. The contribution these resources make to revenue provides a measure of the level of criticality of information resources that can be used as a guide for protection efforts.

Sensitivity will generally be a more subjective call. The unintended disclosure of sensitive information can have many ramifications that may be difficult to determine with any precision. The data owner is typically the best source of for determining the potential consequences of "data leakage" and is normally the individual determining the classification level for data. The classification level will subsequently provide the basis for protection efforts and access control. Most organizations will use three or four sensitivity and criticality classifications such as confidential, internal use and public.

For most organizations, asset classification poses a daunting task, but one that must be undertaken for existing information if security governance is to be effective, efficient and relevant. Done correctly, classification prevents the cost of overprotecting unimportant information and the risk of underprotecting high-value information. It must also be considered that this task grows exponentially more onerous over time, unless addressed. Concurrently, policies, standards and processes must be developed to mandate classification moving forward and

prevent the problem from getting worse. Overclassification is a serious problem for classification implementation. This can be particularly troublesome in organizations with a blame culture where mistakes are not well tolerated. In these cases, data owners may err on the side of caution by marking data more sensitive or critical than warranted, negating the potential benefits of classification. There is a practical solution for organizations where IT operates on a chargeback system and IT services are a cost to the business units. In these cases, additional charges for higher classifications will tend to counteract the tendency to overclassify.

Another essential part of setting the goals for information security that must be considered is to ensure that all assets have a defined owner and defined accountability. This can be done effectively using RACI charts as discussed in section 1.3 Roles and Responsibilities.

In summary, it will not be possible to develop a cost-effective security strategy that is aligned with business requirements prior to:
- Defining business requirements for information security
- Determining the objectives of information security that will satisfy the requirements
- Locating and identifying information assets and resources
- Valuating information assets and resources
- Classifying information assets as to criticality and sensitivity
- Implementing a process to ensure that all assets have a defined owner

1.8.2 DEFINING OBJECTIVES

If an information security strategy is the basis for a plan of action to achieve security objectives, it is necessary to define those objectives. Defining long-term objectives in terms of a desired state of security is necessary for a number of reasons. Without a well-articulated vision of desired outcomes for a security program, it will not be possible to develop a meaningful strategy.

It is often said that if you don't know where you're going, you won't know what direction to go or when you've arrived. Without a strategy, it is not possible to develop a meaningful plan of action and the organization will continue to implement ad hoc tactical point solutions with nothing to provide overall integration. The resulting nonintegrated systems will be increasingly difficult to manage and will become ever more costly and difficult or impossible to secure.

Unfortunately, many organizations do not allocate adequate resources to address these issues until a major incident occurs. Experience has shown that these incidents end up far more costly than if they had been addressed properly in the beginning. A typical example is the recent breach of a major retailer resulting in the theft of 46 million customer credit card records that is estimated to ultimately cost the organization several hundred million US dollars to resolve. The numerous security governance, management and practices flaws subsequently uncovered could have been remediated at a fraction of the expected losses.

Many objectives are stated in terms of addressing risk. Information security strategy objectives should also be stated in terms of specific goals directly aimed at supporting business activities. Some risk mitigation, such as virus and other malware protection, should apply to the organization generally. Such protection is generally not considered a specific business enabler; rather, it supports the overall health of the organization by reducing adverse impacts that hinder all activities of the organization.

A review of the organization's strategic business plans is likely to uncover opportunities for information security activities that can be directly supportive of, or enabling, a particular avenue of business by reducing risk, losses and potential operational disruptions. Other sources of information to guide security efforts can include change management activities, audit reports and steering committee discussions. As an example, the implementation of a public key infrastructure (PKI) can enable high-value transactions between trusted trading partners or customers. Deploying virtual private networks (VPNs) may provide the sales force with secure remote connectivity, enabling strong protection for sensitive information. In other words, information security can enable business activities that would otherwise be too risky to undertake or, as frequently happens, are undertaken with the hope that nothing goes wrong.

Business Linkages

In determining strategic objectives, it is essential to ensure there are direct linkages to specific business activities and goals. These linkages can start from the perspective of the specific objectives of a particular line of business. A review and analysis of all the elements of a particular product line can illustrate this approach.

Consider an organization that manufactures breakfast cereal. The raw materials come into the plant on a just-in-time basis via railcar. The grains are dumped into hoppers that feed the various processing machinery. The finished cereals are packaged and moved to a warehouse in a continuous, highly automated process in a matter of hours.

This relatively straightforward process relies on numerous information flows subject to a failure of availability, confidentiality or integrity. Any breakdown or significant disruption in the systems that support the supply chain side (e.g., ordering, tracking or payments) is likely to cause a disruption in manufacturing. All the automated processing activities in the plant are tied to data processing and information flows. Any interruption or corruption of data would cause a failure of the process, ranging from stopping manufacturing to delivering defective product. To be effective, information security must understand and take into consideration all the information streams that are critical to ensuring continuous operations. Obviously, anything that can affect the integrity or availability of the information needed in this continuous and interdependent process will be a problem. The linkage to the business, in this case, is the ability of an effective security program to prevent disruptions to the information systems essential to production.

The analysis of the foregoing example might look at the dozens of discrete pieces of information handled and processed in this manufacturing operation. Investigation into the history of the process may reveal past failures that are informative of weaknesses in the system. Most system failures are due to human error, and analysis may show that it is possible to reduce errors by either additional or better controls or redundancy for more reliable automated processes.

Typical errors include entry mistakes that might be improved by range checking or other technical processes or instrumentation failure resulting in incorrect processing. Procedural changes or an entry validation process may be required to address entry errors. Instrumentation failures might be addressed by periodic self-test or instrumentation redundancy. Information security concerns also include controlling access to production systems to prevent unauthorized access, incident response and other aspects of risk management.

The development and analysis of business linkages can uncover information security issues at the operational level that can visibly improve the value of information security by making business processes more robust, reducing errors and improving productivity.

Improved business linkage on an ongoing basis can be one of the beneficial outcomes of an information security steering group if high-level representatives of the major departments and business units are included. Linkages may also be established by regular meetings with business owners for discussion regarding security-related issues. This may also provide an opportunity to educate business process owners on potential benefits that security may provide for their operation.

1.8.3 THE DESIRED STATE

The term "desired state" is used to denote a complete snapshot of all relevant conditions at a particular point in the future. For a robust picture, it must include principles, policies and frameworks; processes; organizational structures; culture, ethics and behavior; information; services, infrastructure and applications; and people, skills and competencies.

Defining a state of security in purely quantitative terms is not possible. Consequently, a desired state of security must, to some extent, be defined qualitatively in terms of attributes, characteristics and outcomes. According to COBIT, it can include high-level objectives such as: "Protecting the interests of those relying on information, and the processes, systems and communications that handle, store and deliver the information, from harm resulting from failures of availability, confidentiality and integrity." This statement, while perhaps useful in stating intent and scope, provides little clarity in defining processes or objectives.

Qualitative elements such as desired outcomes should be defined as precisely as possible to provide guidance to strategy development. For example, if specific regulatory compliance is a desired outcome, a significant number of technical and process requirements become apparent. If characteristics include a nonthreatening compliance enforcement approach consistent with the organization's culture, strategy development will have limits on the types of enforcement methods to consider.

A number of useful approaches are available to provide a framework to achieve a well-defined desired state for security. These, and perhaps others, should be evaluated to determine which provides the best form, fit and function for the organization. It may be useful to combine several different standards and frameworks to provide a multidimensional view into the desired state. See **figure 1.8**.

Several of the most accepted approaches are described in the following sections.

COBIT

COBIT 5 provides a comprehensive framework for the governance and management of enterprise IT and extensively addresses IT security, governance, risk and information security in general. Because many aspects of information security involve IT and related activities, it can serve as a framework for determining the desired state for effective information security. *COBIT® 5 for Information Security* builds on the COBIT 5 framework and focuses on information security, providing detailed and practical guidance for information security professionals and other stakeholders.

COBIT 5 is based on five key principles shown in **figure 1.9** for governance and management of enterprise IT and information assets. These are particularly useful and applicable to information security governance:

- **Principle 1: Meeting Stakeholder Needs**—Enterprises exist to create value for their stakeholders by maintaining a balance between the realization of benefits and the optimization of risk and use of resources.
- **Principle 2: Covering the Enterprise End-to-end**—COBIT 5 integrates governance of enterprise IT into enterprise governance:
  - It covers all functions and processes within the enterprise; COBIT 5 does not focus on only the "IT function," but treats information and related technologies as assets that need to be dealt with just like any other asset by everyone in the enterprise.
  - It considers all IT-related governance and management enablers to be enterprisewide and end-to-end (i.e., inclusive of everything and everyone—internal and external—that is relevant to governance and management of enterprise information and related IT).
- **Principle 3: Applying a Single, Integrated Framework**—There are many IT-related standards and good practices, each providing guidance on a subset of IT and information security activities. At a high level, COBIT 5 aligns with other relevant standards and frameworks such as the ISO 27000 series.
- Principle 4: Enabling a Holistic Approach—Efficient and effective governance and management of enterprise IT require a holistic approach, taking into account a number of interacting components. COBIT 5 defines a set of enablers that are broadly defined as anything that can help to achieve the objectives of the enterprise. The COBIT 5 framework defines seven categories of enablers:
  - Principles, policies and frameworks
  - Processes
  - Organizational structures
  - Culture, ethics and behavior
  - Information
  - Services, infrastructure and applications
  - People, skills and competencies
- **Principle 5: Separating Governance From Management**—The COBIT 5 framework makes a clear distinction between governance and management.
  - **Governance** ensures that stakeholder needs, conditions and options are evaluated to determine balanced, agreed-on enterprise objectives to be achieved; settting

direction through prioritization and decision making; and monitoring performance and compliance against agreed-on direction and objectives. In most enterprises, this is the responsibility of the board of directors. Specific responsibilities may be delegated to special organizational structures at an appropriate level.

- **Management** plans, builds, runs and monitors activities in alignment with the direction set by the governance body to achieve the enterprise objectives. In most enterprises, this is the responsibility of the senior management under the leadership of the CEO.

Figure 1.9—COBIT 5 Principles

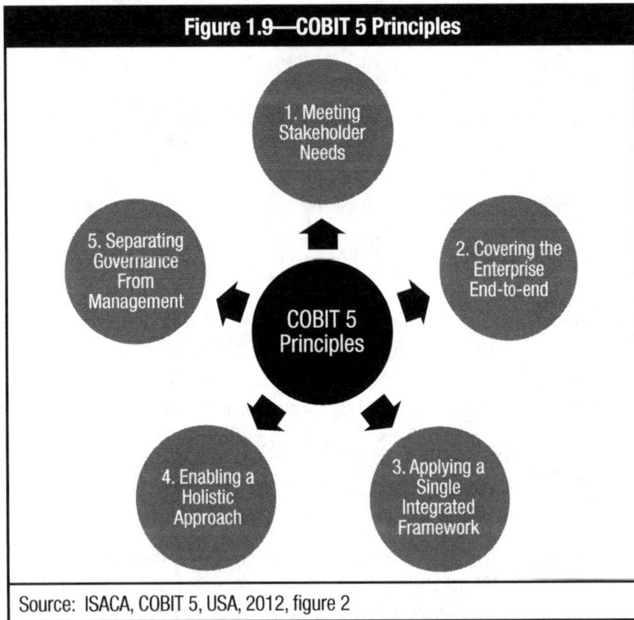

Source: ISACA, COBIT 5, USA, 2012, figure 2

These five principles embodied in COBIT 5 are designed to enable the enterprise to build an effective governance and management framework that optimizes information and technology investment and use for the benefit of stakeholders.

COBIT 5 Process Assessment Model

The COBIT 5 Process Assessment Model (PAM) is a tool that can be used to assess the current state and define a future desired state for information security. The COBIT 5 PAM conforms to ISO/IEC 15504-2 requirements for a PAM and can be used as the basis for conducting an assessment of the capability of each COBIT 5 process, resulting in a 0 to 5 level of maturity.

The process dimension uses COBIT 5 as the process reference model. COBIT 5 provides definitions of processes in a life cycle (the process reference model), together with an architecture describing the relationships among the processes.

The COBIT 5 process reference model is composed of 37 processes describing a life cycle for governance and management of enterprise IT, as shown in **figure 1.10**.

Capability Maturity Model Integration

The Capability Maturity Model Integration (CMMI®) is a capability improvement framework that provides guidance for organizations to elevate performance. CMMI helps organizations benchmark their

Figure 1.10—Overview of the Process Assessment Model

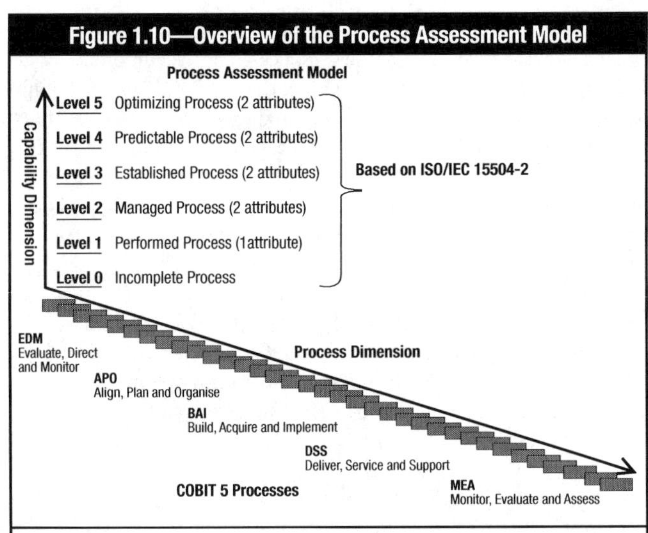

Source: ISACA, *COBIT Process Assessment Model (PAM): Using COBIT 5*, USA, 2013, figure 2

capabilities and build maturity by comparing their operations to good practices and identifying performance gaps. CMMI has five maturity levels; each level builds on the previous for continuous improvement (**figure 1.11**). Organizations that work toward higher maturity levels can advance capabilities and promote more effective processes. Organizations using CMMI to build capability often achieve increased quality, better customer satisfaction, improved employee retention and improved profitability.

Figure 1.11—Characteristics of the Maturity Levels

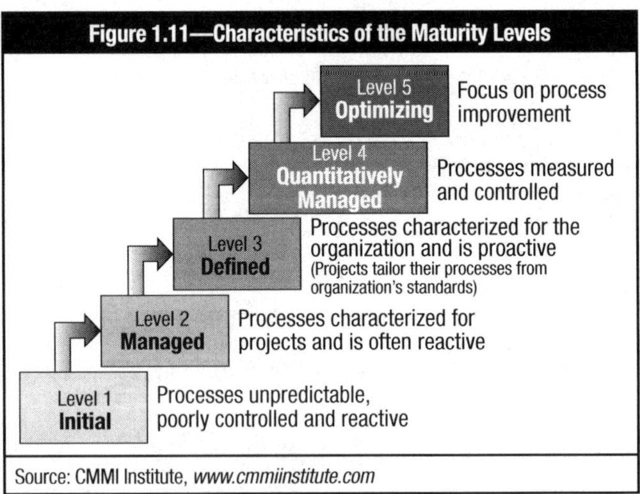

Source: CMMI Institute, *www.cmmiinstitute.com*

Balanced Scorecard

According to the Balanced Scorecard Institute:

> *The balanced scorecard is a management system (not only a measurement system) that enables organizations to clarify their vision and strategy, and translate them into action. It provides feedback around both the internal business processes and external outcomes in order to continuously improve strategic performance and results. When fully deployed, the balanced scorecard transforms strategic planning from an academic exercise into the nerve center of an enterprise.*

The balanced scorecard, as shown in **figure 1.12**, uses four perspectives, develops metrics, collects data and analyzes the data relative to each of these perspectives:
- Learning and growth
- Business process
- Customer
- Financial

Figure 1.12—Balanced Scorecard Dimensions

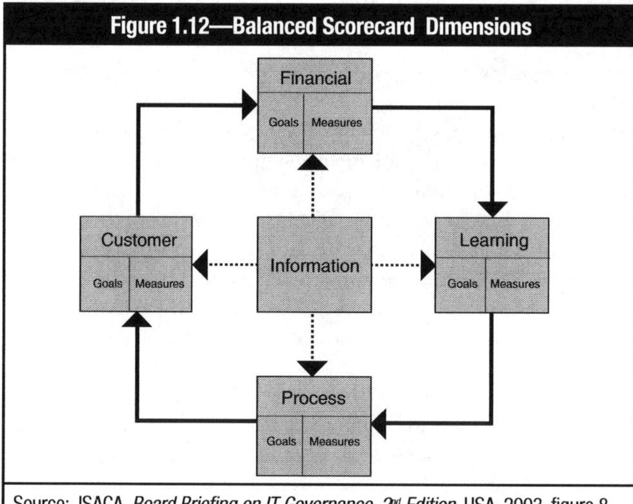

Source: ISACA, *Board Briefing on IT Governance, 2nd Edition*, USA, 2003, figure 8

Architectural Approaches

Enterprise information security architecture (EISA) is a subset of enterprise architecture. An architecture framework can be described as a foundational structure, or set of structures, that can be used for developing a broad range of different architectures, including business process architecture—sometimes referred to as the contextual architecture as well as the more traditional conceptual, logical, physical, functional and operational architectures.

Many methodologies have evolved, including process models, frameworks and ad hoc approaches favored by some consultancies. This evolution occurred as it became evident that an architectural perspective limited to IT was inadequate to address business design and development of security requirements. A number of architectural approaches now provide linkages to and design of the business side of information protection. Architectural approaches that are inclusive of business processes that may be appropriate for defining the desired state of security include (but are not necessarily limited to) framework models such as: COBIT 5, the Open Group Architecture Framework (TOGAF), the Zachman Enterprise Architecture Framework, and the Extended Enterprise Architecture Framework (E2AF). These models can serve to define most or all of the desired state of security, provided they are properly used to reflect and implement the organizational security strategy.

The architecture should describe a method for designing a target or desired state of the enterprise in terms of a set of building blocks, and for showing how the building blocks fit together. The target architecture is referred to as the reference architecture and serves to set the longer-term objectives for technical, systems and process design.

ISO/IEC 27000 Series

To ensure that all relevant elements of security are addressed in an organizational security strategy, the 14 areas of the ISO/IEC 27001:2013 standard can provide a useful framework to gauge comprehensiveness. Organizational policies and standards must be created that can track directly to each element of the standard. It is the standard on which an organization may choose to be certified and assessed (which is increasingly the case globally), while 27002:2013 is the Code of Practice for information security management that supports implementation to comply with the standard.

The 14 security control clauses of ISO/IEC 27001:2013 are:
- A.5: Information security policies
- A.6: Organization of information security
- A.7: Human resource security (controls that are applied before, during or after employment)
- A.8: Asset management
- A.9: Access control
- A.10: Cryptography
- A.11: Physical and environmental security
- A.12: Operations security
- A.13: Communications security
- A.14: System acquisition, development and maintenance
- A.15: Supplier relationships
- A.16: Information security incident management
- A.17: Information security aspects of business continuity management
- A.18: Compliance (with internal requirements, such as policies, and with external requirements, such as laws)

Whereas ISO/IEC 27001:2013 Annex A refers to 114 "controls," they are just sections in ISO/IEC 27002:2013, many of which propose multiple security controls. ISO/IEC 27002:2013 suggests literally hundreds of good practice information security control measures that organizations should consider to satisfy the stated control objectives.

Like ISO/IEC 27001:2013, ISO/IEC 27002:2013 does not mandate specific controls but leaves it to the users to select and implement controls that suit them, using a risk assessment process to identify the most appropriate controls for their specific requirements. They are also free to select controls not listed in the standard, as long as their control objectives are satisfied. The ISO/IEC standard is treated as a generic controls checklist—a menu from which organizations select their own set.

Other Approaches

Other approaches and methods exist that may be useful, such as other ISO standards on quality (ISO 9001:2015), the Six Sigma approach to quality and business management, publications from NIST and Information Security Forum (ISF), and the US Federal Information Security Modernization Act (FISMA). Some of these focus more on management processes and quality management than on strategic security objectives. However, the argument can be made that, if the objective of a security strategy is to fully implement relevant components of ISO/IEC 27001:2013 and 27002:2013, most or all security requirements are likely to be met. That would probably be a needlessly expensive approach, and the standards suggest that they should be carefully tailored

to the specific requirements of the adopting organization. Other methodologies will undoubtedly emerge in the future that may prove to be more effective than those mentioned. The ones outlined are not meant to be exhaustive, merely some of the more widely accepted approaches to arrive at well-defined information security objectives.

It may be useful to employ a combination of methods to describe the desired state to assist in communications with others and as a way to cross-check the objectives to make certain all relevant elements are considered. For example, some combination of COBIT control objectives, CMMI, balanced scorecard and an appropriate architectural model would make a powerful combination. While it may seem overkill, each approach presents a different viewpoint that, in combination, is likely to make certain that no significant aspect is overlooked. Since it is unlikely that an effective security program will develop from a faulty strategy, this may be a prudent approach.

1.8.4 RISK OBJECTIVES

A major input into defining the desired state is the organization's approach to risk, its risk appetite and risk tolerance—that is, what management considers acceptable risk and tolerable deviations from acceptable risk levels. This is another critical step, since defined acceptable risk devolves into the control objectives or other risk mitigation measures employed as well as the criteria by which risk can be evaluated for acceptability. The defining of control objectives is instrumental in determining the type, nature and extent of controls and countermeasures the organization employs to manage risk. **Figure 1.13** presents the relationship among risk, control measures and the cost of controls.

Figure 1.13—Optimizing Risk Costs

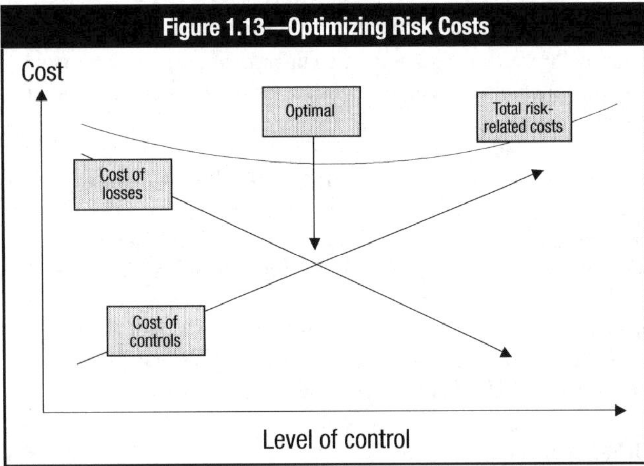

Without a reasonably clear determination of acceptable risk, it is difficult to determine whether information security is meeting its objectives and the appropriate level of resources has been deployed.

Risk is a complex subject with numerous known and unknown variables and is often difficult to ascertain with any precision. (See chapter 2.)

Operational risk management is always a trade-off: If there is a risk associated with taking a particular course of action, there is also a risk of not doing so. Furthermore, individual risk interacts in complex ways, and mitigating one risk almost certainly increases or creates at least one other risk in response.

Risk always carries a cost, whether controlled or not. Risk cost can be expressed as ALE (i.e., the amount of potential loss times the likelihood of occurrence, showing the optimal level of control). **Figure 1.13** illustrates the balance of the cost of controls against the cost of losses, showing the optimal level of control.

The acceptability of some risk can be quantified by using the business continuity approach of developing recovery time objectives (RTOs). Using a summary approach to determining RTOs may provide adequate input for strategy development. This can be an informal determination by business process owners of the amount of time critical systems can be inoperative without serious business consequences. This, in turn, provides the basis for approximating costs of achieving recovery. If this is considered too high, iteration of the process will arrive at an acceptable recovery time at an acceptable cost. This may be considered the acceptable risk.

Acceptable risk can also be approximated by examining the amount of business interruption insurance the organization carries, the amount of the deductible and its cost (i.e., annual premiums). For example, if the organization has US $1 million of insurance with a $10,000 deductible at an annual cost of $50,000, the argument can be made that it is reasonable to spend up to $50,000 in control activities to reduce a $1 million risk to an acceptable risk level of $10,000 (i.e., the residual risk).

Developing the right strategy objectives usually needs to be an iterative approach based on analysis of costs to achieve the desired state and achieve acceptable risk levels. It is likely that lowering the level of acceptable risk will be more costly. However, the approach to achieving the desired state will have a significant bearing on costs as well.

For example, risk may exist because of certain practices that are not necessary or useful to the organization, or are detrimental to its operation. This could include practices that might be considered discriminatory or contrary to law, posing the risk of a lawsuit. Such practices, when examined, may have resulted from outmoded attitudes or approaches that can be efficiently changed at a low cost, resulting in elimination or mitigation of the risk. In other cases, it may be possible to reengineer certain processes or provide better architectures to reduce inherent risk. In other words, the approach to addressing or treating specific risk has a significant impact on costs.

The information security manager must understand that technical controls (e.g., firewalls, intrusion detection systems [IDSs]) are merely one dimension to be considered. Physical, process and procedural controls may be more effective and less costly. In most organizations, process risk poses the greatest hazard and technical controls are unlikely to adequately compensate for poor management or faulty processes.

Once objectives have been clearly defined, there will be a number of ways to architect solutions that will vary significantly in costs and complexity. Whichever process is used, the objective is to define, in meaningful, concrete terms, the desired overall state of security at some future point.

1.9 DETERMINING THE CURRENT STATE OF SECURITY

A current-state evaluation of information security must be determined using the same methodologies or combination of methodologies employed to determine strategy objectives, or the desired state. In other words, whatever combination of methodologies, such as COBIT, CMMI or the balanced scorecard, is used to define the desired state must also be used to determine the current state. This provides an apples-to-apples comparison of the two, providing the basis for a gap analysis that will delineate what is needed to achieve the objectives.

Using these same methodologies periodically can also provide the metrics on progress toward meeting the objectives.

1.9.1 CURRENT RISK

The current state of risk must also be assessed through a comprehensive risk assessment. Just as risk objectives must be determined as a part of the desired state, so must the current state of risk be determined to provide the basis for a gap analysis of what risk exists and to what extent risk must be addressed by the strategy. A full risk assessment includes threat, vulnerability and impact analyses, which individually will provide useful information in building a strategy. Because risk can be addressed in different ways, such as altering risky behavior, developing countermeasures to threats, reducing vulnerabilities or developing controls, this information will provide the basis for determining the most cost-effective strategy to address risk and developing remediation budgets.

Existing controls must also be inventoried, tested and evaluated to determine the extent to which they meet the desired state objectives for risk mitigation. This will provide the basis for deciding whether these controls are sufficient; need to be strengthened, modified or replaced; or additional controls added.

Additional periodic assessments will serve to provide the needed metrics to determine progress. Numerous approaches exist to assess risk. Some of the most commonly used methods include *COBIT® 5 for Risk*, NIST SP 800-30, ISO/IEC 27005 and Operationally Critical Threat, Asset, and Vulnerability Evaluation® (OCTAVE®).

Business Impact Analysis

The current-state evaluation will also include a thorough BIA of critical systems and processes to help round out the current-state picture. Because the ultimate objective of security is to increase stakeholder value by providing business process assurance and minimizing the impacts of adverse events, an impact analysis provides some of the information needed to develop an effective strategy as well as input for asset classification based on business value. The difference between acceptable levels of impact and current level of potential impacts must be addressed by the strategy.

1.10 INFORMATION SECURITY STRATEGY DEVELOPMENT

With the information developed in the previous section, a meaningful security strategy can be developed—a strategy to move from the current state to the desired state. Knowing where one is and where one is going provides the essential starting point for strategy development; it provides the framework for creating a road map. The road map is essentially the specific steps that must be taken to implement the strategy.

A set of information security objectives—coupled with available processes, methods, tools and techniques—creates the means to construct a security strategy. A good security strategy should address and mitigate risk while providing an acceptable level of compliance with the legal, contractual and statutory requirements of the business. It should provide demonstrable support for the business objectives of the organization and maximize value to the stakeholders. The security strategy also needs to address how the organization will embed good security practices into every business process and area of the business.

Often, those charged with developing a security strategy think in terms of controls as the means to implement security. Controls, while important, are not the only element available to the strategist. In some cases, reengineering a process can reduce or eliminate a risk without the need for controls. Potential impacts may be mitigated by architectural modifications rather than controls. It should also be considered that, in some cases, mitigating risk can reduce business opportunities to the extent of being counterproductive. It is also essential for risk mitigation activities to consider impacts on productivity so a careful balance can be struck between security and operational efficiency/effectiveness. This can generally be accomplished by performing a cost-benefit analysis of proposed control activities.

Ultimately, the goal of security is business process assurance, regardless of the business (i.e., ensuring the continued operation of the business by supporting and minimizing disruptions to operations at an acceptable cost). This, in turn, serves to protect value and value creation. While the business of a government agency may not result directly in profits, it is still in the business of providing cost-effective services to its constituency and must still protect the assets for which it has custodial care. Whatever the business, its primary operational goal is to maximize the success of business processes and minimize impediments to those processes.

1.10.1 ELEMENTS OF A STRATEGY

What should go into a security strategy? The starting point and the destination have been defined. The next consideration must be what resources are available and what constraints must be considered when developing the road map. The resources are the mechanisms that will be used to achieve various parts of the strategy, bound by the constraints.

Road Map

The typical road map to achieve a defined, secure desired state includes people, processes, technologies and other resources. It serves to map the routes and steps that must be taken to navigate to the objectives of the strategy.

The interactions and relationships among the various elements of a strategy are likely to be complex. As a consequence, it is prudent to consider the initial stages of developing a security architecture such as those discussed in section 1.11.2 Enterprise Information Security Architecture(s). Architectures can provide a structured approach to defining business drivers, resource relationships and process flows. An architecture can also help ensure that contextual and conceptual elements such as business drivers and consequences are considered in the strategy development stage.

Achieving the desired state is usually a long-term goal consisting of a series of projects and initiatives. Like most large, complex projects, it is necessary to break it down into a series of shorter-term projects that can be accomplished in a reasonable time period given the inevitable resource constraints and budget cycles. The entire road map can, and should, be charted while understanding that there is no steady state for information security and some objectives will need to be modified over time. Some objectives—such as attaining a particular maturity level, reengineering high-risk processes or achieving specific control objectives—may not require modification.

Shorter-term projects aligned with the long-range objectives can serve to provide checkpoints and opportunities for midcourse corrections. They can also provide metrics to validate the overall strategy.

For example, one long-term objective defined in the strategy may be data classification according to sensitivity and criticality. Because of the sheer magnitude of the effort required in a large organization, it is likely to require a number of years to accomplish. The strategy may include the requirement to determine that 25 percent will be targeted for completion each fiscal year using a variety of tactical approaches. A second component of the strategy might be to create policies and standards that preclude the practices that gave rise to the problem to begin with, so it does not get worse while the remediation process is underway.

Development of a strategy to achieve long-term objectives and the road map to get there, coupled with shorter-term intermediate goals, will provide the basis for sound policy and standards development in support of the effort.

1.10.2 STRATEGY RESOURCES AND CONSTRAINTS—OVERVIEW

The following subsections describe the typical resources for implementing an information security strategy and some of the constraints that must be considered.

Note that COBIT 5 defines enablers as factors that individually and collectively influence whether something will work—in this case, governance and management of information security and enterprise IT. Enablers are driven by the goals cascade (i.e., higher-level goals define what the different enablers should achieve).

The COBIT 5 framework describes seven categories of enablers (**figure 1.14**):

- **Principles, policies and frameworks** are the vehicle to translate the desired behavior into practical guidance for day-to-day management.
- **Processes** describe an organized set of practices and activities to achieve certain objectives and produce a set of outputs in support of achieving overall goals.
- **Organizational structures** are the key decision-making entities in an enterprise.
- **Culture, ethics and behavior** of individuals and of the enterprise are very often underestimated as a success factor in governance and management activities.
- **Information** is pervasive throughout any organization and includes all information produced and used by the enterprise. Information is required for keeping the organization running and well governed, but at the operational level, information is very often the key product of the enterprise itself.
- **Services, infrastructure and applications** include the infrastructure, technology and applications that provide the enterprise with information technology processing and services.
- **People, skills and competencies** are required for successful completion of all activities and for making correct decisions and taking corrective actions.

Enablers can function as resources as well as constraints and must be considered from both perspectives.

Figure 1.14—COBIT 5 Enterprise Enablers

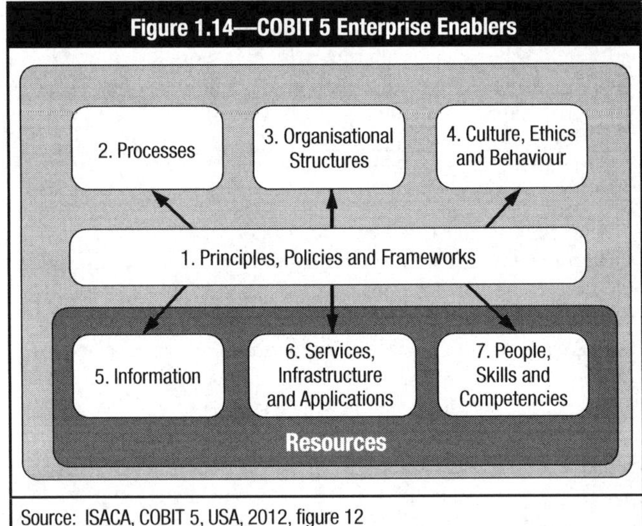

Source: ISACA, COBIT 5, USA, 2012, figure 12

Resources

The resources available to the organization need to be enumerated and considered when developing a security strategy. To the extent possible, the strategy should use existing resources to maximize utilization of existing assets and capabilities.

These resources can be considered the mechanisms, processes and systems that are available, in some optimal mix, to achieve the desired state of security over time. They include:

- Policies
- Standards
- Procedures

- Guidelines
- Architecture(s)
- Controls—physical, technical, procedural
- Countermeasures
- Layered defenses
- Technologies
- Personnel security
- Organizational structure
- Roles and responsibilities
- Skills
- Training
- Awareness and education
- Audits
- Compliance enforcement
- Threat assessment
- Vulnerability assessment
- BIA
- Risk analysis
- Resource dependency analysis
- Third-party service providers
- Other organizational support and assurance providers
- Facilities
- Environmental security

Constraints

There are also a number of constraints that must be considered when developing a security strategy and subsequent action plan. Constraints typically include:

- **Legal**—Laws and regulatory requirements
- **Physical**—Capacity, space, environmental constraints
- **Ethics**—Appropriate, reasonable and customary
- **Culture**—Both inside and outside the organization
- **Costs**—Time, money
- **Personnel**—Resistance to change, resentment against new constraints
- **Organizational structure**—How decisions are made and by whom, turf protection
- **Resources**—Capital, technology, people
- **Capabilities**—Knowledge, training, skills, expertise
- **Time**—Window of opportunity, mandated compliance
- **Risk appetite**—Threats, vulnerabilities, impacts

Some of the constraints, such as ethics and culture, may have been dealt with in developing the desired state. Others will undoubtedly arise as a consequence of developing the road map and action plan.

1.11 STRATEGY RESOURCES

There are typically numerous resources available to develop an information security strategy, but they will vary with the organization. The information security manager must determine what resources are available and be aware that there may be cultural, financial or other reasons that certain options may be precluded, such as management reluctance to change or modify policies. This section covers some of the most essential concepts of information security and constitutes essential knowledge for the CISM candidate.

1.11.1 POLICIES AND STANDARDS

There is broad range of interpretation of policy, standards, procedures and guidelines. The definitions used in this document are in agreement with the major standards bodies and should be adopted to preclude miscommunication. Policies and standards are considered tools of governance and management, respectively, and procedures and guidelines the purview of operations. For clarity, the four are defined in the following subsections.

Policies

Policies are the high-level statements of management intent, expectations and direction. Well-developed high-level policies in a mature organization can remain fairly static for extended periods.

An example of an effective high-level policy statement on access control could be: *Information resources shall be controlled in a manner that effectively precludes unauthorized access.*

Policies can be considered the "constitution" of security governance and must be clearly aligned with and support the strategic security objectives of the organization.

Standards

Standards, in this context, are the metrics, allowable boundaries or the process used to determine whether procedures, processes or systems meet policy requirements. Metrics are noted in the sense that a procedure either complies with the standard or it does not. Boundaries are set in terms of allowable limits on processes, people and technologies.

A standard for passwords used for access control could be: *Passwords for medium- and low-security domains shall be composed of no fewer than eight characters consisting of a mixture of upper- and lowercase letters and at least one number and one punctuation mark.*

The standard for access control for employees on the premises can include password composition requirements, minimum and maximum password length, frequency of password changes, and rules for reuse. Generally, a standard must provide sufficient parameters or boundaries that a procedure or practice can be unambiguously determined to meet the requirements of the relevant policy. Standards must be carefully crafted to provide only the necessary limits to ensure security while maximizing procedural options.

Standards must change as requirements and technologies change. Multiple standards will usually exist for each policy, depending on the security domain or classification level. For example, the password standard would be more restrictive when accessing high-security domains.

Procedures

Procedures are the responsibility of operations, including security operations, but they are included here for clarity. Procedures must be unambiguous and include all necessary steps to accomplish specific tasks. They must define expected outcomes, displays and required conditions precedent to execution. Procedures must also contain the steps required when unexpected results occur.

Procedures must be clear and unambiguous, and terms must be exact. For example, the words "must" and "shall" are used for any task that is mandatory. The word "should" must be used to mean a preferred action that is not mandatory. The terms "may" or "can" must be used only to denote a purely discretionary action. Discretionary tasks should appear in procedures only where necessary, since they dilute the message of the procedure.

Procedures for passwords include the detailed steps required for setting up password accounts and for changing or resetting passwords.

Guidelines

Guidelines for executing procedures are also the responsibility of operations. Guidelines should contain information that will be helpful in executing the procedures. This can include clarification of policies and standards, dependencies, suggestions and examples, narrative clarifying the procedures, background information that may be useful, and tools that can be used. Guidelines can be useful in many other circumstances as well, but they are considered here in the context of information security governance.

1.11.2 ENTERPRISE INFORMATION SECURITY ARCHITECTURE(S)

While still not implemented by many organizations, an EISA can be a powerful development, implementation and integration tool for the development and implementation of a strategy. However, its effectiveness depends on it being an integral part of enterprise architecture (EA) to help ensure that proper controls are implemented and integrated throughout the organization's infrastructure, processes and technologies.

The failure of organizations to embrace the notion of security architecture appears to have several identifiable causes. One is that such projects are expensive and time-consuming, and there is little or no understanding or appreciation at most organizational levels for the necessity or the potential benefits.

It may also be that there are few competent security architects with sufficiently broad and deep experience to address the wide range of issues necessary to ensure a reasonable degree of success. This lack of architecture over time has resulted in functionally less security integration and increasing vulnerability across the enterprise at the same time that technical security has seen significant improvement. This lack of integration contributes to the increasing difficulty in managing enterprise security efforts effectively.

A number of architectural frameworks and processes currently exist and some of the more prevalent ones are referenced below. Information security architecture is also covered more extensively in section 3.8 Information Security Infrastructure and Architecture.

A number of these approaches are similar and have evolved from the development of enterprise architecture. For example, the Zachman framework approach of developing a who, what, why, where, when and how matrix is shared by SABSA and E2AF. See **figure 1.15**.

The approach for EA, including security, that has gained ground during the past decade is TOGAF, which addresses the following four interrelated areas of specialization called architecture domains:

- Business architecture, which defines the business strategy, governance, organization and key business processes of the organization
- Applications architecture, which provides a blueprint for the individual application systems to be deployed, the interactions among the application systems, and their relationships to the core business processes of the organization with the frameworks for services to be exposed as business functions for integration
- Data architecture, which describes the structure of an organization's logical and physical data assets and the associated data management resources
- Technical architecture, or technology architecture, which describes the hardware, software and network infrastructure needed to support the deployment of core, mission-critical applications

Figure 1.15—SABSA Security Architecture Matrix

	Assets (What)	Motivation (Why)	Process (How)	People (Who)	Location (Where)	Time (When)
Contextual	The Business	Business Risk Model	Business Process Model	Business Organization and Relationships	Business Geography	Business Time Dependencies
Conceptual	Business Attributes Profile	Control Objectives	Security Strategies and Architectural Layering	Security Entity Model and Trust Framework	Security Domain Model	Security-related Lifetimes and Deadlines
Logical	Business Information Model	Security Policies	Security Services	Entity Schema and Privilege Profiles	Security Domain Definitions and Associations	Security Processing Cycle
Physical	Business Data Model	Security Rules, Practices and Procedures	Security Mechanisms	Users, Applications and the User Interface	Platform and Network Infrastructure	Control Structure Execution
Component	Detailed Data Structures	Security Standards	Security Products and Tools	Identities, Functions, Actions and ACLs	Processes, Nodes, Addresses and Protocols	Security Step Timing and Sequencing
Operational	Assurance of Operational Continuity	Operational Risk Management	Security Service Management and Support	Application and User Management and Support	Security of Sites, Networks and Platforms	Security Operations Schedule

Source: Copyright SABSA Institute, *www.sabsa.org*. Reproduced with permission.

The objectives of the various approaches are essentially the same. The framework details the organization, roles, entities and relationships that exist, or should exist, to perform a set of business processes. The framework should provide a rigorous taxonomy that clearly identifies what processes a business performs and detailed information about how those processes are executed and secured. The end product is a set of artifacts that describe, in varying degrees of detail, exactly what and how a business operates and what security controls are required. COBIT 5 provides extensive resources and tools to achieve these objectives and has seen wide implementation globally.

COBIT 5 moves from an IT-centric approach to a holistic architectural framework. COBIT 5 consolidates and integrates the COBIT 4.1, Val IT™ 2.0 and Risk IT frameworks, and draws from ISACA's IT Assurance Framework™ (ITAF™) and BMIS. It aligns with frameworks and standards such as Information Technology Infrastructure Library (ITIL), ISO, Project Management Body of Knowledge (PMBOK), PRINCE2 and TOGAF.

The major drivers for the development of COBIT 5 for information security include:

- The need to describe information security in an enterprise context, including:
  - The full end-to-end business and IT functional responsibilities of information security
  - All aspects that lead to effective governance and management of information security, such as organizational structures, policies and culture
  - The relationship and link of information security to enterprise objectives
- An ever-increasing need for the enterprise to:
  - Maintain information risk at an acceptable level and protect information against unauthorized disclosure, unauthorized or inadvertent modifications, and possible intrusions
  - Ensure that services and systems are continuously available to internal and external stakeholders, leading to user satisfaction with IT engagement and services
  - Comply with the growing number of relevant laws and regulations as well as contractual requirements and internal policies on information and systems security and protection, and provide transparency on the level of compliance
  - Achieve all of the above while containing the cost of IT services and technology protection

The choice of approaches may be limited by an existing organizational standard, but if one does not exist, the choice should be made based on form, fit and function. In other words, a particular approach may be more consistent with existing organizational practices or may be more suitable for a particular situation. The various approaches may also entail considerably greater efforts and resources. Some are more oriented to or limited to technical architectures and will not be well suited for governance purposes.

While a specific security architecture may be of considerable benefit, it is essential that it be guided by, and tightly integrated with, the overall enterprise architecture. The development of a current enterprise architecture, such as TOGAF (**figure 1.16**), will address security as an essential component of the overall

Figure 1.16—The Structure of the TOGAF Document

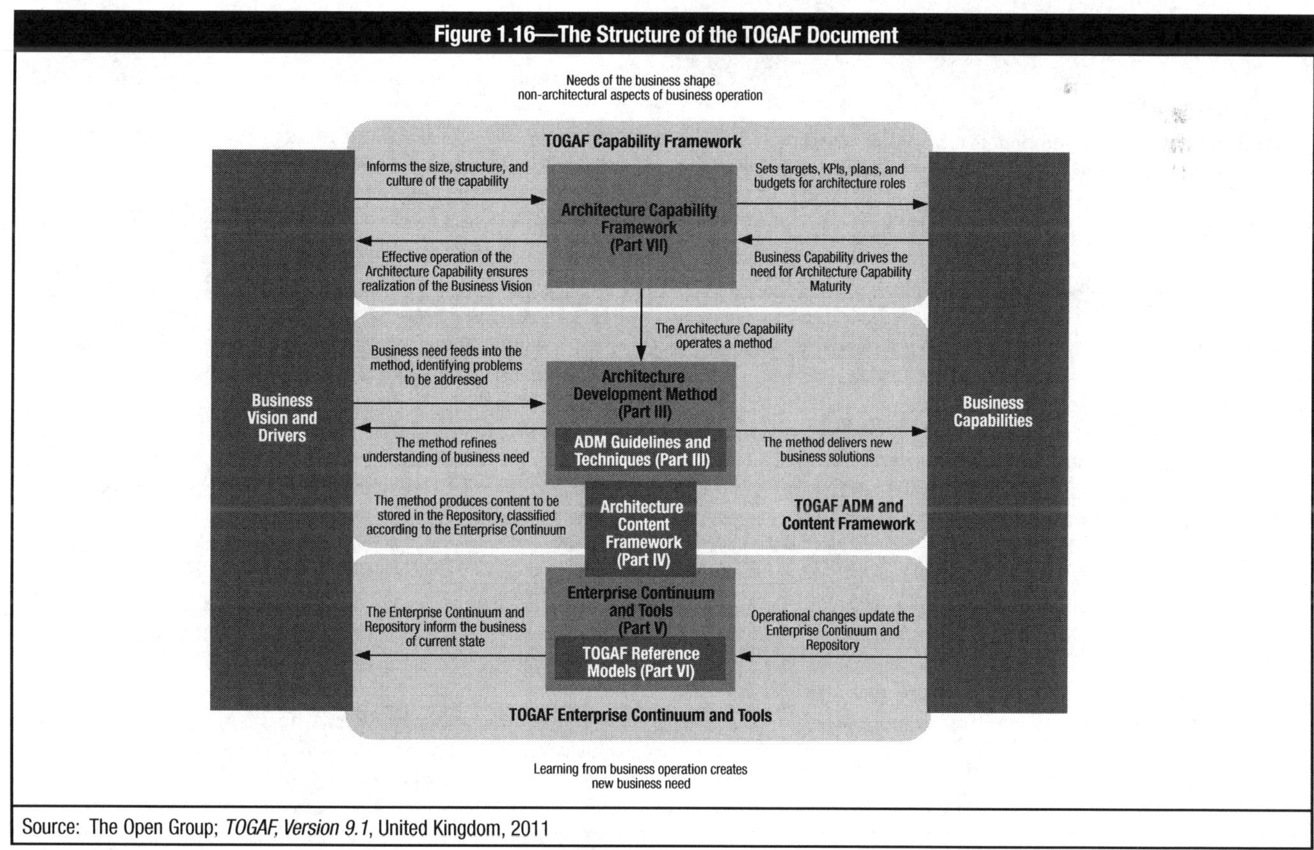

Source: The Open Group; *TOGAF, Version 9.1*, United Kingdom, 2011

design and, in most cases, will be the preferred approach to ensure effective integration.

Alternative Enterprise Architecture Frameworks

In addition to those mentioned, other approaches to enterprise and security architecture include the following (the choice of an architectural approach should be based on factors such as form, fit, function and, in some organizations, mandate):

- AGATE French Délégation Générale pour l'Armement Atelier de Gestion de l'ArchiTEcture des systèmes d'information et de communication
- Integrated Architecture Framework of Capgemini
- Interoperable Delivery (of European government services to public) Administrations, Business and Citizens (IDABC)
- Method for an Integrated Knowledge Environment (MIKE2.0), which includes an enterprise architecture framework called the Strategic Architecture for the Federated Enterprise (SAFE)
- Model-driven Architecture (MDA) of the Object Management Group
- NIH Enterprise Architecture Framework
- OBASHI business and IT methodology and framework (OBASHI)
- Open Security Architecture
- SAP Enterprise Architecture Framework, an extension of TOGAF, to better support commercial off-the-shelf programs and service-oriented architecture
- Service-Oriented Modeling Framework (SOMF)
- United Kingdom Ministry of Defence Architecture Framework (MODAF)
- United States Department of Defense Architectural Framework (DoDAF)
- United States Office of Management and Budget Federal Enterprise Architecture (FEA)

1.11.3 CONTROLS

Controls are the primary components to consider when developing an information security strategy. Controls can be physical, technical or procedural. The choice of controls must be based on a number of considerations including ensuring their effectiveness, their cost or potential restriction to business activities, and their optimal form of control.

Extensive discussion of controls, their usage, implementation and enforcement is found in section 2.7.21 Controls.

IT Controls

COBIT focuses on IT controls, which constitute the majority of controls required in many organizations although appropriate procedural and administrative controls must also be considered. Arguably, COBIT is one of the most developed and comprehensive approaches to determining control objectives for IT and subsequent implementation and management.

COBIT defines control objectives as "a statement of the desired result or purpose to be achieved by implementing control procedures in a particular IT activity."

Non-IT Controls

The information security manager must be aware that information security controls must be developed for non-IT-related information processes as well. This will include secure marking, handling and storage requirements for physical information and considerations for handling and preventing social engineering. Environmental controls must also be taken into account so otherwise secure systems are not subject to simply being stolen, as has occurred in some well-publicized cases.

Countermeasures

Countermeasures are the protection measures that directly reduce a vulnerability or a threat. Countermeasures can simply be considered targeted controls.

Countermeasures to threats should be considered from a strategic perspective. They can be passive or active but, in many cases, they may be more effective and less constricting than general controls. A countermeasure might consist of making the most sensitive information accessible only from a separate subnet, not externally available, or it might consist of changing to an inherently more secure operating system or preventing unsecured systems from connecting to the network.

Layered Defenses

Layering defenses, or defense in depth, is an important concept in designing an effective information security strategy or architecture. The layers must be designed so that the cause of failure of one layer does not also cause failure of the next layer.

The number of layers needed will be a function of asset sensitivity and criticality as well as the reliability of the defenses and the degree of exposure. Excessive reliance on a single control is likely to create a false sense of confidence. For example, a company that depends solely on a firewall can still be subject to numerous attack methodologies. A further defense may be the creation through education and awareness of a "human firewall," which can constitute a critical layer of defense. Segmenting the network can constitute another defensive layer.

1.11.4 TECHNOLOGIES

The past few decades have seen the development of numerous security technologies to address the ever-growing threats to information resources. Technology is one of the cornerstones of an effective security strategy. The information security manager must be familiar with how these technologies can serve as controls in achieving the desired state of security. Technology, however, cannot compensate for management, cultural or operational deficiencies, and the information security manager is cautioned to not place excessive reliance on these mechanisms. As **figure 1.17** demonstrates, to achieve effective defenses against security incidents, a combination of policies, standards and procedures must come together with technology.

There are a number of technologies with security mechanisms that can play a critical role in the success of an organization's security strategy. These technologies are discussed in more detail in chapter 3. Given the ongoing and rapid development of technology in this area, the prudent information security manager will utilize available resources to stay current on the latest developments.

Figure 1.17—Defense in Depth by Function

Defenses Against System Compromise	Policies, Standards, Procedures, Technology
Prevention	Authentication
	Authorization
	Encryption
	Firewalls
	Data labeling/handling/retention
	Management
	Physical security
	Intrusion prevention
	Virus scanning
	Personnel security
	Awareness and training
Containment	Authorization
	Data privacy
	Firewalls/security domains
	Network segmentation
	Physical security
Detection/notification	Monitoring
	Measurements/metrics
	Auditing/logging
	Honeypots
	Intrusion detection
	Virus detection
Reaction	Incident response
	Policy/procedure change
	Additional security mechanisms
	New/better controls
Evidence collection/ event tracking	Auditing/logging
	Management/monitoring
	Nonrepudiation
	Forensics
Recovery/restoration	Backups/restoration
	Failover/remote sites
	Business continuity/disaster recovery planning

Source: Krag Brotby

1.11.5 PERSONNEL

Personnel security is an important area of the information security strategy that must be considered as a preventive means of securing an organization. Because the most costly and damaging compromises are usually the result of insider activities, whether intentional or unintentional, the first line of defense is to try to ensure the trustworthiness and integrity of new and existing personnel. Limited background checks can provide indicators of negative characteristics, but the extent of these checks may be constrained by privacy and other laws, particularly in European Union nations.

In addition, the extent and nature of background investigations should be relevant and proportional to the sensitivity and criticality of the requirements of the position held. An extensive background investigation of a receptionist might be considered an unwarranted privacy intrusion, for example. Privacy regulations of the relevant jurisdiction must be considered because they vary greatly in different countries. Nevertheless, consideration must be given to controls aimed at preventing the employment of personnel likely to harm the organization and providing ongoing intelligence indicative of emerging or potential problems with existing staff.

Methods of tracking incidents of theft should be developed, and these events should be investigated and tracked when feasible. The appearance of what may be considered minor events may be indications of a more serious situation. It may also be an indicator of personnel involved in illegal or improper activities.

If the organization's policy is that email is not private and may be inspected by the company, and employees have been properly made aware of this policy, it may be appropriate to consider monitoring email of personnel who have been identified as potential problems. Legal protections vary on this type of monitoring and it is the responsibility of the information security officer to understand the legal requirements of the jurisdiction involved.

It may also be prudent to develop an investigation and background-checking policy and standards. These should be reviewed by the organization's legal and HR departments. These policies should also be reviewed by senior management for consistency with the organization's culture and governance approach.

1.11.6 ORGANIZATIONAL STRUCTURE

The structure of the organization will have a significant effect on developing an information security strategy. A flexible and evolving structure is likely to be useful in developing a security strategy. In more constrained structures, efforts at developing a strategy may be perceived by various factions as a threat to their autonomy or authority. This has often been the case as the information security function has risen in importance and position in the organizational structure.

Although it was sufficient in the past for information security to report to the CIO, that structure has become inadequate in effectively addressing increased risk, growing losses and sophistication of attackers. It also often results in a conflict of interest. As a result, information security increasingly has C-level management responsibilities and reports to the COO, CEO or the board of directors. There are several reasons why this is the case. One is that the broad requirements of information security are outside the scope and focus of the typical CIO. Another reason is the inherent conflict of interest resulting from information security being primarily a regulatory activity, while the CIO is often focused on cost and performance of IT. Information security, due to efforts to ensure security, is often perceived as a

constraint on IT operations. CIOs and IT departments are usually under pressure to increase performance and cut costs. Security is often the victim of these pressures. Finally, it must be considered that for information security to be effective, it must be aligned with the business rather than technology.

Centralized and Decentralized Approaches to Coordinating Information Security

An organization's cultural makeup will affect many aspects of the strategy, including the determination of whether the security organization is most effective using a centralized or decentralized approach. While many benefits can be achieved through the centralization and standardization of security, often the structure of an organization makes this an ineffective approach.

Multinational companies that choose a centralized approach need to carefully consider different local legal requirements in each country where they have a presence. For example, some countries may not allow business data to be stored or processed outside of their national boundaries; some governments may collect taxes such as a withholding tax for the software or hardware used by the entities within their jurisdiction, regardless of where the software or hardware is physically located.

One example of a distributed approach is an organization that has grown through acquisitions and operates more as independent entities rather than as a single company. In this situation, it is not unusual for separate IT groups to exist along with different software and hardware. In this example, it would not be unusual for different security organizations to exist with separate approaches, policies and procedures. In these situations, there may still be benefits to creating a single set of overall enterprisewide security policies and then deal with local differences through local standards and procedures.

A decentralized security process has advantages in that security administrators are normally closer to the users and understand local issues better. Often, they can respond more quickly to requests for changes to access rights or security incidents.

However, there are also disadvantages. For example, the quality of service may vary by location, based on the level of training the local staff possesses and the degree to which they may be encumbered with other unrelated duties.

There may be different approaches and techniques used for security depending on whether a centralized or decentralized approach is taken, but the overall responsibilities and objectives of security will not change. They still must:

- Be closely aligned with the business objectives
- Be sponsored and approved by senior management
- Have monitoring in place
- Have reporting and crisis management in place
- Have organizational continuance procedures
- Have risk management in place
- Have appropriate security awareness and training programs

1.11.7 EMPLOYEE ROLES AND RESPONSIBILITIES

With the many tasks employees must complete, it is important that the strategy includes a mechanism that defines all security roles and responsibilities, as discussed in section 1.3 Roles and Responsibilities, and incorporates them in employee job descriptions. Ultimately, if employees are compensated based on their adherence to meeting their job responsibilities, there is a better chance of achieving security governance objectives. An employee's annual job performance and objectives can include security-related measurements.

The information security manager should work with the personnel director to define security roles and responsibilities. The related competencies required for each job position should also be defined and documented.

1.11.8 SKILLS

The skills required to implement a security strategy are an important consideration. Choosing a strategy that uses skills already available is likely to be a more cost-effective option, but at times, skills may need to be developed or the required activities outsourced. A skills inventory is important to determine the resources available in developing a security strategy. Proficiency testing may be useful to determine if the requisite skills are available or can be achieved through training.

1.11.9 AWARENESS AND EDUCATION

Training, education and awareness are vital in the overall strategy because security is often weakest at the end-user level. It is also here that one should consider the need for the development of methods and processes that enable the policies, standards and procedures to be more easily followed, implemented and monitored. A recurring security awareness program aimed at end users reinforces the importance of information security and is now required by law in some jurisdictions for a number of sectors.

In most organizations, evidence indicates that the majority of personnel are not aware of security policies and standards, even where they do exist. Awareness and training programs can provide for widespread acknowledgment that security is important to the organization. Because security relies heavily on individual compliance, a robust security awareness program must be in place and must be considered in strategy development.

Broadening and deepening the appropriate skills of security personnel through training can do a great deal to improve an organization's overall security effectiveness. The challenge is to determine what the "appropriate" skills are and how they need to be improved to effectively protect the organization against the seemingly endless array of security risk.

Finding employees with the necessary combination of security skills to be effective in an ever-changing, diverse environment and complex regulatory climate can be difficult and expensive. One way some organizations attempt to ensure that all needed skills are available is to hire overqualified people. The concern with this approach is that these individuals are costly to acquire and maintain and, failing to be challenged, are often dissatisfied with the position. This can lead to excessive employee turnover, unproductive attitudes or substandard performance.

Training for new or existing security personnel to equip them with the skills needed to meet specific existing and emerging security requirements can be more cost-effective. This is a strong argument for an ongoing program of training targeted to the needs of the organization and, at the same time, providing a career path for employees based on continuous improvement.

For training to be an effective option, it must be targeted to specific systems, processes and policies and to the organization's unique and specific way of doing business and its unique security context. Anything less is likely to be inadequate, and anything more wastes resources. Properly executed, this approach can seamlessly integrate into existing programs and initiatives, shore up areas of deficiency, and align security processes with business processes.

1.11.10 AUDITS

Audits—both internal and external—are one of the main processes used to determine information security deficiencies from a controls and compliance standpoint and are one of the essential resources in strategy development.

Internal audits in larger organizations are performed by an internal audit department, generally reporting to either a CRO or to an audit committee of the board of directors. In most cases, a subunit of internal audit or an independent IT audit group will focus on information technology resources. In smaller organizations, internal IT audits or reviews may be performed by the information security manager or may be delegated to an information security officer. Typically, the focus is on policy compliance of people, processes and technology.

External audits are most often conducted by the finance department and often do not find their way to information security. Because these audits can provide powerful monitoring tools for the information security manager, it is important to ensure that the security department has access to this information. As with other departments, it is important for the information security manager to develop a good working relationship with the finance department in order to facilitate the flow of information that is essential to effective security management.

As regulatory oversight has increased, many organizations are required to file various audit and other reports with regulatory agencies. Many of these reports have information security implications that can provide useful intelligence and monitoring information for the information security manager. For example, under compliance provisions of the US Sarbanes-Oxley Act, the financial controls of public companies are required to be tested annually within 90 days of reporting to the US Securities and Exchange Commission (SEC). It is important that the results of such tests are available to the information security manager and should be required as part of strategy considerations.

1.11.11 COMPLIANCE ENFORCEMENT

Security violations are an ongoing concern for information security managers, and it is important to develop procedures for handling them as a part of developing the strategy. Senior management buy-in and backing for these procedures are critical, especially in the area of enforcement. Security managers often find that the greatest compliance problems arise with management. If there is a lack of commitment and compliance in management ranks, it may be difficult, if not impossible, to enforce compliance across an organization.

The most effective approach to compliance in an organization where openness and trust are valued and promoted by management is likely to be a system of self-reporting and voluntary compliance based on the understanding that security is clearly in everyone's best interest. This approach usually also requires that these processes be carefully thought out, clearly communicated and cooperatively implemented. Determining how to accomplish this is an element of strategy.

An important consideration for compliance enforcement is to prioritize compliance requirements. In a typical organization, there are some policies and procedures that are more critical than others and, from a security perspective, may require 100 percent compliance (e.g., the operation of a nuclear power plant). Because it may not be practical, possible or cost-effective to monitor all compliance equally, prioritization of compliance requirements should focus on areas of greatest risk and impact.

1.11.12 THREAT ASSESSMENT

While threat assessment is performed as a part of overall risk assessment, it is an important element for strategic consideration by itself. One reason is that risk treatment options should consider the most cost-effective approach for addressing any particular risk. For example, mitigation measures can address threat directly, they can reduce or eliminate vulnerabilities, or they can address and compensate for impacts. The most cost-effective choice is facilitated by separately analyzing threat and vulnerability, as well as exposure and impact.

Another reason for developing a threat profile is that the strategy should consider viable threats regardless of whether a current vulnerability is known to exist. This is a proactive approach and takes into account that all vulnerabilities cannot be known and new ones that may result in an unacceptable risk are introduced continuously.

In addition, policy development should map to a threat profile. This is because, in a broad sense, threats are relatively constant (e.g., fire, flood, earthquake, malware, theft, fraud, attacks, mistakes), whereas vulnerabilities change frequently as a result of changes in business, processes, technology and personnel. As a part of policy development, existing policies should be reviewed since, especially in older organizations, there may well exist a number of policies that are no longer relevant and do not address any conceivable threat or rational requirement. See section 2.7.10 Threats.

1.11.13 VULNERABILITY ASSESSMENT

In most organizations, technical vulnerability assessments using automated scans are common but by themselves are of limited value for security strategy development. Comprehensive vulnerability assessments that include physical elements—such as procedures, practices, technologies, facilities, service level agreements, and legal and contractual requirement—are essential. Processes and facilities are frequently the most vulnerable

components; yet, because of the greater difficulty in performing assessments on them, they are also the least frequently done. The process of developing a strategy will offer opportunities to address many of these vulnerabilities in a prudent, proactive approach. Even if no known or apparent threats exist for weaknesses discovered during these assessments, cost-effective opportunities to address systemic weaknesses during strategy development should be considered. See section 2.7.11 Vulnerabilities.

1.11.14 RISK ASSESSMENT AND MANAGEMENT

While both threat and vulnerability assessments can be useful in their own right, in considering the elements of security strategy, identifying and assessing the overall risk to the organization is also required. While threats and vulnerabilities that pose no risk to the organization may not be immediately significant, the ever-changing risk landscape makes it likely that this will not continue to be the case. In any event, to the extent that strategy development can address threats and vulnerabilities, it should do so as a matter of good practice.

Formally assessing risk is accomplished by first determining the viable threats to information resources that an organization faces. This includes physical and environmental threats (e.g., flood, fire, earthquake, pandemic) as well as technology threats (e.g., malicious software, system failure, internal and external attacks).

The next consideration is the likelihood that these threats will materialize and their probable magnitude. This is the risk identification phase of risk assessment. For example, it may be highly likely in some areas that an earthquake will occur, but it may be that, on average, it will not be severe enough to cause damage. In this case, the more infrequent major earthquakes that historically have caused damage should be considered. Some threats may be extreme, but too rare to warrant consideration (e.g., a comet strike) and can be ignored.

The next step is to determine the extent of organizational weaknesses and exposure to these threats. The combination of the frequency and magnitude and the extent of the organization's vulnerability will determine the relative level of risk. By using this information to calculate the resulting probable ALE—considering frequency, magnitude and exposure—management is in a position to decide on acceptability.

Both the frequency of occurrence as well as probable magnitude can be indicated by the extent to which these threats have materialized in the past and by the experiences of peer organizations. The consideration of frequency is important, even if the magnitude is moderate. While a single event may constitute an acceptable risk, if it is a frequent occurrence, the aggregate risk may not be acceptable. For example, an organization may accept a US $10,000 loss once a year, but that level of loss every day may not be acceptable. See section 2.7.4 Risk Assessment and Management Approaches.

1.11.15 INSURANCE

Strategy resources include the option of addressing some risk with insurance. Generally, the types of risk suitable for insuring are rare, high-impact events such as floods, hurricanes, fire, embezzlement or liability lawsuits. The most common types of insurance that can be considered include first-party, third-party and fidelity bonds. First-party insurance covers the organization in the event of damage from most sources and can include business interruption, direct loss and recovery costs. Third-party insurance deals with potential liability to third parties and generally includes defense against lawsuits and covers damages up to predetermined limits. Fidelity insurance or bonding involves protection against employee or agent theft or embezzlement.

1.11.16 BUSINESS IMPACT ANALYSIS

For management, business impact is the bottom line of risk. A BIA is an exercise that determines the consequences of losing the support of any resource to an organization and is a part of the risk assessment process. The consequences are typically reduced to financial impacts. Risk that cannot result in an appreciable impact is not important. BIAs are an important component of developing a strategy that addresses potential adverse impacts to the organization. A BIA must also be considered as a requirement to determine the criticality and sensitivity of systems and information. As such, it will provide the basis for developing an approach to information classification and addressing business continuity requirements. See section 2.8.2 Impact Assessment and Analysis.

1.11.17 RESOURCE DEPENDENCY ANALYSIS

Business dependency analysis is similar to resource dependency analysis, but is less granular and includes the elements that would be considered in a BCP, including such things as staples and paper clips. Resource dependency is similar to disaster recovery planning (DRP) and considers the systems, hardware and software required to perform specific organizational functions. Resource dependency can provide another perspective on the criticality of information resources. It can, to some extent, be used instead of an impact analysis to ensure that the strategy considers resources critical to business operations. The analysis is based on determining the resources (such as systems, software, connectivity) and dependencies (such as input processes and data repositories) required by operations critical to the organization.

1.11.18 OUTSOURCED SERVICES

Outsourcing is common, both onshore and offshore, as companies focus on core competencies and ways to cut costs. From an information security point of view, these arrangements can present risk that may be difficult to quantify and potentially difficult to mitigate. Typically, both the resources and skills of the outsourced functions are lost to the organization, which itself will present risk. Providers may operate on different standards and can be difficult to control. The security strategy should consider outsourced security services carefully to ensure that they either are not a critical single point of failure or there is a viable backup plan in the event of service provider failure.

Increasing use of cloud services presents a host of new and often obscure set of risk in addition to a range of potential benefits. Cloud benefits and risk are discussed further in section 3.11.10 Cloud Computing.

Risk posed by outsourcing can also materialize as the result of mergers and acquisitions. Typically, significant differences in culture, systems, technology and operations between the parties present a host of security challenges that must be identified and addressed. Often, in these situations, security is an afterthought and the security manager must strive to gain a presence in these activities and assess the risk for management consideration.

1.11.19 OTHER ORGANIZATIONAL SUPPORT AND ASSURANCE PROVIDERS

When developing a security strategy, there usually are a number of support and assurance service providers within an organization that should be considered a part of information security resources. These can include a variety of departments such as legal, compliance, audit, procurement, insurance, disaster recovery, physical security, training, project office and human resources. Other departments or groups, such as change management or quality assurance, also have elements of assurance as a part of their operation.

Typically, these assurance functions are not well integrated. Strategic considerations include approaches to ensure that these functions operate seamlessly to prevent gaps that may lead to security compromises. It will also serve to reduce or eliminate duplication of efforts and help prevent various groups working at cross-purposes (i.e., prevent the assurance activities of one group from undermining by the activities of another).

1.12 STRATEGY CONSTRAINTS

Numerous constraints must be considered when developing a security strategy. They will set the boundaries for the options available to the information security manager and should be thoroughly defined and understood before initiating strategy development.

1.12.1 LEGAL AND REGULATORY REQUIREMENTS

There are a number of legal and regulatory issues affecting information security that must be considered in developing a strategy. Information security is inevitably intertwined with questions of privacy; intellectual property; and contractual, civil and criminal law. Any effort to design and implement an effective information security strategy must be built on a solid understanding of the pertinent legal requirements and restrictions. Different regions in a global organization may be governed by conflicting legislation. An example of this is in the area of privacy legislation, where different cultures place different degrees of importance on privacy.

In some countries, thorough background checks may be performed on new employees, checks that are illegal under laws in other countries. To address these situations, the global organization may need to establish different security strategies for each regional division, or it can base policy on the most restrictive requirements to be consistent across the enterprise.

There are also a number of legal and regulatory issues associated with Internet business, global transmissions and transborder data flows (e.g., privacy, tax laws and tariffs, data import/export restrictions, restrictions on cryptography, warranties, patents, copyrights, trade secrets, national security). These vary depending upon the organization's location, and result in constraints and boundaries on security strategies. Research into these areas must be undertaken in conjunction with legal and regulatory departments as well as any areas of the business that may be affected.

From the perspective of the information security manager, regulatory compliance should be treated as any other risk and the extent of compliance is ultimately a business decision that must be made by senior management with input as to risk and potential impact.

The strategy must also take into consideration that personnel safety is a priority and the subject of regulations in many jurisdictions.

Requirements for Content and Retention of Business Records

There are two main aspects an information security strategy must take into consideration regarding the content and retention of business records and compliance:
- The business requirements for business records
- The legal and regulatory requirements for records

Business requirements may exceed the legal and regulatory requirements imposed by applicable legislating bodies due to the nature of the organization's business. Some organizations have business needs requiring access to data that are 10 to 20 years old, or more. This can include, for example, customer records, patient records and engineering information. As a rule, the retention strategy and subsequent policy must, at a minimum, meet the legal requirements in the applicable jurisdiction and industry.

Depending upon an organization's location and industry, regulatory bodies have requirements that an organization must comply with, including legal, medical and tax records.

Regulations such as Sarbanes-Oxley have imposed various mandatory retention requirements for various types and categories of information, regardless of the storage medium. The strategy will require the information security manager to stay current with these requirements and ensure compliance. Also to be considered are the requirements of any lawful preservation order, which would require an organization or individual to retain specific data when required by law enforcement or other authorities. It is also generally the case that archived information must be indexed sufficiently to be located and retrieved.

E-discovery

Increasingly, civil and criminal actions rely on evidence obtained from email and other electronic communications in response to a production request or subpoena. If information has been archived without being classified and cataloged, retrieving the required material can be an arduous and expensive task, as has occurred in a number of high-profile cases. The increasing prevalence of this type of discovery also contributes to the need for careful consideration of a retention policy that limits the length of time that certain kinds of information such as email are retained within the bounds of legal requirements. Generally, the best option is to have a policy that requires destruction of any data not required to be retained by law or for specific business reasons.

1.12.2 PHYSICAL

There are a variety of physical and environmental factors that may influence or constrain an information security strategy. The obvious ones include such matters as capacity, space, environmental hazards and availability of infrastructure.

Others constraints that have, on occasion, been ignored include a data center that a major oil company operation placed in a basement known to flood periodically. The security strategy should make certain that provisions are made for the consideration of environmental hazards and adequate infrastructure capacity. There also needs to be consideration for physical requirements for recovery in the case of a disaster.

The strategy must include a requirement of consideration for personnel and resource safety.

1.12.3 ETHICS

The perception an organization's customers and the public at large have of the organization's ethical behavior can have a major impact on the organization and affect its value. These perceptions are often influenced by location and culture, and an effective strategy will include ethical considerations in the areas of its operations.

1.12.4 CULTURE

Internal culture of the organization must be taken into account in developing a security strategy. The culture in which the organization operates must also be considered. A strategy that is at odds with cultural norms may encounter resistance and may make successful implementation difficult.

1.12.5 ORGANIZATIONAL STRUCTURE

Organizational structure will have a significant impact on how a governance strategy can be devised and implemented. Often, various assurance functions exist in silos that have different reporting structures and authority. Cooperation among these functions is important and typically requires senior management buy-in and involvement.

1.12.6 COSTS

The development and implementation of a strategy consumes resources, including time and money. Obviously, the strategy needs to consider the most cost-effective way it can be implemented.

Organizations often justify spending based on a project's value. With security projects, however, the control of specific risk or compliance with regulations are typically the primary drivers.

Generally, from a business perspective, a cost-benefit or other financial analysis is the most accepted approach to justifying expenditures and should be considered when developing a strategy.

A traditional, but speculative, approach is to consider the value of avoiding specific risk by estimating the potential losses incurred by a specific event and multiplying by the probability of it occurring in a given year. This results in an ALE. Thus, the cost of the controls required to preclude such an event can be compared to the ALE to determine the ROI.

Many practitioners believe that ROI is not a good approach to justifying security programs. This is considered especially true for programs implemented for the purposes of regulatory compliance. For example, under Sarbanes-Oxley, enhanced penalties consisting of long sentences in federal prisons and large fines for senior executives are prescribed for some violations. The ROI on programs to prevent such penalties can be difficult to quantify although a subjective value from the affected executives should be readily obtainable.

Technological advances, such as single sign-on and user access provisioning technologies and procedures, have resulted in savings in time and cost over traditional manual administration techniques, which may provide a reasonable basis for ROI calculations. There are a number of examples that compare the costs of traditional processes against the newer procedures, and these can be used in developing a business case.

1.12.7 PERSONNEL

A security strategy must consider what resistance may be encountered during implementation. Resistance to significant changes, as well as possible resentment against new constraints possibly viewed as making tasks more difficult or time-consuming, should be expected.

1.12.8 RESOURCES

An effective strategy must consider available budgets; the TCO of new or additional technologies; and the manpower requirements of design, implementation, operation, maintenance and eventual decommissioning. Typically, the TCO must be developed for the full life cycle of technologies, processes and personnel.

1.12.9 CAPABILITIES

The resources available to implement a strategy should include the known capabilities of the organization, including expertise and skills. A strategy that relies on demonstrated capabilities is more likely to succeed than one that does not.

1.12.10 TIME

Time is a major constraint in developing and implementing a strategy. There may be compliance deadlines that must be met or support for certain strategic operations such as a merger that must be accommodated. There may be windows of opportunity for particular business activities that mandate specific timelines for implementation of certain strategies.

1.12.11 RISK ACCEPTANCE AND TOLERANCE

The level of acceptable risk and the risk tolerance of the organization play a major role in developing an information security strategy. While they are difficult to measure, there are a variety of methods to arrive at useful approximations. One method is to develop RTOs for critical systems by performing a BIA. The shorter the times decided by appropriate managers, the greater the cost and the lower the risk appetite. The relationship of the cost of downtime compared to the cost of recovery provides a quantifiable basis for determining acceptable risk. If the organization has business continuity insurance, the deductible (the amount of loss before insurance pays a claim) is a good quantifiable indication of acceptable risk. Obviously, the cost of protection should never exceed the benefit derived.

1.13 ACTION PLAN TO IMPLEMENT STRATEGY

Implementing an information strategy will typically require a number of projects or initiatives. An analysis of the gaps between the current state and the desired state for each defined metric identifies the requirements and priorities needed for an overall plan or road map to achieve the objectives and close the gaps.

1.13.1 GAP ANALYSIS—BASIS FOR AN ACTION PLAN

A gap analysis is required for various components of the strategy previously discussed, such as maturity levels, each control objective, and each risk and impact objective. The analysis will identify the steps needed to move from the current state to the desired state to achieve the defined objectives. This exercise may need to be repeated annually, or more frequently, to provide performance and goal metrics and information needed for possible midcourse corrections in response to changing environments or other factors. A typical approach to gap analysis is to work backward from the endpoint to the current state and determine the intermediate steps need to accomplish the objectives.

CMMI or other methods can be used to assess the gap between the current and desired state. Some typical areas that should be assessed and/or ensured include:

- A security strategy with senior management acceptance and support
- A security strategy intrinsically linked with business objectives
- Security policies that are complete and consistent with strategy
- Complete standards for all relevant, consistently maintained policies
- Complete and accurate procedures for all important operations
- Clear assignment of roles and responsibilities
- An organizational structure ensuring appropriate authority for information security management without inherent conflicts of interest
- Information assets that have been identified and classified as to criticality and sensitivity
- Effective controls that have been designed, implemented and maintained
- Effective security metrics and monitoring processes in place
- Effective compliance and enforcement processes
- Tested and functional incident and emergency response capabilities
- Tested business continuity/disaster recovery plan
- Appropriate security approvals in change management processes
- Risk that is properly identified, evaluated, communicated and managed
- Adequate security awareness and training of all users
- The development and delivery of activities that can positively influence security orientation of culture and behavior of staff
- Regulatory and legal issues that are understood and addressed
- Addressing of security issues with third-party service providers
- The timely resolution of noncompliance issues and other variances

1.13.2 POLICY DEVELOPMENT

One of the most important aspects of the action plan to execute the strategy is to create or modify policies and standards as needed. Policies are the constitution of governance, standards are the law. Policies must capture the intent, expectations and direction of management. As a strategy evolves, it is vital that supporting policies are developed to articulate the strategy. For example, if the objective is to become ISO/IEC 27001:2013-compliant over a three-year period, then the strategy must consider which elements are addressed first, what resources are allocated, how the elements of the standard can be accomplished and so forth. The road map should show the steps and the sequence, dependencies and milestones. The action plan is essentially a project plan to implement the strategy following the road map.

If the objective is ISO/IEC 27001:2013 compliance, each of the relevant 14 domains and major subsections must be the subject of one or more policies. In practice, this can be effectively accomplished with about two dozen specific policies for even large institutions. Each of the policies is likely to have a number of supporting standards typically divided by security domains. In other words, a set of standards for a high-security domain will be more stringent than the standards for a low-security domain. Other standards may need to be developed for different business units, depending on their activities and regulatory requirements.

The completed strategy provides the basis for creation or modification of existing policies. The policies should be directly traceable to strategy elements. If they are not, either the strategy is incomplete or the policy is incorrect. It should be apparent that a policy that contradicts the strategy is counterproductive. The strategy is the statement of intent, expectations and direction of management. The policies must, in turn, be consistent with and support the intent and direction of the strategy.

Most organizations today have some information security policies. Typically, they have evolved over time, are usually created in response to a security problem or regulations, and are often inconsistent and sometimes contradictory. These policies generally have no relationship to a security strategy (if one exists) and only a coincidental relation to business activities.

Policies are one of the primary elements of governance. They must be properly created, accepted and validated by the board and senior management and broadly communicated throughout the organization. There may be occasions where subpolicies must be created to address unique situations separate from the bulk of the organization. An example is where a separate part of the organization is performing highly classified military work. Policies that reflect the specific security requirements for classified military work may exist as a separate set.

There are several attributes of good policies that should be considered:

- Security policies should be an articulation of a well-defined information security strategy and capture the intent, expectations and direction of management.
- Each policy should state only one general security mandate.
- Policies must be clear and easily understood by all affected parties.

- Policies should rarely be more than a few sentences long.
- There should rarely be a reason to have more than two dozen policies.

Most organizations have created security policies prior to developing a security strategy. Indeed, most organizations still have not developed a security strategy. In many cases, policy development has not followed the approach defined above and has been ad hoc in a variety of formats. Often, these policies have been written to include standards and procedures in lengthy, detailed documents compiled in large, dusty volumes relegated to the stock room.

In many cases, especially in smaller organizations, effective practices have been developed that may not be reflected in written policies. Existing practices that adequately address security requirements may usefully serve as the basis for policy and standards development. This approach minimizes organizational disruptions, communications of new policies, and resistance to new or unfamiliar constraints.

1.13.3 STANDARDS DEVELOPMENT

Standards are powerful security management tools. They set the permissible bounds for procedures and practices of technology and systems, and for people and events. Properly implemented, they are the law to the constitution of policy. They provide the measuring stick for policy compliance and a sound basis for audits. They govern the creation of procedures and guidelines. Standards set the security baselines, reflect acceptable risk and control objectives, and are the criteria for evaluating risk as being acceptable.

Standards are the predominant tool of implementing effective security governance and must be owned by the information security manager. They must be carefully crafted to provide only necessary and meaningful boundaries without unnecessarily restricting procedural options. Standards serve to interpret policies and must reflect the intent of policy. Standards must be unambiguous, consistent, and precise as to scope and audience. Standards must exist for the creation of additional standards and policies regarding format, content and required approvals.

Standards must be disseminated to those governed by them as well as those impacted. Review and modification processes must be developed as well.

Exception processes must be developed for standards not readily attainable for technological or other reasons. A process for implementing mitigating controls must be developed for out-of-compliance situations.

1.13.4 TRAINING AND AWARENESS

An effective action plan to implement a security strategy must consider an ongoing program of security awareness and training. In most organizations, evidence indicates that the majority of personnel are not aware of security policies and standards, even where they do exist. To ensure awareness of new or modified policies, all impacted personnel must be trained appropriately so they can see the connection between the policies and standards and their daily tasks. This information should be tailored to individual groups to ensure that it is relevant and must be presented in terms that are clear and understood by the intended audience. For example, presenting new standards on hardening servers is not likely to be meaningful to the shipping department. In addition to providing information to those affected by changes, it is important to ensure that staff members involved in the various aspects of implementing the strategy are also appropriately trained. This includes understanding the objectives of the strategy (KGIs), the processes that will be used and performance metrics for the various activities (KPIs), and the critical success factors (CSFs), which are the elements or events that must occur to achieve the KGIs.

See section 3.10.2 Security Awareness Training and Education for additional discussion on training and awareness.

1.13.5 ACTION PLAN METRICS

The plan of action to implement the strategy will require methods to monitor and measure progress and the achievement of milestones. As with any project plan, progress and costs must be monitored on an ongoing basis to determine conformance with the plan and to allow for midcourse corrections on a timely basis. There are likely to be a variety of near-term goals, each requiring resources and a plan of action for achievement.

There are a number of approaches that can be used for ongoing monitoring and measurement of progress. One or more of the methods used to determine the current state can be used on a regular basis to determine and chart how the current state has changed. For example, a balanced scorecard might be used effectively, by itself, as an ongoing means of tracking progress. Another commonly used approach is to use the CMMI to define both the current state and the objectives. The PAM provides a basis for performing ongoing gap analysis to determine progress toward achieving the goals.

In addition, each plan of action will benefit from developing an appropriate set of KPIs, defining CSFs and setting agreed-upon KGIs. For example, the plan of action to achieve regulatory compliance for Sarbanes-Oxley may require, among other things:
- A detailed analysis by competent legal personnel to determine regulatory requirements for affected business units
- Knowledge of the current state of compliance
- Definition by management of the required state of compliance

Key Goal Indicators

Defining clear objectives and achieving consensus on the goals are essential to developing meaningful metrics. For this particular plan, the key goals could include:
- Achieving Sarbanes-Oxley controls testing compliance mandates
- Completing independent controls testing compliance validation and attestation
- Preparing the required statement of control effectiveness

Sarbanes-Oxley requires that, for organizations publicly traded in the US, all financial controls must be tested for effectiveness within 90 days of reporting. The results of testing must be signed by the CEO and CFO and must be attested to by the organization's auditors. The results must then be included in the organization's public filings to the SEC.

Critical Success Factors

To achieve Sarbanes-Oxley compliance, certain steps must be accomplished to successfully meet the required objectives, including:
- Identifying, categorizing and defining controls
- Defining appropriate tests to determine effectiveness
- Committing resources to accomplish required testing

Large organizations have hundreds (or more) of controls that have been developed over a period of time. In many cases, these controls are *ad hoc* and have not been subject to formal processes. It will be necessary to identify control processes, procedures, structures and technologies so an appropriate testing regime can be developed. Determining the necessary resources and testing procedures will be critical to accomplishing the required tests.

Key Performance Indicators

Indicators of the key or critical performance factors necessary to achieve the objectives include:
- Control effectiveness testing plans
- Progress in controls effectiveness testing
- Results of testing control effectiveness

For management to track progress in the testing effort, appropriate testing plans must be developed that are consistent with the defined goals and encompassing the CSFs. Because of the limited time (90 days) available to perform the required tests, management will need reports on the progress and results of testing.

General Metrics Considerations

Considerations for information security metrics include ensuring that what is being measured is, in fact, relevant. Because security is difficult to measure in any objective sense, relatively meaningless metrics are often used simply because they are readily available. Metrics serve only one purpose: providing the information necessary for making decisions. It is, therefore, critical to understand what decisions must be made and who makes them and then find ways of supplying that information in an accurate and timely fashion. Different metrics are more or less useful for different parts of the organization, and should be determined in collaboration with business process owners and management.

Metrics will generally fall into one of three categories—strategic, tactical and operational. Senior management is typically not interested in detailed technical metrics, such as the number of virus attacks thwarted or password resets, but in information of a strategic nature (e.g., significant emerging risks that may impact the achievement of business objectives). While technical metrics may be of significance to the IT security manager, senior management typically wants a summary of information important from a management perspective—information that typically excludes detailed technical data. This includes:
- Progress according to plan and budget
- Significant changes in risk and possible impacts to business objectives
- Results of disaster recovery testing
- Audit results
- Regulatory compliance status

The information security manager may want more detailed tactical information, including:
- Policy compliance metrics
- Significant process, system or other changes that may affect the risk profile
- Patch management status
- Exceptions and variances to policy or standards

In organizations that have an IT security manager, it is likely that most technical security data available can be useful. This includes:
- Vulnerability scan results
- Server configuration standards compliance
- IDS monitoring results
- Firewall log analysis

Useful information security management metrics are often difficult to design and implement. Typically, the focus is on IT vulnerabilities obtained using automated scans without knowing whether a threat exists, the extent of exposure or whether there is a potential impact. From an information security management and strategic perspective, this information is of little value.

This often results in the collection of vast amounts of data to try to ensure that nothing significant is overlooked. The result can be that the sheer volume of data makes it difficult to see the big picture, and efforts should be made to develop processes to distill technical data into information needed to manage effectively.

Improvements in overall monitoring can be achieved by careful analysis of available metrics to determine their relevancy. For example, it may be interesting to know how many packets were dropped by the firewalls, but this sheds little light on risk to the organization or potential impacts. It may be information the IT department finds useful, but it is of no value to information security management. On the other hand, knowing the amount of time it takes to recover critical services after a major incident is likely to be extremely useful to all parties.

Metrics design and monitoring activities should take into consideration:
- What is important to manage information security operations
- IT security management requirements
- The needs of business process owners
- What senior management wants to know

Communication with each of the constituencies may be helpful in determining the kinds of security reports they would find useful. Reporting processes can then be devised that provide each group with the security information they require to make informed security-related decisions.

1.13.6 ACTION PLAN INTERMEDIATE GOALS

For most organizations, a variety of specific near-term goals that align with the overall information security strategy can readily be defined once the overall strategy has been completed. Based on the BIA determination of business-critical resources and the state

of security as determined by the foregoing CMMI gap analysis, prioritization of remedial activities should be straightforward.

If the objectives of the security strategy are to achieve CMMI level 4 (Quantitatively Managed) compliance or certification, then an example of a near-term action (or tactical) plan may state:

During the next 12 months:
- *Each business unit must identify current applications in use*
- *Twenty-five percent of all stored information must be reviewed to determine ownership, criticality and sensitivity*
- *Each business unit will complete a BIA for information resources to identify critical resources*
- *Business units must achieve regulatory compliance*
- *All security roles and responsibilities must be defined*
- *A process will be developed to ensure business process linkages*
- *A comprehensive risk assessment must be performed for each business unit*
- *All users must be educated on an acceptable use policy*
- *All policies must be reviewed and revised as necessary for consistency with strategic security objectives*
- *Standards must exist for all policies*

Near-term goals and milestones are required as part of the action plans; however, the long-term desired state needs to be clearly defined to maximize potential synergies and ensure that short- or intermediate-term action plans are ultimately aligned with the end goals. For example, a tactical solution that needs to be replaced because it will not integrate into the overall plan is likely to be more costly than one that will integrate.

It is important that the strategy and long-range plan serve to integrate near-term tactical activities. This counters the tendency to implement point solutions that are typical of the firefighting, crisis mode of operation in which many security departments find themselves. As many security managers have discovered, numerous unintegrated solutions, implemented in response to a series of crises over a period of years, become increasingly costly and difficult to manage.

1.14 INFORMATION SECURITY PROGRAM OBJECTIVES

Implementing the strategy with an action plan will result in an information security program. The program is, essentially, the project plan to implement and establish ongoing management of some part or parts of the strategy.

The objective of the information security program is to protect the interests of those relying on information and the processes, systems and communications that handle, store and deliver the information, from harm resulting from failures of availability, confidentiality and integrity.

While emerging definitions are adding concepts such as information usefulness and possession (the latter to cope with theft, deception and fraud), the networked economy certainly has added the need for trust and accountability in electronic transactions.

For most organizations, the security objective is met when:
- Information is available and usable when required, and the systems that provide it can appropriately resist attacks (availability).
- Information is observed by or disclosed to only those who have a right to know (confidentiality).
- Information is protected against unauthorized modification (integrity).
- Business transactions, as well as information exchanges between enterprise locations or with partners, can be trusted (authenticity and nonrepudiation).

The relative priority and significance of availability, confidentiality, integrity, authenticity and nonrepudiation vary according to the data within the information system and the business context in which they are used. For example, integrity is especially important relative to management information due to the impact that information has on critical strategy-related decisions. Confidentiality may be the most important based on regulatory or legal requirements regarding personal, financial or medical information or to protect trade secrets.

It is important to understand that these concepts apply equally to electronic systems as well as physical systems. Confidentiality, for example, is as much at risk from social engineering or "dumpster diving" as it is from a successful external attack. Additionally, integrity of information can be compromised at least as easily from forged physical inputs to the system as from electronic compromise.

It must also be considered that many significant losses occur from insiders as well as from external attacks. The result is that controls used to detect anomalies and ensure integrity of systems must be equally concerned with nontechnical attacks by insiders.

1.15 CASE STUDY

The CISO of a large manufacturing company with three years' experience began to realize that the information security function was having less influence and information risk recommendations were increasingly being ignored by the organization's business units. Despite continuing efforts to communicate with managers from the various corporate departments and business units and a positively regarded security training and awareness program, she was having little success in getting management buy-in for implementing needed controls.

The decreasing importance of information security became apparent when she was not invited to a series of meetings concerning significant changes in company operations. During subsequent budget meetings, several executives (including the CEO) said they were unsure of the value that information security contributes to the company and suggested the budget might be better allocated elsewhere. In addition, the CISO has noticed that she has not been invited to a board meeting to present updates regarding the information security program in more than a year.

The company is publicly held and has been in operation for nearly 90 years. It has been profitable for all but a few years of its existence, although profits were down during the recent worldwide recession. The company has approximately 9,000 employees and a few hundred long-term contractors. It makes a wide variety of small hand and power tools and is known internationally for the wide variety of high-quality products it makes.

The CEO is also the chairman of the board of directors and appoints people to the board on the basis of their willingness to give approval for his initiatives with little delay. Consequently, the board provides little oversight and guidance to the business, and although the audit committee of the board is increasingly concerned about the unmitigated risks the organization faces, it has little influence on the CEO. Due to the CEO's meager appreciation for information security, the CISO is worried that her job is at risk unless she can improve management's perception of the value of information security.

An additional complication is the recent recession, which saw a drop in the sales of manufactured goods. Competitors were selling goods below cost to reduce inventories and increase cash flow. As a result, the CEO has seen this increase in competition and reduction in profits as further justification to reduce expenses not directly correlated with revenue production.

Business units are the backbone of this company. Except for sharing the machinery and equipment used in manufacturing, each is a fiercely independent silo with the mission of being as profitable as possible. Business unit managers are highly valued and are placed at a fairly senior level in the organizational chart; all of them report to the COO. Each business unit often faces short deadlines to beat the competition to market with new products. As a result, business unit administrators must carefully plan their use of machinery and equipment within each plant because several business units often need machinery and equipment at the same time. Additionally, the machinery and equipment must be available on a 24/7 basis. All machinery and equipment are controlled by process automation software on computers that run in control rooms. This software is highly distributed to the point that any kind of network disruption or outage is likely to adversely affect production.

The CISO has come to realize that the information security practice set up three years ago is not adequate to deal with current ever-growing threats. Although she made efforts to adjust the program according to changes that occurred within the organization, it is clear that some areas were not sufficiently considered. One of these areas is information security metrics relevant to management's concerns. Presently, the following types of metrics are used to measure the success of the information security program:

- Percentage of Windows systems that run antivirus software that is updated daily
- Percentage of incoming network traffic that is evaluated by a firewall on an exterior gateway
- Percentage of network traffic that is monitored by an IDS

1.15 CASE STUDY *(CONT.)*

- Percentage of systems that are patched within the required time period
- Percentage of systems that are in compliance with baseline security standards
- Percentage of spam that is deleted by the company's spam wall
- The number of vulnerabilities uncovered during automated scans

She realizes that metrics she considered sufficient three years ago when the program was implemented are clearly not sufficient to be useful to management in their decision-making process or adequately persuasive to generate support for the information security program. She determines that while the current metrics have been useful operationally, they are not relevant or meaningful at a tactical and strategic level and this is the underlying cause of management failing to understand the value of the information security program and the need for a number of improvements that would increase budget requirements.

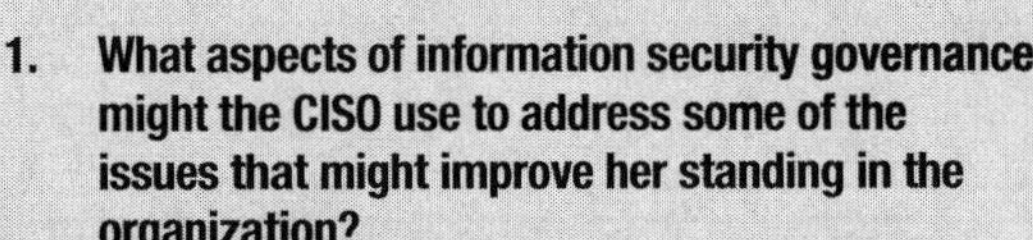

1. **What aspects of information security governance might the CISO use to address some of the issues that might improve her standing in the organization?**

2. **In what major ways and areas can information security help the business in reaching its goals?**

3. **How adequate is the current set of information security metrics in terms of the company's business goals and directions?**

4. **How can the current set of metrics be improved? Why?**

Answers on page 71.

CHAPTER 1 ANSWER KEY

KNOWLEDGE CHECK: ROLES AND RESPONSIBILITIES (PAGE 37)

Note: *COBIT 5: Enabling Processes* provided the basis for this activity.

	Information Security Manager	Board of Directors	Chief Information Officer	Chief Executive Officer	Business Process Owner
Define the target IT capabilities.	C		R	A	I
Conduct a gap analysis.	R		A		R
Define the strategic plan and road map.	C		A	C	
Communicate the IT strategy and direction.	I	I	R	R	I

KNOWLEDGE CHECK: METRICS (PAGE 44)

1. Which of the above metrics are SMART?
   Answer: A and B

2. Which of the above metrics could be considered strategic?
   Answer: B, C and D

3. Which of the above metrics are related to performance management?
   Answer: A, C, D and E

4. How could the metrics that are not SMART be rewritten to be SMART?
   Suggested revisions:
   C. Percentage of critical business systems that are patched within seven days from the patch being released
   D. Percentage of network incidents that are responded to in the time allowed over a three-month period
   E. Number of information security activities that are at CMMI level 3 (Defined) or above

CASE STUDY ANSWERS (PAGE 70)

Question 1: The following aspects could help the CISO improve her standing in the organization:

- Obtaining management commitment: Obtaining management commitment in this situation would help the CISO explain the importance of information security to the organization and help her to influence the board of directors to consider information security in its decisions.
- Business case development: Developing a business case is one key tool in obtaining management commitment. It provides the information required for an organization to decide whether a project should proceed. The business case can be used to prove the worth of IT and information security initiatives in the organization and provide concrete justification for the costs.
- Information security strategy development: An information security strategy can help to ensure strategic alignment, value delivery and resource optimization to the business. It will help to define the direction of information security in terms of the business, thus ensuring that it has proven benefit to the organization.
- Information security program metrics: Information security program metrics can be used to justify that the program is meeting its defined goals and providing value to the business.

Question 2: Information security can help this business reach its goals by implementing controls that provide assurance of achieving the enterprise's aforementioned goals and ensuring that security objectives are aligned with organizational goals and business strategy.

Question 3: The current set of information security metrics does not tie well to the business's goals. The metrics are not very meaningful in the context of how information security supports the business strategy; they simply measure arbitrary factors that may not accurately portray the information security function. This set of metrics seems to be based on discussions with the CEO, who, as it has been seen, does not have a sufficient understanding of the information security function and how it can align to meet business objectives.

Question 4: The current set of metrics could be improved by applying SMART concepts to them. They should be specific, measurable, attainable, relevant and timely. They should tie into the information security program objectives as they support the objectives of the business. Some sample objectives include:

- Percentage of manufacturing processes per month that are uninterrupted by security problems
- Downtime for computers used in computer-aided engineering and computer-aided manufacturing due to security problems, supported by a quantification of impact in financial terms
- Impact on transactions with third-party entities (e.g., shippers and suppliers) that have been affected by security problems
- Mean time per security incident from time of onset to incident closure
- Monthly cost of interruptions in manufacturing processes due to security problems
- Number and cost of security incidents or events that could have been prevented if certain controls were implemented
- Number and nature of security objectives not achieved due to funding constraints or lack of business unit cooperation
- The ratio of affirmative media reports compared to negative reports
- Maturity level of organization's information security activities

Chapter 2:

Information Risk Management

Section One: Overview

Section Two: Content

erview

dge base that the information nd to appropriately apply risk principles and practices to an organization's information security program.

DOMAIN DEFINITION

Manage information risk to an acceptable level based on risk appetite to meet organizational goals and objectives.

LEARNING OBJECTIVES

The objective of this domain is to ensure that the CISM candidate:

- Understands the importance of risk management as a tool for meeting business needs and developing a security management program to support these needs
- Understands ways to identify, rank and respond to risk in a way that is appropriate as defined by organizational directives
- Assesses the appropriateness and effectiveness of information security controls
- Reports on information security risk effectively

CISM EXAM REFERENCE

This domain represents 30 percent of the CISM examination (approximately 45 questions).

TASK AND KNOWLEDGE STATEMENTS

TASK STATEMENTS

There are nine tasks within this domain that a CISM candidate must know how to perform:

T2.1 Establish and/or maintain a process for information asset classification to ensure that measures taken to protect assets are proportional to their business value.

T2.2 Identify legal, regulatory, organizational and other applicable requirements to manage the risk of noncompliance to acceptable levels.

T2.3 Ensure that risk assessments, vulnerability assessments and threat analyses are conducted consistently, and at appropriate times, to identify and assess risk to the organization's information.

T2.4 Identify, recommend or implement appropriate risk treatment/response options to manage risk to acceptable levels based on organizational risk appetite.

T2.5 Determine whether information security controls are appropriate and effectively manage risk to an acceptable level.

T2.6 Facilitate the integration of information risk management into business and IT processes (e.g., systems development, procurement, project management) to enable a consistent and comprehensive information risk management program across the organization.

T2.7 Monitor for internal and external factors (e.g., threat landscape, cybersecurity, geopolitical, regulatory change) that may require reassessment of risk to ensure that changes to existing or new risk scenarios are identified and managed appropriately.

T2.8 Report noncompliance and other changes in information risk to facilitate the risk management decision-making process.

T2.9 Ensure that information security risk is reported to senior management to support an understanding of potential impact on the organizational goals and objectives.

KNOWLEDGE STATEMENTS

There are 19 knowledge statements within the information risk management domain:

K2.1 Knowledge of methods to establish an information asset classification model consistent with business objectives

K2.2 Knowledge of considerations for assigning ownership of information assets and risk

K2.3 Knowledge of methods to identify and evaluate the impact of internal or external events on information assets and the business

K2.4 Knowledge of methods used to monitor internal or external risk factors

K2.5 Knowledge of information asset valuation methodologies

K2.6 Knowledge of legal, regulatory, organizational and other requirements related to information security

K2.7 Knowledge of reputable, reliable and timely sources of information regarding emerging information security threats and vulnerabilities

K2.8 Knowledge of events that may require risk reassessments and changes to information security program elements

K2.9 Knowledge of information threats, vulnerabilities and exposures (including cybersecurity) and their evolving nature

K2.10 Knowledge of risk assessment and analysis methodologies

K2.11 Knowledge of methods used to prioritize risk scenarios and risk treatment/response options

K2.12 Knowledge of risk reporting requirements (e.g., frequency, audience, content)

K2.13 Knowledge of risk treatment/response options (avoid, mitigate, accept or transfer) and methods to apply them

K2.14 Knowledge of control baselines and standards and their relationships to risk assessments

K2.15 Knowledge of information security controls and the methods to analyze their effectiveness

K2.16 Knowledge of gap analysis techniques as related to information security

K2.17 Knowledge of techniques for integrating information security risk management into business and IT processes

K2.18 Knowledge of compliance reporting requirements and processes

K2.19 Knowledge of cost/benefit analysis to assess risk treatment options

RELATIONSHIP OF TASK TO KNOWLEDGE STATEMENTS

The task statements are what the CISM candidate is expected to know how to perform. The knowledge statements delineate each of the areas in which the CISM candidate must have a good understanding to perform the tasks. The task and knowledge statements are mapped, insofar as it is possible to do so. Note that although there is often overlap, each task statement will generally map to several knowledge statements.

Task and Knowledge Statements Mapping			
Task Statement		**Knowledge Statements**	
T2.1	Establish and/or maintain a process for information asset classification to ensure that measures taken to protect assets are proportional to their business value.	K2.1	Knowledge of methods to establish an information asset classification model consistent with business objectives
		K2.2	Knowledge of considerations for assigning ownership of information assets and risk
		K2.3	Knowledge of methods to identify and evaluate the impact of internal or external events on information assets and the business
		K2.5	Knowledge of information asset valuation methodologies
		K2.7	Knowledge of reputable, reliable and timely sources of information regarding emerging information security threats and vulnerabilities
		K2.8	Knowledge of events that may require risk reassessments and changes to information security program elements
T2.2	Identify legal, regulatory, organizational and other applicable requirements to manage the risk of noncompliance to acceptable levels.	K2.1	Knowledge of methods to establish an information asset classification model consistent with business objectives
		K2.4	Knowledge of methods used to monitor internal or external risk factors
		K2.6	Knowledge of legal, regulatory, organizational and other requirements related to information security
T2.3	Ensure that risk assessments, vulnerability assessments and threat analyses are conducted consistently, and at appropriate times, to identify and assess risk to the organization's information.	K2.7	Knowledge of reputable, reliable and timely sources of information regarding emerging information security threats and vulnerabilities
		K2.8	Knowledge of events that may require risk reassessments and changes to information security program elements
		K2.9	Knowledge of information threats, vulnerabilities and exposures (including cybersecurity) and their evolving nature
		K2.10	Knowledge of risk assessment and analysis methodologies
		K2.14	Knowledge of control baselines and standards and their relationships to risk assessments
		K2.16	Knowledge of gap analysis techniques as related to information security
T2.4	Identify, recommend or implement appropriate risk treatment/response options to manage risk to acceptable levels based on organizational risk appetite.	K2.10	Knowledge of risk assessment and analysis methodologies
		K2.11	Knowledge of methods used to prioritize risk scenarios and risk treatment/response options
		K2.13	Knowledge of risk treatment/response options (avoid, mitigate, accept or transfer) and methods to apply them
		K2.19	Knowledge of cost/benefit analysis to assess risk treatment options
T2.5	Determine whether information security controls are appropriate and effectively manage risk to an acceptable level.	K2.11	Knowledge of methods used to prioritize risk scenarios and risk treatment/response options
		K2.12	Knowledge of risk reporting requirements (e.g., frequency, audience, content)
		K2.13	Knowledge of risk treatment/response options (avoid, mitigate, accept or transfer) and methods to apply them
		K2.14	Knowledge of control baselines and standards and their relationships to risk assessments
		K2.15	Knowledge of information security controls and the methods to analyze their effectiveness
		K2.16	Knowledge of gap analysis techniques as related to information security
		K2.17	Knowledge of techniques for integrating information security risk management into business and IT processes
		K2.19	Knowledge of cost/benefit analysis to assess risk treatment options
T2.6	Facilitate the integration of information risk management into business and IT processes (e.g., systems development, procurement, project management) to enable a consistent and comprehensive information risk management program across the organization.	K2.17	Knowledge of techniques for integrating information security risk management into business and IT processes
		K2.18	Knowledge of compliance reporting requirements and processes

Task and Knowledge Statements Mapping *(cont.)*	
Task Statement	**Knowledge Statements**
T2.7 Monitor for internal and external factors (e.g., threat landscape, cybersecurity, geopolitical, regulatory change) that may require reassessment of risk to ensure that changes to existing or new risk scenarios are identified and managed appropriately.	K2.3 Knowledge of methods to identify and evaluate the impact of internal or external events on information assets and the business K2.4 Knowledge of methods used to monitor internal or external risk factors K2.6 Knowledge of legal, regulatory, organizational and other requirements related to information security K2.7 Knowledge of reputable, reliable and timely sources of information regarding emerging information security threats and vulnerabilities K2.8 Knowledge of events that may require risk reassessments and changes to information security program elements K2.9 Knowledge of information threats, vulnerabilities and exposures (including cybersecurity) and their evolving nature K2.10 Knowledge of risk assessment and analysis methodologies K2.11 Knowledge of methods used to prioritize risk scenarios and risk treatment/response options K2.13 Knowledge of risk treatment/response options (avoid, mitigate, accept or transfer) and methods to apply them
T2.8 Report noncompliance and other changes in information risk to facilitate the risk management decision-making process.	K2.6 Knowledge of legal, regulatory, organizational and other requirements related to information security K2.7 Knowledge of reputable, reliable and timely sources of information regarding emerging information security threats and vulnerabilities K2.8 Knowledge of events that may require risk reassessments and changes to information security program elements K2.12 Knowledge of risk reporting requirements (e.g., frequency, audience, content) K2.18 Knowledge of compliance reporting requirements and processes
T2.9 Ensure that information security risk is reported to senior management to support an understanding of potential impact on the organizational goals and objectives.	K2.12 Knowledge of risk reporting requirements (e.g., frequency, audience, content) K2.18 Knowledge of compliance reporting requirements and processes

TASK STATEMENT REFERENCE GUIDE

The following section contains the task statements a CISM candidate is expected to know how to accomplish mapped to the areas in the review manual with information that support the execution of the task. The references in the manual focus on the knowledge the information security manager must know to accomplish the tasks and successfully negotiate the exam.

Task Statement Reference Guide	
Task Statement	**Reference in Manual**
T2.1 Establish and/or maintain a process for information asset classification to ensure that measures taken to protect assets are proportional to their business value.	1.8.1 The Goal 2.3.1 Ensure Asset Identification, Classification and Ownership 2.8 Information Asset Classification
T2.2 Identify legal, regulatory, organizational and other applicable requirements to manage the risk of noncompliance to acceptable levels.	2.5.3 Defining the External Environment 2.5.4 Defining the Internal Environment 2.7.22 Legal and Regulatory Requirements
T2.3 Ensure that risk assessments, vulnerability assessments and threat analyses are conducted consistently, and at appropriate times, to identify and assess risk to the organization's information.	2.7 Risk Assessment
T2.4 Identify, recommend or implement appropriate risk treatment/response options to manage risk to acceptable levels based on organizational risk appetite.	2.7.18 Risk Treatment (Response) Options
T2.5 Determine whether information security controls are appropriate and effectively manage risk to an acceptable level.	2.12 Security Control Baseline 3.12 Controls and Countermeasures 3.12.10 Control Testing and Modification
T2.6 Facilitate the integration of information risk management into business and IT processes (e.g., systems development, procurement, project management) to enable a consistent and comprehensive information risk management program across the organization.	2.11 Risk Management Integration With Life Cycle Processes

Task Statement Reference Guide			
Task Statement		**Reference in Manual**	
T2.7	Monitor for internal and external factors (e.g., threat landscape, cybersecurity, geopolitical, regulatory change) that may require reassessment of risk to ensure that changes to existing or new risk scenarios are identified and managed appropriately.	2.13 3.13	Risk Monitoring and Communication Security Program Metrics and Monitoring
T2.8	Report noncompliance and other changes in information risk to facilitate the risk management decision-making process.	2.13.3	Reporting Significant Changes in Risk
T2.9	Ensure that information security risk is reported to senior management to support an understanding of potential impact on the organizational goals and objectives.	2.13.3	Reporting Significant Changes in Risk

SUGGESTED RESOURCES FOR FURTHER STUDY

Brotby, W. Krag; *Information Security Management Metrics: A Definitive Guide to Effective Security Monitoring and Measurement*, Auerbach Publications, USA, 2009

International Electrotechnical Commission (IEC), *IEC 31010:2009 Risk management—Risk assessment techniques*, Switzerland, 2009

International Organization for Standardization (ISO), *ISO/IEC 27005:2011 Information technology—Security techniques—Information security risk management*, Switzerland, 2011

ISO, *ISO 31000:2009 Risk management—Principles and guidelines*, Switzerland, 2009

ISACA, *COBIT® 5 for Risk*, USA, 2013, *www.isaca.org/cobit*

ISACA, *Implementing and Continually Improving IT Governance*, USA, 2010

The Institute of Internal Auditors Australia, *HB 158-2010 Delivering assurance based on ISO 31000:2009—Risk management—Principles and guidelines,* Australia, 2010

National Institute of Standards and Technology (NIST), *NIST Special Publication 800-39 Managing Information Security Risk: Organization, Mission, and Information System View*, USA, 2011

NIST, *NIST Special Publication 800-30, Revision 1 Guide for Conducting Risk Assessments*, USA, 201

Note: Publications in bold are stocked in the ISACA Bookstore.

SELF-ASSESSMENT QUESTIONS

CISM exam questions are developed with the intent of measuring and testing practical knowledge in information security management. All questions are multiple choice and are designed for one best answer. Every CISM question has a stem (question) and four options (answer choices). The candidate is asked to choose the correct or best answer from the options. The stem may be in the form of a question or incomplete statement. In some instances, a scenario or a description problem may also be included. These questions normally include a description of a situation and require the candidate to answer two or more questions based on the information provided. Many times a CISM examination question will require the candidate to choose the most likely or best answer.

In every case, the candidate is required to read the question carefully, eliminate known incorrect answers and then make the best choice possible. Knowing the format in which questions are asked, and how to study to gain knowledge of what will be tested, will go a long way toward answering them correctly.

2-1 The overall objective of risk management is to:

A. eliminate all vulnerabilities, if possible.
B. reduce risk to the lowest possible level.
C. manage risk to an acceptable level.
D. implement effective countermeasures.

2-2 The information security manager should treat regulatory compliance as:

A. an organizational mandate.
B. a risk management priority.
C. a purely operational issue.
D. another risk to be managed.

2-3 To address changes in risk, an effective risk management program should:

A. ensure that continuous monitoring processes are in place.
B. establish proper security baselines for all information resources.
C. implement a complete data classification process.
D. change security policies on a timely basis to address changing risk.

2-4 Information classification is important to properly manage risk **PRIMARILY** because:

A. it ensures accountability for information resources as required by roles and responsibilities.
B. it is a legal requirement under various regulations.
C. it ensures adequate protection of assets commensurate with the degree of risk.
D. asset protection can then be based on the potential consequences of compromise.

2-5 Vulnerabilities discovered during an assessment should be:

A. handled as a risk, even though there is no threat.
B. prioritized for remediation solely based on impact.
C. a basis for analyzing the effectiveness of controls.
D. evaluated for threat, impact and cost of mitigation.

2-6 Indemnity agreements can be used to:

A. ensure an agreed-upon level of service.
B. reduce impacts on organizational resources.
C. transfer responsibility to a third party.
D. provide an effective countermeasure to threats.

2-7 Residual risk can be determined by:

A. assessing remaining vulnerabilities.
B. performing a threat analysis.
C. conducting a risk assessment.
D. implementing risk transfer.

2-8 Data owners are **PRIMARILY** responsible for creating risk mitigation strategies to address which of the following areas?

A. Platform security
B. Entitlement changes
C. Intrusion detection
D. Antivirus controls

2-9 A risk analysis should:

A. limit the scope to a benchmark of similar companies.
B. assume an equal degree of protection for all assets.
C. address the potential size and likelihood of loss.
D. give more weight to the likelihood vs. the size of the loss.

2-10 Which of the following is the **FIRST** step in selecting the appropriate controls to be implemented in a new business application?

A. Business impact analysis (BIA)
B. Cost-benefit analysis
C. Return on investment (ROI) analysis
D. Risk assessment

ANSWERS TO SELF-ASSESSMENT QUESTIONS

2-1 A. It is not possible to eliminate all vulnerabilities, and vulnerabilities that have no impact or exposure do not require the expenditure of resources to remediate or achieve elimination.
B. Reduction of risk to the lowest level generally is more costly and unnecessarily restrictive and the goal is to achieve control objectives that will result in acceptable levels of risk.
C. The objective of risk management is managing risk to a level acceptable to the organization.
D. Countermeasures are a targeted control and a part of the risk treatment process to achieve the overall objective of risk management.

2-2 A. A regulatory requirement, whether mandated by the organization, should be treated as any other risk.
B. A regulatory requirement is a priority to the extent the level of risk compares to other risk. Priorities for risk management are usually based on probability and impact—the more likely the compromise and the more severe the consequences, the higher the priority.
C. Regulatory compliance is not just an operational issue, but a management issue.
D. There are numerous regulations that may affect an organization. Priority will be a management decision based on those regulations with the greatest level of enforcement (risk) and the most severe sanctions (consequences or impact) in addition to the cost of compliance (mitigation), just as with any other risk. In some cases, management may decide that the cost of potential sanctions will be less than the cost of compliance. While it is generally preferable to be as compliant as reasonably possible, the extent of regulatory compliance is a management decision, not a security decision. All risk must be prioritized, and compliance may not pose the greatest risk.

2-3 **A. Risk changes as threats, vulnerabilities or potential impacts change over time. The risk management program must have processes in place to monitor those changes and modify countermeasures, as appropriate, to maintain acceptable levels of residual risk.**
B. Security baselines are set to achieve acceptable levels of particular levels of identified risk. Baselines do not address changes in risk and may need to be changed when risk changes.
C. Data classification is based on business value of the data (i.e., the data's sensitivity and/or criticality) and does not address changes in risk.
D. Policies should not require changes as a result of changes in risk. Standards and procedures may need to be changed to address significant changes in risk.

2-4 A. Classification does not ensure accountability.
B. Classification is not generally a legal requirement.
C. Classification is not based on risk.
D. Classification is based on potential impact or consequences of compromise.

2-5 A. Vulnerabilities may not be exposed to potential threats. Also, there may be no threat or possibly little or no impact even if exploited. While threats are always evolving, without additional information, the appropriate treatment cannot be determined.
B. Vulnerabilities should be prioritized for remediation based on probability of compromise (which is affected by the level of exposure), impact and cost of remediation.
C. Vulnerabilities discovered will, to some extent, show whether existing controls are in place to address a potential risk but they do not indicate the controls' effectiveness.
D. Vulnerabilities uncovered should be evaluated and prioritized based on whether there is a credible threat, the impact if the vulnerability is exploited and the cost of mitigation. If there is a potential threat but little or no impact if the vulnerability is exploited, there is little risk, and it may not be cost-effective to address it.

2-6 A. Indemnity agreements are not used to define service levels; these are provided by service level agreements (SLAs).
B. Indemnity agreements serve to reduce financial impacts by providing compensation for adverse events in the scope of the agreement.
C. Legal responsibility cannot be transferred by indemnity agreements or any other instrument.
D. Indemnity agreements are not a countermeasure to threats, but they can be considered a compensating control.

2-7 A. Determining remaining vulnerabilities after countermeasures are in place says nothing about threats; therefore, risk cannot be determined.
B. Risk cannot be determined by threat analysis alone.
C. Regardless of whether risk is residual, it is determined by a risk assessment.
D. Transferring all risk is not relevant to determining residual risk.

2-8 A. Platform security is the responsibility of IT.
B. Data owners are concerned with, and responsible for, who has access to their resources; therefore, they need to be concerned with the strategy of how to mitigate risk of data resource usage.
C. Intrusion detection is the responsibility of IT.
D. Antivirus controls are typically IT security concerns.

2-9 A. A risk analysis would not normally consider the benchmark of similar companies as providing relevant information other than for comparison purposes.
B. Assuming an equal degree of protection would be rational only in the rare event that all assets are similar in sensitivity and criticality.
C. A risk analysis deals with the potential size and likelihood of loss.
D. Because the likelihood determines (on an annualized basis) the size of the loss, both elements must be considered in the calculation.

2-10 A. If the risk is determined to be unacceptable, a business impact analysis (BIA) can be used to determine the level of mitigation necessary.
B. A cost-benefit analysis can be used to determine whether mitigation cost is appropriate, considering the potential impact after a BIA has been performed.
C. Return on investment (ROI) analysis focuses on the business value of the control compared to its cost over time.
D. It is necessary to first consider the risk and determine whether it is acceptable to the organization. Risk assessment can identify threats and vulnerabilities and calculate the risk. Controls are evaluated by comparing the cost of the control against the potential impact if the risk were exploited.

Section Two: Content

2.0 INTRODUCTION

Risk can be defined as the combination of the probability (or likelihood) of an event and its consequences. The probability of an event is the likelihood that a given threat will exploit an exposed vulnerability. If there are no consequences or impact, there is considered to be no risk. Conversely, the greater the consequences or impact, the greater the risk.

Exposure, or the extent to which a vulnerability is exposed, to a threat also factors into the risk equation since the extent of exposure affects the probability of compromise (i.e., less exposure results in less likelihood, or frequency, of compromise, thereby reducing risk). Exposure is also referred to as the attack surface. It is affected by the extent and effectiveness of controls and where a particular device is located within a network (i.e., a server in the middle of the network is likely to be less exposed to attack than one on the perimeter).

Classifying assets according to their business value is an essential part of effective risk management because the greater the value, the greater the potential impact and, therefore, the greater the risk. Business value is a combination of criticality to operations and/or sensitivity, which is a function of the possible damage to the organization resulting from unauthorized disclosure (e.g., strategic plans, customer lists). As it is prudent to protect high-value information assets to a greater extent, the most cost-effective management of information risk can be achieved by allocating protection resources in proportion to business value. The task is to develop a classification schema that optimizes this allocation as well as the criteria by which these resources can be effectively classified. The basis for classification levels is determined by the potential impact to the organization if the asset is compromised.

Organizations operate under a large number of legal and regulatory requirements that vary considerably among jurisdictions. In addition to external requirements, the information security manager must be knowledgeable about internal policies, standards and procedures. From a risk perspective, a number of nonexplicit constraints must also be identified including local and organizational culture and perceptions of ethical behavior that, unless effectively addressed, can result in significant reputational damage and the loss of customer and public trust. One of the responsibilities of the information security manager is to use an appropriate risk identification, analysis and evaluation approach to identify all forms of information-related risk and their potential impacts. Another responsibility is to develop response options so management can make informed decisions on acceptable risk management approaches.

A structured, consistent process for risk assessment is essential if it is to be managed successfully. The first step is to understand the threat landscape and then to determine the vulnerabilities of the organization that makes it susceptible to compromise. Both technical and physical weaknesses must be assessed as well as the exposure of those vulnerabilities to viable threats. The extent to which the identified risk is mitigated by existing controls must also be considered.

Once a risk has been identified, analyzed and evaluated to determine whether it meets the criteria for acceptability, the options for treatment (or response) must be analyzed. The choices available include accept the risk, transfer the risk, cease the activity creating the risk (avoid) or mitigate the risk. The costs and potential consequences for each of the options must be considered. If mitigating controls are selected because the risk exceeds the organization's risk appetite, then the mitigation options and cost-effectiveness must be assessed. In addition, the control risk of the possible risk reduction options must also be determined (i.e., the risk that the control will fail or be inadequate).

Control effectiveness typically changes over time as risk changes. As a consequence, it is essential to periodically test and assess whether existing controls continue to be effective and still meet the control objectives. Testing controls will generally require simulating the conditions the control is designed to address, such as is done in penetration testing.

All organizational activities create some degree of risk. It is essential that an ongoing information security activity understands and analyzes the organization's processes to determine which pose a significant risk to information security. This applies to physical and procedural activities as well as IT processes. Because remediation is generally the responsibility of the individual business units, the information security manager usually serves in an investigatory, monitoring and facilitative role. This requires working with individual organizational units to understand what information they use and develop; the extent to which it is sensitive or critical; how it is handled, stored and processed; what media forms it exists in; and where it is located. Based on this information, the information security manager can suggest appropriate remediation options.

The risk landscape is constantly changing with new threats, new regulations, new attack modalities and evolving vulnerabilities. The information security manager monitors the changing situation to avoid putting the organization at increased risk. Public sources of information on external threats should be reviewed consistently. Internal changes must also be monitored on a continuous basis; some can be tracked through the change control process, but other aspects must be monitored by metrics, reporting, audits or direct observation.

It is also critical to assess risk on a regular or event-driven basis, such as after any incident or security event. Any successful compromise is the result of either a lack of adequate controls or a control failure, which indicates risk was not assessed accurately and must be reassessed.

Management decisions usually take risk into consideration and, as a consequence, the information security manager must report changes in risk to properly inform the decision-making process. For many organizations subject to regulatory oversight, a lack of or inadequate compliance can pose a significant risk. It is important for the information security manager to assess the risk associated with regulatory compliance by staying current with relevant regulatory enforcement actions and the resulting sanctions. Using the standard risk assessment approach, this

information is important to management in determining the appropriate level of compliance, timetable to achieve it and budgetary requirements.

The key elements that must be supplied to management on a consistent basis regarding information security risk are likelihood and impact related to identified risk. Likelihood (or probability) is determined by the risk analysis process and impact by a BIA. This information provides the basis for management to make prudent decisions.

2.1 RISK MANAGEMENT OVERVIEW

Risk management is a process aimed at achieving an optimal balance between realizing opportunities for gain and minimizing vulnerabilities and loss. This is usually accomplished by ensuring that the impact of threats exploiting vulnerabilities is within acceptable limits at an acceptable cost. Risk management is different from managing risk, which is often used synonymously with risk mitigation or risk response.

In practical business terms, risk management means that risk is managed so it does not materially impact the business process in an adverse way and an acceptable level of assurance and predictability to the desired outcomes of any important organizational activity are provided for. Risk is inherent in all activities; typically, a higher level of strategic risk will be taken to attempt to achieve higher returns. However, each organization needs to decide individually what level of risk is appropriate in order to achieve the financial and organizational results it wants to accomplish.

The foundation for effective risk management is a comprehensive risk assessment, based on a solid understanding of the organization's risk universe. It is not possible to devise a relevant risk management program if there is no understanding of the nature and extent of risk to information resources and the potential impact on the organization's activities. The structure of an organization's risk function can be centralized or decentralized. Mature organizations generally manage risk through an enterprise risk management (ERM) group or department to ensure consistency. In some organizations there is a separate IT risk department, whereas others have IT risk as a team under the ERM group. In some cases, the information security management, business resiliency and/or incident management programs may be integrated into an existing technology risk management framework. In others, risk is managed in a decentralized manner in a number of different departments and operational units, necessitating efforts to ensure continuity and integration of risk management activities.

Risk management, the development of a BIA, the creation of an information asset inventory and risk analysis are fundamental prerequisites to developing a meaningful security strategy. Organizations that develop an information security governance program, as detailed in chapter 1, include risk management as an integral part of their overall program. Risk management must be addressed regardless of the state of governance. At a high level, risk management is accomplished by balancing risk exposure against mitigation costs and implementing appropriate controls and countermeasures.

Controls are designed as part of the information risk management framework, which incorporates policies, standards, procedures, practices and organizational structures. It is designed to provide reasonable assurance that business objectives are achieved and undesired events are prevented or detected and addressed. The framework must address people, processes and technology and it encompasses the physical, technical, contractual and procedural aspects of the organization. It must take into consideration the strategic, tactical, administrative and operational components of the organization to be effective.

Countermeasures include any process that serves to counter specific threats and can be considered a targeted control. They can range from modifying architecture or reengineering processes, to reducing or eliminating internal threats to technical vulnerabilities, to creating an awareness program for all employees to target social engineering and promote early recognition and reporting of security incidents.

Because risk management decisions typically have major financial implications and can require changes across the entire organization, it is imperative that senior management is supportive of the process and fully understands and agrees with the results of the program.

Risk management can mean different things to different people in the organization. For example, a business manager might assume threats seldom occur and is not convinced of the ROI for security measures. An IS auditor's view may be that risk management means the prevention of loss, whereas an insurance manager could define it as cost-effective risk financing.

The information security manager should also understand that risk management must operate at multiple levels, including the strategic, management and operational levels. The significance of business experience and business decision making in any risk assessment process should be recognized as important to achieving realistic and successful outcomes from the process. The likelihood and relevance of a particular threat or risk is usually a matter of judgment, so experience is beneficial in arriving at realistic results. A good practice is to report both probable outcomes as well as worst-case situations.

The overall assessment of risk includes three distinct phases:
- Risk identification
- Risk analysis
- Risk evaluation

In risk identification, a set of risk scenarios and possible outcomes are developed based on a thorough assessment of the threat landscape and a vulnerability assessment (both logical and physical, including consideration of exposures) and then determining the probable ways a compromise might occur as well as the possible consequences (see section 2.7.9 Identification of Risk).

Risk analysis consists of a thorough analysis of the identified risk using various techniques and performing BIAs across the enterprise to develop a clear understanding of potential impacts.

Risk evaluation uses the results of the analysis to determine if the risk falls within the acceptable range or it must be mitigated.

The next step is risk response, which can consist of risk mitigation, acceptance, avoidance or transfer.

Depending on the organization and its maturity with respect to risk management, a simple risk management process may have a greater chance of success then a complex one and is likely to be less costly.

2.1.1 THE IMPORTANCE OF RISK MANAGEMENT

The management of risk to information resources is a fundamental function of information security. It provides the rationale and justification for virtually all information security activities. Information security as a discipline exists to manage the risk to confidentiality, integrity and availability of information. Effective risk management is one of the main objectives of governance, as discussed in section 1.1.2 Outcomes of Information Security Governance, as well as the key to managing regulatory requirements. Information security provides a level of assurance for business activities by managing risk to levels acceptable and appropriate to the mission of the business or organization. Without identifying and subsequently analyzing the spectrum of risk for a given business activity, it is not possible to determine the potential cost or impact of a particular event and, consequently, how to determine appropriate mitigation measures.

The effectiveness of risk management depends on the degree to which it is a part of an organization's culture and the extent to which risk management becomes everyone's responsibility.

The design and implementation of the risk management process in the organization will be influenced by the organization's:
- Culture (risk averse or aggressive)
- Mission and objectives
- Organizational structure
- Ability to absorb losses
- Products and services
- Management and operation processes
- Specific organizational practices
- Physical, environmental and regulatory conditions

2.1.2 OUTCOMES OF RISK MANAGEMENT

Effective risk management serves to reduce the incidence of significant adverse impacts on an organization by addressing threats, mitigating exposure, and/or reducing vulnerability or impact. To the extent this is accomplished, risk management provides a level of predictability that supports the organization's ability to operate effectively and profitably.

As stated in chapter 1, one of the outcomes of good governance is effective risk management. This includes executing appropriate measures to mitigate risk and reduce potential impacts on information resources to an acceptable level and providing:
- An understanding of the organization's threat, vulnerability and risk profiles
- An understanding of risk exposure and potential consequences of compromise
- Awareness of risk management priorities based on potential consequences
- An organizational risk mitigation strategy sufficient to achieve acceptable consequences from residual risk
- Organizational acceptance/deference based on an understanding of the potential consequences of residual risk
- Measurable evidence that risk management resources are used in an appropriate and cost-effective manner

2.2 RISK MANAGEMENT STRATEGY

A risk management strategy is the plan to achieve the risk management objectives. Ultimately, those objectives are to achieve an acceptable level of risk across the enterprise resulting in an acceptable level of disruption to the organization's activities.

The acceptable level of risk is a management decision generally based on a variety of factors including:
- The ability of the organization to absorb loss
- Management's risk appetite
- Costs to achieve acceptable risk levels
- Risk-benefit ratios

Acceptable risk determines control objectives, which become the main objectives of the strategy.

To be effective and ensure consistent management of risk across the enterprise, it is essential that the information risk strategy is integrated with the other risk management activities in the organization. This is to prevent gaps in protection and duplication of efforts and will ensure that the various risk management activities are not working at cross-purposes.

The information risk strategy must also be consistent with and integrated into the overall security governance strategy. The information security strategy must be based on the organization's overall objectives and business strategy.

The execution of the plans to achieve the risk management objectives is determined by a number of internal and external factors. Some of the main internal factors to consider include organizational maturity, history, culture, structure and acceptable risk. Various external factors, such as industry sector and legal and regulatory requirements, must be addressed by the strategy as well.

An information risk management strategy must include determining the optimal approach to align processes, technology and behavior. It takes into account all credible risk and the full range of options for its appropriate management. This includes effective processes for risk identification and risk monitoring and metrics. It also includes promoting adequate robustness in defenses to withstand a wide range of viable attacks and resilience sufficient to recover from any conceivable mishap.

2.2.1 RISK COMMUNICATION, RISK AWARENESS AND CONSULTING

For risk management to become part of the organization's culture, it is necessary to communicate and create awareness of the issues across the organization at each step of the risk management process.

Communication should involve all stakeholders and focus on consultation and development of a common understanding of the objectives and requirements of the program. This allows variations in needs and perceptions to be identified and addressed more effectively. A template for risk communication is shown in **figure 2.1**.

2.2.2 RISK AWARENESS

Awareness is a powerful tool in creating the culture, shaping ethics and influencing the behavior of the members of an organization. The risk and security awareness program should include communication of risk and security information, periodic testing as a metric for awareness, and a channel for staff to report risk and security issues. The staff and operational teams of an organization are often the first to be aware of any problems or abnormal activities. Through awareness programs, it is possible to develop a team approach to risk management that enables every member of an organization to identify and report on risk and work to defend systems and networks from attacks. Each member of the team can help identify vulnerabilities, suspicious activity and possible attacks. This may enable a faster response and better containment of a risk when an attack happens. Risk awareness acknowledges that risk is an integral part of the business. This does not imply that all risk is to be avoided or eliminated, but rather that:

- Risk is well understood and known.
- Information risk issues are identifiable.
- Employees recognize that organizational risk can affect them personally.
- The enterprise recognizes and uses the means to manage risk.

A risk awareness program creates an understanding of overall risk, risk factors and the various types of risk that an organization faces. An awareness program should be tailored to the needs of the individual groups within an organization and deliver content suitable for that group. The program should not disclose vulnerabilities or ongoing investigations except where the problem has already been addressed. Using examples and descriptions of the types of attacks and compromises that other organizations have experienced can reinforce the need for diligence and caution when addressing risk.

Awareness education and training can serve to mitigate some of the most significant areas of organizational risk and achieve the most cost-effective improvement in risk and security. This can generally be achieved by educating an organization's staff in required procedures and policy compliance, as well as ensuring that staff can identify and understand the risk that threatens the organization. Training must effectively communicate the risk and its potential impact so staff can understand the justification for what many may see as inconvenient extra steps that risk mitigation and security controls often require.

The information security manager's ability to develop awareness and training programs that will be effective in the environment depends on his/her understanding of the organization's structure and culture as well as the most effective types of communication. Periodically changing risk awareness messages and the means of delivery will help maintain a higher level of risk awareness. Procedural and technical controls can be complex, and it is essential to provide training as needed to ensure that staff understands the procedures and can correctly perform the required steps.

Figure 2.1—Risk Communication Plan

A risk communication plan defines frequency, types and recipients of information about risk. The main purpose of the plan is to reduce the overload of non-relevant information (avoiding the likelihood of 'risk noise').

Life Cycle and Stakeholders: Life Cycle Stage	Internal Stakeholder	External Stakeholder	Description/Stake
Information planning	ERM committee, audit committee		• Ensure that a risk communication plan exists and approve it. • Ensure that risk is communicated efficiently and timely.
Information design	Risk function		• Outline the different aspects to be included in the risk communication plan (e.g., frequencies/types/recipient). • Ensure that risk is communicated efficiently and in a timely manner.
Information build/acquire	Risk function, business process owners/CIO		• Detail the outlined aspects of the risk communication plan. • Ensure the risk communication plan includes key requirements for risk communication.
Information use/operate: store, share, use	Board, executive management, CIO, risk function, business process owners, compliance, internal audit	External audit, regulator	Effective utilisation of the risk communication plan by the risk function ensures availability of correct and concise information about risk to the other stakeholders.
Information monitor	Board, risk committee, audit committee, risk function	External audit	• Timely action is taken on the status of risk and risk capabilities. • The adequacy of the risk communication plan is validated regularly.
Information dispose	Risk function		Ensure that information is disposed of in a timely, secure and appropriate manner.

Source: ISACA, *COBIT® 5 for Risk*, USA, 2013, figure 63, page 152

Awareness of information security policies, standards and procedures by all personnel is essential to achieving effective risk management. Employees cannot be expected to comply with policies or standards they are not aware of or follow procedures they do not understand. The information security manager should use a standardized approach, such as short computer- or paper-based quizzes, to gauge awareness levels. Periodic use of a standardized testing approach provides metrics for awareness trends and training effectiveness. Further training needs can be determined by a skills assessment or a testing approach. Indicators for additional training requirements can come from various sources such as tracking help desk activity, operational errors, security events and audits.

An awareness program for management should highlight the need for management to play a supervisory role in protecting systems and applications from attack. Managers have the responsibility to oversee the actions of their staff and direct compliance with the policies, procedures and practices of the organization.

Awareness training for senior management should highlight liability, need for compliance, due care and due diligence, and the need to create the tone and culture of the organization through policy and good practice. Senior managers may need to be reminded that they are the ones who own the risk and bear the responsibility for determining risk acceptance levels.

2.3 EFFECTIVE INFORMATION RISK MANAGEMENT

Effective information risk management activities must be supported on an ongoing basis by all members of the organization. Identified risk must have an owner and clear accountability to ensure proper management of the risk.

Senior management support, ownership and accountability lend credibility to risk management efforts. Even the best designed and implemented controls will not function as intended if operations are conducted by careless, indifferent or untrained personnel. An organizational culture that includes sound information security practices coupled with senior management commitment to effective risk management is required to achieve the objectives of the program.

In addition, personnel must be aware of the risk the organization faces, understand their responsibilities and be trained in applicable control procedures. Compliance with information security controls must be monitored, tested and enforced on a continuing basis. The information security manager must also consider developing approaches to achieve a level of integration with the typically numerous risk management activities of other parts of the organization. These include legal, facilities, physical security, human resources (HR), audit, and privacy and compliance activities. The objective is stated as assurance process integration, one of the six outcomes identified in chapter 1. Other efforts at promoting integration of risk management (as well as governance and compliance) across the enterprise are promoted by the Open Compliance and Ethics Group (OCEG) and supported by numerous vendors and consultancies.

2.3.1 DEVELOPING A RISK MANAGEMENT PROGRAM

Initial steps in developing a risk management program include establishing:

- Context and purpose of the program
- Scope and charter
- Authority, structure and reporting relationships
- Asset identification, classification and ownership
- Risk management objectives
- The methodology to be used
- The implementation team

Establish Context and Purpose

A primary requirement is to determine the organization's purpose for creating an information risk management program, identify the desired outcomes and define objectives. It might be a limited effort to reduce the impacts of Internet-based attacks and accidents or to ensure compliance with legal or regulatory requirements. If the context and scope of risk management activities are not established, the resulting program may be far broader than expected, and responsibilities will be distributed across several departments. If the organization has one or more existing risk management functions, it is necessary to determine how they will be integrated and how the responsibilities will be divided. This integration maximizes cost-effectiveness, minimizes gaps in protection, and reduces the chances of working at cross-purposes or having the information security manager's risk management activities subverted by inadequate efforts elsewhere in the organization.

Setting the risk management context involves defining the internal and external environment; organizational structure and lines of authority; the process, project or activity; scope; and goals and objectives.

Determining the organization's risk appetite and tolerance is an essential element of risk management. **Risk appetite** is what is considered by management to be an acceptable level of risk, and **risk tolerance** is the acceptable level of deviation from the acceptable risk level. Each organizational unit or department has a different risk appetite it considers acceptable in the pursuit of its business objectives. Risk appetite is a business decision based on a number of criteria—including industry sector, regulatory considerations, ability to absorb loss, reputational concerns, mission and culture—rather than any specific quantitative measures. Typically, the board of directors and executive management set the tone for the risk management program. This results in an organization taking a security posture of being risk-averse or risk-aggressive. The tone at the top influences all aspects of the program, including how well it ultimately functions. As with other aspects of security, a top-down approach is essential for an effective risk management program.

Define Scope and Charter

Because all departments and operational units in the organization have some level of responsibility for managing risk, it is important to clearly define the scope of responsibility and authority that specifically falls to the information security manager and to other stakeholders. This helps prevent gaps in the process, improves overall consistency of risk management efforts

and reduces unnecessary duplication of effort. This effort can be clarified by developing an information risk management RACI chart (see section 1.3 Roles and Responsibilities).

Because most information security activities are in some way related to managing risk, this exercise should map closely to the security manager's job responsibilities. Regardless of the scope of responsibility of the information security manager, the total scope for risk management needs to be defined and the overall objectives determined.

While many parts of the organization are responsible for some aspects of risk management, the main areas of information security typically interface with physical security, operational risk, IT and business management. This relationship can be contentious when goals are not aligned or are openly contradictory and the scope of responsibilities is poorly defined. For example, it is necessary to determine who is responsible for ensuring that sensitive information is not left at print stations, which could result in confidential documents not being shredded before being discarded, which enables the loss of data through dumpster diving. This illustrates just one of the numerous points of intersection of information security, IT security, facilities, physical security and other assurance providers that require clear definition of the areas of respective authority and responsibility.

Define Authority, Structure and Reporting

Authority to take certain actions and make decisions must be clearly defined. Many incidents have resulted in unnecessarily severe consequences as a result of a lack of clarity as to who has the authority to make decisions and take certain actions.

The reporting and authority structure of an organization has an impact on the ability to effectively manage risk. This may be the result of different parts of the organization having different parts of risk management under their control, coupled with conflicting perspectives. This fractured risk management structure will further degrade if reporting goes to different decision makers in various parts of the organization. A lack of clear governance and integration of risk management activities often results in dire consequences, and the information security manager should address such issues by assessing and presenting the risk and making recommendations to senior management.

Ensure Asset Identification, Classification and Ownership

To appropriately scope and prioritize risk management efforts, a complete and accurate information asset register must exist. Although it may be a daunting task, it is necessary to locate all instances of information assets as part of the asset identification process.

In addition, information assets need to be classified in terms of business value or sensitivity and criticality to the organization (see 2.8 Information Asset Classification).

It is also essential to ensure that all information assets have an identified owner with specific responsibilities for managing risk to those assets (i.e., the important underlying concept is that all assets and all identified risk have an owner with defined responsibilities). This will help promote accountability for policy compliance and risk management requirements throughout the organization. Policies requiring asset and risk ownership should be in place, as well as processes established to assign ownership as assets are acquired, transferred or created.

Determine Objectives

Clear objectives and priorities for the information risk management program are essential. While the ultimate objectives, or desired state, may be to manage all risk to acceptable levels, even if that were possible, time and resource limitations will make that unlikely to occur at the same time. As a result, it will be necessary to set priorities for the program and prioritize risk accordingly. In other words, some types of risk cannot be addressed and must be accepted, some can wait, and some should be addressed immediately. Of course, before risk can be prioritized it must be identified, and its likelihood and impact must be determined through risk analysis and impact assessment. Priorities can then be set, starting with types of risk that are determined to have a high likelihood of materializing as well as a high impact, and working down from there.

Determine Methodologies

Many approaches to assessing, analyzing and mitigating risk are used. In many organizations, standard approaches should be used if they are adequate for the purpose. If these practices have not been established or are inadequate, the information security manager should evaluate the available choices and seek to implement those that are the best for the organization.

Designate Program Development Team

Once the scope of information risk management activities is defined and the objectives are clarified, the next step is to designate an individual or team responsible for developing and implementing the information risk management program. While the team is primarily responsible for the risk management plan, a successful program requires the integration of risk management at all levels of the organization. Operations staff and board members (through an oversight or steering committee) should assist the risk management committee in identifying risk, determining acceptable risk levels, developing suitable loss-control and intervention strategies, and determining where the authority and responsibility for various aspects of risk management will reside. Of overriding importance is the need for the risk management program to be properly aligned with the strategy and direction of the business. For this reason, it is vital that participation include representatives from all key business units.

2.3.2 ROLES AND RESPONSIBILITIES

Regarding risk management, the information security manager is typically responsible for developing, collaborating and managing the information risk management program to achieve an acceptable level of risk by meeting the control objectives. The cost of achieving these objectives is always a consideration, and the information security manager must strive for the most cost-effective solutions. The information security manager must also take responsibility for working with other risk management teams and assurance activities in the organization to promote the integration of activities and provide an effective and coordinated level of business process assurance. Effective integration of a typical organization's numerous risk management activities helps prevent gaps in protection and inconsistent risk management effectiveness, reduces duplication of efforts and the possibility of working at cross-purposes, and generally provides the most cost-effective implementations.

2.4 INFORMATION RISK MANAGEMENT CONCEPTS

The information security manager must have a broad understanding of a number of concepts fundamental to security and risk management, including technical, strategic, tactical, administrative and operational elements. Some of the main concepts are discussed in the following section.

2.4.1 CONCEPTS

The information security manager should be familiar with a number of concepts and their use in the context of risk management. Some of the main concepts include:

- Acceptable interruption window (AIW)
- Business impact analysis (BIA)
- Controls
- Countermeasures
- Criticality
- Exposure
- Frequency
- Impacts
- Information asset classification
- Key risk indicators (KRIs)
- Maximum tolerable outage (MTO)
- Proximity
- Recovery time objectives (RTOs)
- Recovery point objectives (RPOs)
- Redundancy
- Resource valuation
- Risk
- Risk analysis
- Risk appetite
- Risk assessment
- Risk evaluation
- Risk identification
- Risk metrics/trends
- Risk reporting
- Risk tolerance
- Risk treatment and response
- Sensitivity
- Service delivery objectives (SDOs)
- Threats
- Velocity
- Volatility
- Vulnerabilities

Other risk management aspects related to information security that should be understood include:

- Business continuity/disaster recovery
- Business process reengineering
- Enterprise and security architectures
- IT and information security governance
- Policies, standards and procedures
- Project management timelines and complexity
- Service level agreements (SLAs)
- System robustness and resilience
- Systems life cycle management

2.4.2 TECHNOLOGIES

There are also a variety of information security technologies and technical concepts that are important for the information security manager to understand thoroughly as they relate to risk management. Some of these include:

- Antispam devices
- Antivirus/malware, including spyware/adware
- Application security measures
- Cloud services, deployment and risk
- Encryption and public key infrastructure (PKI)
- Environmental controls
- Intrusion detection/prevention
- Logical access controls
- Network access controls
- Network and Internet protocols
- Physical security measures
- Platform security
- Routers, firewalls and other network components (e.g., bridges, gateways)
- Telecommunications and voice-over Internet protocol (VoIP)
- Wireless security and protocols
- Virtualization

In addition, while personnel and facilities security may not be part of a risk management program, these are areas of risk that need to be considered as a part of risk management. The information security manager must be aware of and factor in personnel issues and personnel security controls as well as environmental and facilities controls as a part of risk assessment and management activities.

2.5 IMPLEMENTING RISK MANAGEMENT

As a part of planning a risk management program, the information security manager must identify all other organizational risk management activities and seek to integrate these functions or leverage the activities within the context of the information security program. Larger organizations usually have a risk management function that deals with activities typically related to physical risk. Financial institutions typically have a department dealing with credit risk. Other departments or roles, such as HR and privacy officers, and compliance functions, such as audit, typically are involved in managing risk within the organization. To be effective, mechanisms should be put in place to ensure good communication with other risk management and assurance functions. This is to ensure that effective information risk management is not bypassed or subverted by the lack of effective processes in other domains. It also prevents duplication of effort and minimizes gaps in assurance functions that can adversely affect information protection activities as well as other areas of operational and business risk.

2.5.1 THE RISK MANAGEMENT PROCESS

Risk management consists of a series of processes that take into account the end-to-end requirements of identifying, analyzing, evaluating and maintaining risk at acceptable levels. These include weighing policy alternatives in consultation with interested parties, considering risk assessment and other factors, and selecting appropriate prevention and control options that have acceptable costs and impacts on the organization's ability to operate efficiently.

Risk management usually consists of the following processes:

- **Establish scope and boundaries**—Process for the establishment of global parameters for the performance of risk management within an organization. Both internal and external factors have to be taken into account to provide context.
- **Identify information assets and valuation**—An information asset inventory and valuation process to determine assets at risk and potential impacts of compromise
- **Perform risk assessment**—A process consisting of risk identification, analysis and evaluation, including:
  - Identifying viable threats, vulnerabilities and exposures
  - Analyzing the level of risk and potential impact
  - Evaluating whether the risk meets the criteria for acceptance
- **Determine risk treatment or response**—Process of selecting strategies to deal with identified risk that exceeds the acceptable level. Risk treatment strategies are avoiding, by cessation of risky activities; mitigating, by developing and implementing controls; transferring risk to a third party, which could be inside or outside the organization; and accepting. Risk is usually accepted if there is no cost-effective way to mitigate it, there is little exposure or potential impact, or it is simply not feasible to address it effectively.
- **Accept residual risk**—The decision and approval by management to accept the remaining risk after the treatment process, if needed, is concluded (i.e., the risk may be accepted after evaluation shows it is within acceptable limits or if no effective treatment option is available)
- **Communicate about and monitor risk**—A process to exchange and share information related to risk, as well as reviewing the effectiveness of the whole risk management process. Communication of risk is usually performed among decision makers and other stakeholders inside and outside the organization. Through communication and monitoring, the scope, boundaries, evaluated risk and action plans remain relevant and updated.

The risk response workflow is shown in **figure 2.2**. Note that the concepts of avoid, reduce, share and accept are often referred to using different terms, but the meaning is similar:

- Terminate the activity (avoid).
- Reduce the risk (mitigate).
- Transfer the risk (share).
- Retain the risk (accept).

Developing a systematic, analytical and continuous risk management process as shown in **figure 3.3** is critical to any successful security program and must be implemented as a formal process. Determining the correct or appropriate level of security is dependent on the potential risk that an organization faces, the potential impact, and the organization's ability to accept or otherwise mitigate risk. Risk is unique to each organization.

Figure 2.2—Risk Response Workflow

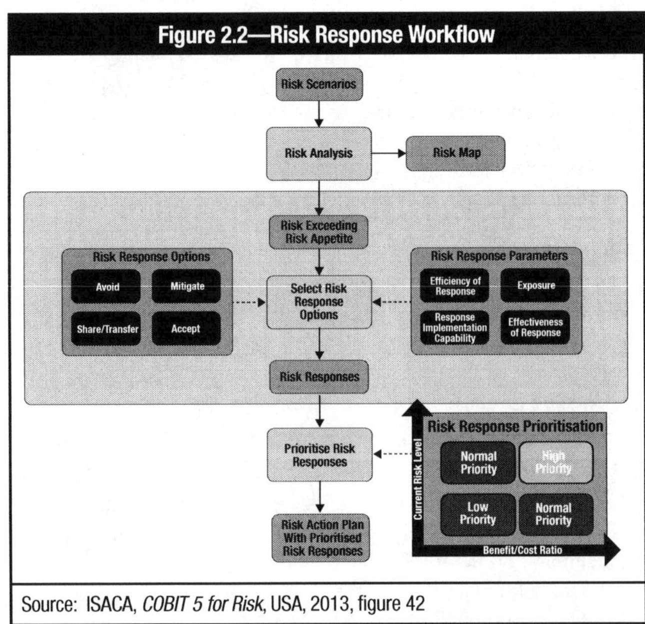

Source: ISACA, *COBIT 5 for Risk*, USA, 2013, figure 42

Figure 2.3—Continuous Risk Management Steps

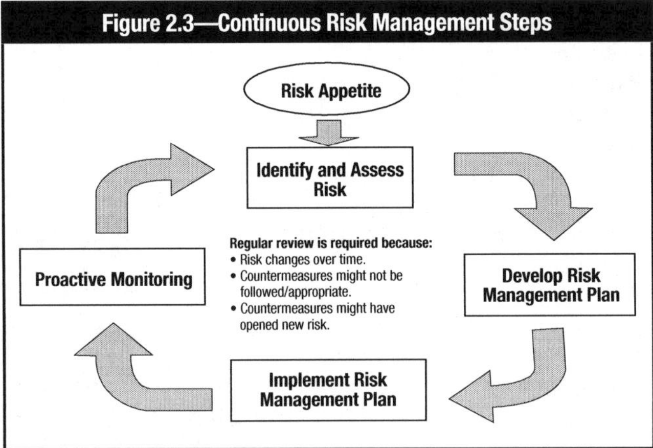

The information security manager should set up a regular, formal process in which risk assessments are performed at the organizational, system and application levels, including business processes both logical and physical. Ensuring that there are measurements (metrics) in place to assess the risk and the effectiveness of security measures is part of the information security manager's ongoing responsibility. The effectiveness of security measures or controls should be assessed against control objectives and/or security baselines. The information security manager should also explore and recommend to asset owners continuous manual and automated techniques to monitor the organization's risk. This risk assessment process is important because it focuses the organization's security activities on issues that have the greatest likelihood, impact and significance.

Overgeneralization by applying risk factors across industries or regions should be avoided. Furthermore, the effectiveness of an information security program is frequently challenged by organizational, technological and business/operational change. These changes can, to a large extent, be monitored through links to all change management processes in the organization. Risk management should be a continuous and dynamic process to ensure that changing threats and vulnerabilities are addressed in a timely manner.

In addition, processes must be developed to monitor the status of security controls and countermeasures to determine their ongoing effectiveness. Controls usually degrade over time and are subject to failure, mandating ongoing control monitoring and periodic testing.

2.5.2 DEFINING A RISK MANAGEMENT FRAMEWORK

To develop an organization's systematic risk management program, a reference model should be used and adapted to the circumstances of the organization. The reference model will reflect the desired state discussed in chapter 1 (i.e., a snapshot of a future state that meets all risk management objectives). Several publications/standards are available to provide guidance on information technology and security risk management approaches. Examples include:

- COBIT 5
- *ISO 31000:2009 Risk management—Principles and guidelines*
- *IEC 31010:2009 Risk management—Risk assessment techniques*
- *National Institute of Standards and Technology (NIST) Special Publication 800-39 Managing Information Security Risk: Organization, Mission and Information System View*
- *HB 158-2010 Delivering assurance based on ISO 31000:2009—Risk management—Principles and guidelines*
- *ISO/IEC 27005:2011 Information technology—Security techniques—Information security risk management*

The standards referenced in the list have similar risk management requirements, including:

- **Policy**—An organization's senior management/executive leadership needs to define and document its policy for risk management, including objectives for, and its commitment to, risk management. The policy must be relevant to the organization's strategic context, goals, objectives and the nature of its business. Management should ensure that this policy is understood, and that standards are developed, implemented and maintained at all levels in the organization.
- **Planning and resourcing**—Responsibility, authority, accountability and interrelationships of personnel who perform and verify work affecting risk management must be defined and documented (see section 1.3 Roles and Responsibilities). The organization must identify resource requirements and facilitate the implementation of risk management programs, through the assignment of trained personnel for ongoing management of work activities and the verification activities for internal review.
- **Implementation program**—The organization must define the steps required to implement an effective risk management system.
- **Management review**—Executive management must ensure periodic review of the risk management system to ensure its continuing stability and effectiveness in satisfying requirements of the program. Records of such reviews must be maintained.
- **Risk management process**—Risk management can be applied at both strategic and tactical levels in the organization—products/services, business/IT processes, projects, decisions, applications and platforms. The organization must prioritize individual risk treatment according to the organization's business objectives, risk appetite and regulatory environment for the given industry.
- **Risk management documentation**—For each stage of the process, adequate records must be kept that are sufficient to satisfy an independent audit.

By establishing the framework for risk management, the basic parameters within which risk must be managed are defined, including the criteria for acceptable risk and control objectives. Consequently, the scope for the rest of the risk management process is also set. It includes the definition of basic assumptions for the organization's external and internal environment and the overall objectives of the risk management process and activities. Although the definition of scope and framework are fundamental for the establishment of risk management, they are independent from the particular structure of the management process, methods and tools to be used for the implementation.

In order to define an efficient framework it is important to:

- Understand the background of the organization and its risk (e.g., its core processes, valuable assets, competitive areas)
- Evaluate existing risk management activities and criteria for acceptable risk levels
- Develop a structure and process for the development of risk management initiatives and controls sufficient to achieve acceptable risk levels (control objectives)

This approach is useful for:

- Clarifying and gaining common understanding of the organizational objectives
- Identifying the environment in which these objectives are set
- Specifying the main scope and objectives for risk management, applicable restrictions, or specific conditions and the outcomes required
- Developing a set of criteria against which the risk will be measured
- Defining a set of key elements for structuring the risk identification and assessment process

The risk management program must be integrated with the organization's overall management system and, whenever possible, must adapt various elements—such as policies, organizational processes, accountability, resources and communication methods—to its specific needs.

Many organizations' existing management practices and processes include elements of risk management and many organizations have already adopted a formal risk management process for particular types of risk or circumstances. These should be critically reviewed and assessed to determine whether they meet current control objectives and mitigation and other requirements (e.g., efficiency, user acceptance, cultural compatibility, regulatory requirements).

2.5.3 DEFINING THE EXTERNAL ENVIRONMENT

Defining the external environment includes specifying the environment in which the organization operates. The external environment typically includes:

- The local market; the business; and the competitive, financial and political environments
- The law and regulatory environment
- Social and cultural conditions
- External stakeholders

It is also important that both the perceptions and values of the various stakeholders and any externally generated threats and/or opportunities are properly evaluated and taken into consideration.

2.5.4 DEFINING THE INTERNAL ENVIRONMENT

Key areas that must be evaluated to provide a comprehensive view of the organization's internal environment include:
- Key business drivers (e.g., market indicators, competitive advances, product attractiveness)
- The organization's strengths, weaknesses, opportunities and threats
- Internal stakeholders
- Organization structure and culture
- Assets in terms of resources (i.e., people, systems, processes, capital)
- Goals and objectives, and the strategies already in place to achieve them

2.5.5 DETERMINING THE RISK MANAGEMENT CONTEXT

In business terms, the process of managing risk must provide a balance between costs and benefits. The context is the scope of risk management activities and the environment in which risk management operates, including the organizational structure and culture.

Determining the risk management context involves defining the:
- Range of the organization and the processes or activities to be assessed
- Full scope of the risk management activities
- Roles and responsibilities of various parts of the organization participating in the risk management process
- Organizational culture in terms of risk-averseness or -aggressiveness

The criteria by which risk will be evaluated have to be determined and agreed upon. Deciding whether risk treatment is required is usually based on operational, technical, financial, regulatory, legal, social or environmental criteria, or combinations thereof. The criteria should be in line with the scope and qualitative analysis of the organization's internal policies and procedures and must support its goals and objectives.

Important criteria to be considered are:
- Impact—The kinds of consequences that will be considered
- Likelihood—The probability of compromise
- The rules that will determine whether the risk level is such that further treatment activities are required

These criteria may need to change during later phases of the risk management process as a result of changing circumstances or as a consequence of the risk assessment and evaluation process itself.

2.5.6 GAP ANALYSIS

Gap analysis in the context of risk management refers to determining the gap between existing controls and control objectives. Control objectives based on acceptable risk should have been defined as a consequence of developing information security governance and strategy as discussed in chapter 1.

Control objectives should lead to controls that achieve acceptable risk, which, in turn, serve to set the information security baseline.

Control objectives may change as a part of risk management activities as exposures, business objectives or regulations change, potentially creating a gap between the existing control and its revised objectives. Periodically analyzing the gap between controls and their objectives should be a standard practice. This is normally accomplished as a part of the process of controls testing for effectiveness. To the extent that effectiveness of controls fails to meet the control objective and achieve acceptable risk, the controls may need to be modified, redesigned or supplemented with additional control activities. For more information on the relationship between controls and control objectives, see section 3.12 Controls and Countermeasures.

2.5.7 OTHER ORGANIZATIONAL SUPPORT

The information security industry provides many subscription services that an information security manager can integrate into an information security program. These services help to leverage expertise of external service providers without actually assigning them responsibility for executing any part of the security program. There are:
- **Good practices published by organizations**—Organizations such as ISACA, the SANS Institute and the International Information Systems Security Certification Consortium (ISC)²® can be valuable sources of comparison data with which to evaluate an information security program.
- **Security networking roundtables**—These organizations gather information security professionals from similar industries to discuss topics of common interest. Some are free and are sponsored by a security technology vendor. An information security manager may consider restricting attendance to vendor-sponsored forums if the organization is already a client of the sponsoring vendor so as to not feel pressure to purchase a vendor's product or service. Others roundtables are membership fee-based.
- **Security news organizations**—A number of organizations publish daily or weekly news relevant to risk management and information security generally (e.g., *Computer World*, SANS, *Tech Target, CIO Magazine).*
- **Security-related studies**—Various organizations, including PricewaterhouseCoopers (PwC), Ernst and Young (EY), Verizon, Symantec and Ponemon, provide annual studies regarding a variety of security-related matters.
- **Security training organizations**—These institutions provide classes on technical topics in information security such as vulnerability analysis and platform security configuration strategies.
- **Vulnerability alerting services**—These services allow information security managers to maintain a list of technologies in use at their organization and receive news with respect to vulnerabilities found in any technology on the list.

The field of information security changes as technology advances, and the threat landscape changes as cybercrime evolves and new laws and regulations are enacted. The information security manager should have frequent interaction with sources of new information about security products, services, threats, vulnerabilities, regulations, laws and management techniques.

2.6 RISK ASSESSMENT AND ANALYSIS METHODOLOGIES

A variety of methodologies for assessing risk are available, and some are more complex than others. Whichever approach is used, the outcome should be similar. The choice of methodologies, unless mandated by management or a regulatory body, should be based on the best fit for the organization.

Using the COBIT 5 approach to risk assessment, which aligns with ISO/IEC 27005:2011, assessment includes identification, analysis and evaluation (**figure 2.4**).

Figure 2.4—Information Security Risk Management Process

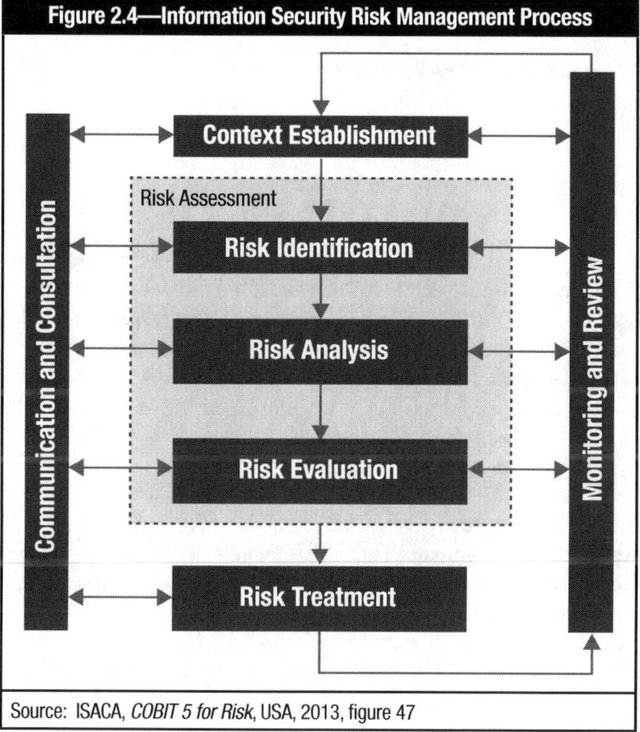

Source: ISACA, *COBIT 5 for Risk*, USA, 2013, figure 47

No single best approach to the selection of a methodology for conducting a risk assessment exists; however, the results must meet the goals and objectives of the organization in identifying the relative risk rating of assets and processes critical to the business. It is also essential to identify as much significant risk as possible. An approach to achieving this is to develop risk scenarios.

Risk assessment—coupled with either a BIA or information asset classification process to determine criticality and/or sensitivity and subsequent analysis—is used as a basis for identifying appropriate and cost-effective controls or countermeasures to mitigate the identified risk. It should be noted that sensitivity and/or criticality is often stated as business value.

Most risk assessment approaches have three phases. Each is discussed in detail in subsequent sections. These include:

- **Risk identification** is the process of using risk scenarios to determine the range and nature of risk to the organization.
- **Risk analysis** is the process of combining the vulnerability information gathered during an assessment and the threat information gathered from other sources to determine risk of compromise both in terms of frequency and potential magnitude. Analysis may include statistical computations such as VAR, ALE or ROSI to gain greater insight into risk distribution (i.e., the maximum risk and probable risk and potential consequences [impact]).
- **Risk evaluation** is the process of comparing the results of the risk analysis against established criteria for acceptability, impact, likelihood and the need for further treatment. Deciding whether risk treatment is required is usually based on operational, technical, financial, regulatory, legal, social or environmental criteria, or combinations of criteria. The criteria should be in line with the scope and qualitative analysis of the organization's internal policies and procedures and must support its goals and objectives.

2.7 RISK ASSESSMENT

Figure 2.5 is an example of a standard approach to risk assessment where the first step is to locate and identify information assets and then to determine asset valuation. Without an accurate inventory of assets, it is difficult to know what vulnerabilities exist that could be exploited by threats. It is also important to determine the relative business value of the assets because the potential impact of the loss of the asset (consequences) is a component of determining risk as well as the basis for classification.

Assuming an asset is of significant value, the next step is to determine what vulnerabilities to loss or damage exist. If vulnerabilities exist, an assessment of viable threats must be performed. If the asset has value as well as vulnerabilities

Figure 2.5—Risk Analysis Framework

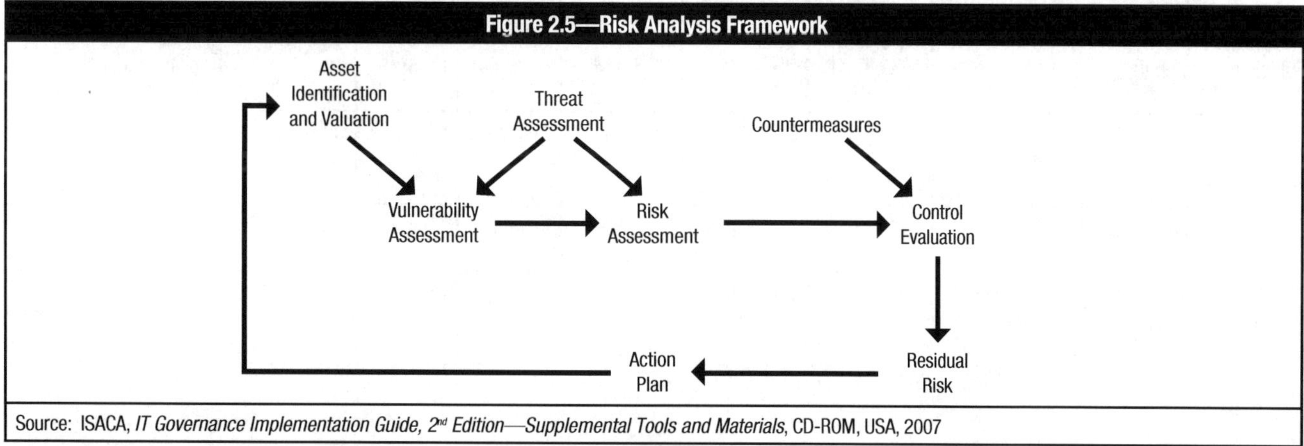

Source: ISACA, *IT Governance Implementation Guide, 2nd Edition—Supplemental Tools and Materials*, CD-ROM, USA, 2007

susceptible to viable threats, then there is a risk. To clarify, it is obvious that vulnerabilities for which no threats exist pose no risk (although threat is dynamic and may arise at any time). Another way to look at it is that the greater the value, and the greater the number and degree of vulnerabilities coupled with a greater number of viable threats, the greater the risk of loss.

2.7.1 INFORMATION ASSET IDENTIFICATION AND VALUATION

The first step in risk assessment is to locate and inventory all information assets and determine their relative or approximate business value (i.e., criticality or sensitivity). This is necessary because the business value is a part of the risk determination (i.e., risk exists only if there is likelihood of impact and consequences [risk = likelihood × consequences]). In addition, it is not possible to determine vulnerability and the extent of exposure of unidentified assets. The business value will also be needed for the purposes of classification and to provide the justification for the level of protection required.

The process of valuation, which consists of relating all values in a common financial form, is straightforward for some assets. Hardware can be easily valued based on replacement costs. The value of information, in some cases, is the cost of recreating or restoring it or it may be based on the contribution to generating revenues. In other cases, the value is related to the consequential costs and possible regulatory sanctions stemming from the exposure of confidential information or the loss of trade secrets.

The impact or consequences of a breach of personally identifiable information (PII) can be regulatory sanctions; in addition, the individuals suffering identity theft losses may file lawsuits for damages and class-action lawsuits on behalf of thousands of victims may arise. Another consequence is possible reputational damage, often resulting in loss of share value. Clearly, in this case, the valuation cannot be based on the intrinsic value of the information, which may be low or zero. Rather, valuation must be based on the total range of potential losses and other impacts.

Marketing materials are another type of information with no intrinsic value but which can, nevertheless, create unintended liabilities and, therefore, risk that must be considered. Inaccurate representations of products or services or information leading to wrong investor decisions can result in significant losses as a consequence of various legal actions. Therefore, a prudent organization must consider ensuring systematic review and control of information to manage the risk of potential liability created by publicly released information.

Categories of typical information assets that must be assigned a value and protected include, but are not limited to:
- Proprietary information and processes of all types, including information that can damage the organization
- Current financial records and future projections
- Acquisition or merger plans
- Strategic marketing plans
- Trade secrets
- Patent-related information
- PII

2.7.2 INFORMATION ASSET VALUATION STRATEGIES

Companies typically find resource valuation problematic and do not undertake the effort. Often companies do not have an accurate list of their information assets, and the effort to inventory and categorize these assets can appear to be a daunting task. Another reason is that even when assets are known, it is often difficult to place an exact value on assets such as PII or trade secrets, although a high degree of accuracy is not necessary.

In most cases, effective resource valuation is best based on loss scenarios. Information can be classified and put into a matrix with each loss scenario to make a complex problem more manageable and understandable. See **figure 2.6**.

The accuracy of the valuation is not as critical as having an approach to prioritize efforts. Values within the same order of magnitude as the actual loss (should it occur) are sufficient for planning purposes. Media reports contain many well-documented loss scenarios and loss amounts with which to base a valuation.

Figure 2.6—Matrix of Loss Scenarios

Scenario	Type of Data	Size of Loss	Reputation Loss	Lawsuit Loss	Fines/ Reg Loss	Market Loss	Expected Loss per year	Notes
Hacker steals data, publicly blackmails company.	Customer data	1K records 10K records	US $1M US $20M	US $1M US $10M	US $1 US $35M	US $1M US $5M	US $10M	Regulatory loss of ability to make acquisitions for 1 year
Employee steals data, sells data to competitors.	Strategic plan	3-year plan	Minimal	Minimal	Minimal	US $20M	US $2M	Competitor duplicates new products, brings them to market faster.
Contractor steals data, sells data to hackers.	Employee data	10K records	US $5M	US $10M	Minimal	Minimal	US $200,000	
Backup tapes and data are found in garbage, makes front-page news.	Customer data	10M records	US $20M	US $20M	US $10M	US $5M	US $200,000	

2.7.3 INFORMATION ASSET VALUATION METHODOLOGIES

Information asset valuation methodologies use many different variables, including the level of technical complexity and the level of potential direct and consequential financial loss. Quantitative valuation methodologies are generally the most precise but can be quite complex once actual and downstream impacts have been analyzed.

Another valuation methodology is qualitative in nature, where an independent decision is made based on business knowledge, executive management directives, historical perspectives, business goals and environmental factors. There are situations in which quantitative data are not available and this alternative method is the only option. Many information systems managers use a combination of techniques. In some cases, simply assigning value based on a subjective scale of low, medium and high may be satisfactory.

The most straightforward approach is the monetary value that represents the purchase price, replacement cost or book value, if that is representative of the importance to the organization. If it is not, other approaches must be considered. If it is an asset that directly or indirectly generates revenue, a computed value such as net present value (NPV) may be a reasonable approach.

Another approach is to consider value-add or other more intangible values. For example, an e-commerce application and server may have hardware and software costs of only US $50,000, but are an essential component in generating millions in revenue every month. In this situation, value may be computed in terms of revenue generation or the financial impact for any unanticipated downtime.

Intangible assets are usually comprised of intellectual property such as trade secrets, patents and copyrights, knowledge management, brand reputation, corporate culture, customer loyalty and trust, and innovation. Intangible assets that may prove difficult to quantify are the organization's reputation and consumer trust. Although a hacking incident itself may not create any direct losses, customers may leave due to lack of confidence in the organization, especially if there are strong competitors. Auditors may represent intangible assets under the heading of "goodwill."

For example, when customers' personal credit data are stolen, the organization could incur the costs of notifying a large number of people about the incident. Additionally, this type of incident could result in potential costs related to legal defense; for example, those affected may file a lawsuit. In the case where credit cards are compromised, the affected banks typically insist on recovering the costs of reissuing new cards in addition to any fraud costs they had to cover. This typically is in the range of US $10 or more for each card.

In a publicly traded company, intangible assets represent the difference between the tangible assets as recorded in the financials and the company's market capitalization value. For example, a company with a US $5 billion market capitalization has US $1 billion in tangible assets and US $4 billion in intangible assets. Therefore, 80 percent of the company's value (US $4 billion) is made up of intangible assets. It is obvious that most of these intangible assets fall under the purview of information security for the purposes of protecting them and preserving their value.

In addition, the information security manager needs to be aware of ongoing changes in the organization and should alter the use of valuation methodologies to best meet the needs as a result of these changes. If quantitative data are outdated and cannot be updated in a reasonable time frame, it may be desirable to use qualitative data either in place of or to augment the quantitative data.

While a detailed discussion on methods for establishing intangible asset values is beyond the scope of this manual, it is important for the information security manager to understand valuation approaches and the necessity for this activity.

2.7.4 RISK ASSESSMENT AND MANAGEMENT APPROACHES

Numerous risk management models and assessment approaches are available to the information security manager. The approach selected should be determined by the best form, fit and function. The approaches include COBIT, OCTAVE®, NIST 800-39, HB 158-2010, ISO/IEC 31000, ITIL® and CRAMM. Other approaches—such as factor analysis of information risk (FAIR), risk factor analysis and VAR—may be more suitable, depending on the organization and the specific requirements.

While the particulars and terms may differ somewhat, all the approaches cover essentially the same processes as shown in **figure 2.2**. In the COBIT 5 approach, risk scenarios are the process for identifying risk, followed by analysis. Risk evaluation is the next step to determine whether risk exceeds acceptable levels. The three steps are the components of a risk assessment. Risk response, or risk treatment, is conditioned by the criteria in the risk response parameters. This is followed by response (treatment) priorities based on a cost-benefit and risk-level analysis with high cost-benefit and high likelihood/high impact taking the highest priority.

> **Note:** Information security managers should have broad knowledge of the existence of various methodologies to determine the most suitable approach or combination of approaches for their organization. Specific approaches will not be tested in the CISM examination.

The risk assessment approach developed by NIST is demonstrated in the next section as an example of a well-developed, comprehensive methodology.

2.7.5 NIST RISK ASSESSMENT METHODOLOGY

The risk assessment methodology encompasses nine primary steps:

- Step 1—System (or general domain) characterization
- Step 2—Threat identification
- Step 3—Vulnerability identification
- Step 4—Control analysis
- Step 5—Likelihood determination
- Step 6—Impact analysis
- Step 7—Risk determination
- Step 8—Control recommendations
- Step 9—Results documentation

Steps 2, 3, 4 and 6 can be conducted in parallel after step 1 has been completed.

Figure 2.7 depicts these steps and the inputs to and outputs from each step.

Figure 2.7—NIST Risk Assessment Methodology

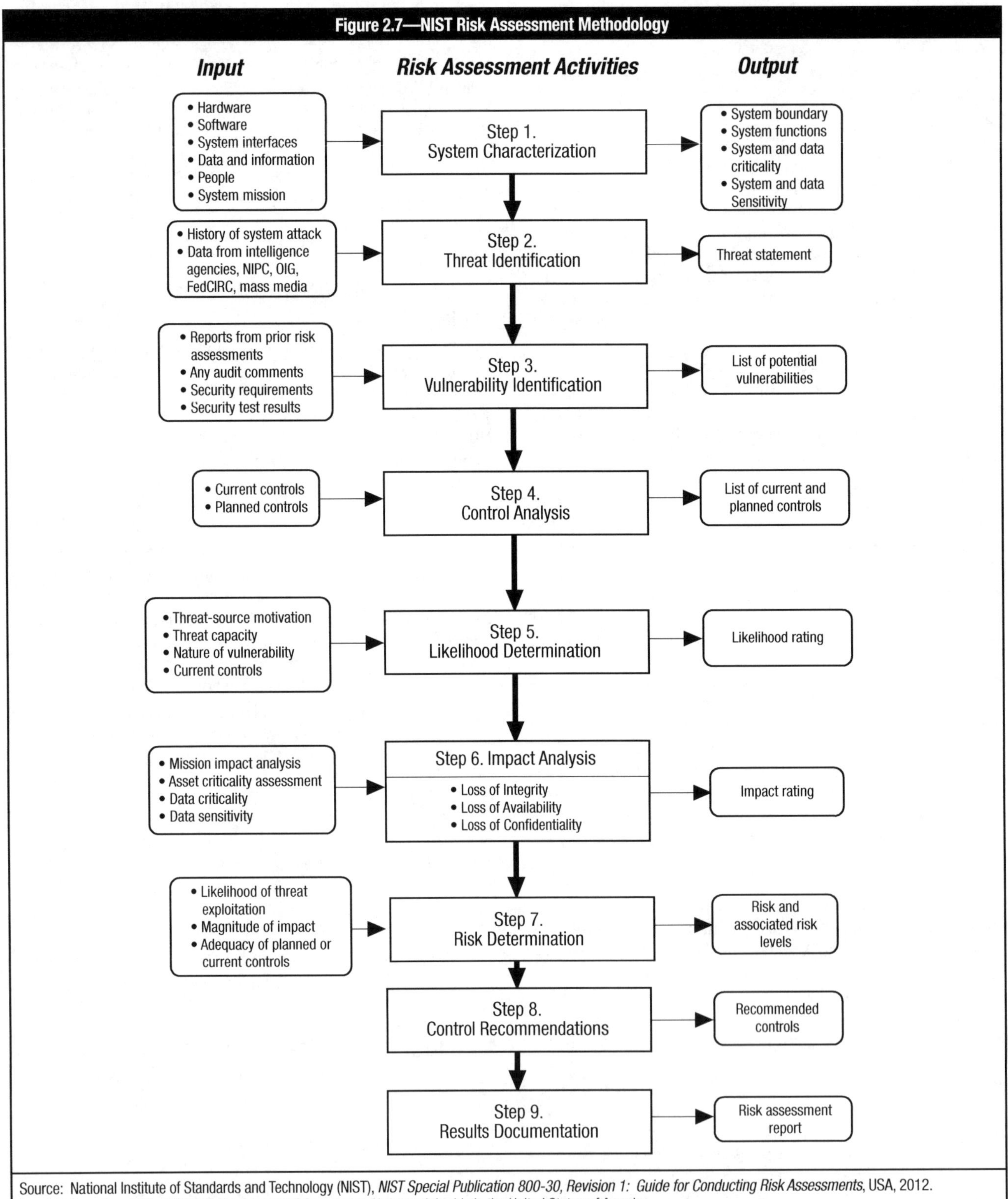

Source: National Institute of Standards and Technology (NIST), *NIST Special Publication 800-30, Revision 1: Guide for Conducting Risk Assessments*, USA, 2012. Reprinted courtesy of the NIST, US Department of Commerce. Not copyrightable in the United States of America.

2.7.6 ISO/IEC PROCESS STEPS

Figure 2.8 shows the ISO/IEC 27005 process steps for risk assessment and their relation to COBIT 5.

2.7.7 AGGREGATED AND CASCADING RISK

Aggregated risk exists where a particular threat affects a large number of minor vulnerabilities that, in the aggregate, can have a significant impact. Another possibility is that a large number of threats can simultaneously affect a number of minor vulnerabilities, resulting in a large aggregate risk. In this situation, it is possible for risk that is individually acceptable to have a catastrophic impact collectively.

Cascading risk can also manifest unacceptable impacts as a result of one failure leading to a chain reaction of failures. An example of this occurred on the east coast of the United States when a failure at a small power utility in the Midwest caused a cascade of failures across the power grid, ultimately encompassing most of the northeastern quarter of the United States. Similarly, to the extent that portions of the enterprise's IT and other operations have closely coupled dependencies, the information security manager must consider how any particular failures or combinations of failures will affect dependent systems.

2.7.8 OTHER RISK ASSESSMENT APPROACHES

Developments over the past few decades have resulted in notable improvements in defining the bounds of probable risk. However, few of these directly address information risk effectively. Some of these developments are starting to see adoption in the information security field and it is likely that the use of more sophisticated techniques and methods will occur in the coming years. Some of these methods are described in the following section.

Factor Analysis of Information Risk

A promising approach for decomposing risk and understanding its components is FAIR. The approach offers a reasoned, detailed analysis process that is designed primarily as a complement to other assessment approaches with the objective of increasing accuracy (**figure 2.9**).

FAIR provides a reasoned and logical framework for the following:
- **A taxonomy** of the factors that make up information risk. This taxonomy provides a foundational understanding of information risk, without which one could not reasonably do the rest. It also provides a set of standard definitions for the necessary terms.
- **A method for measuring** the factors that drive information risk, including threat event frequency, vulnerability and loss.
- **A computational engine** that derives risk by mathematically simulating the relationships among the measured factors.

Figure 2.8—ISO/IEC 27005 Process Steps Covered by *COBIT 5 for Risk*

ISO/IEC 27005 Process Step	Important Concepts of the Component	*COBIT 5 for Risk* Coverage
Context Establishment	This process step includes: • Setting the basic criteria necessary for information security risk management (ISRM) • Defining the scope and boundaries • Establishing an appropriate organisation operating the ISRM	This process step is included in the enabler dimension goal. More specifically, contextual quality: the extent to which outcomes of the enabler are fit for purpose given the context in which they operate.
Risk Assessment	Risk assessment determines the value of the information assets, identifies the applicable threats and vulnerabilities that exist (or could exist), identifies the existing controls and their effect on the risk identified, determines the potential consequences, and finally, prioritises the derived risk and ranks it against the risk evaluation criteria set in the context establishment. This process step consists of the following activities: risk identification, risk analysis and risk evaluation.	Appendix B.3 describes risk assessment as one item of the Information enabler.
Risk Identification	Risk identification includes the identification of: • Assets • Threats • Existing Controls • Vulnerabilities • Consequences The output of this process is a list of incident scenarios with their consequences related to assets and business processes.	The sequence used in ISO/IEC 27005 to identify risk is partly aligned to the *COBIT 5 for Risk* approach. The identification of risk comprises the following elements in *COBIT 5 for Risk*: • Control • Value • Threat condition that impose a noteworthy level of IT risk *COBIT 5 for Risk* also uses scenario development for identifying risk. Key attributes of potential and known risk events are stored in a repository. Attributes may include name, description, owner, expected/actual frequency, potential/actual magnitude, potential/actual business impact, disposition, etc.

Figure 2.8—ISO/IEC 27005 Process Steps Covered by *COBIT 5 for Risk* (cont.)

ISO/IEC 27005 Process Step	Important Concepts of the Component	*COBIT 5 for Risk* Coverage
Risk Analysis	The risk estimation process step includes the following important concepts: • Assessment of consequences • Assessment of incident likelihoods • Determination of level of risk	Risk analysis is the process whereby frequency and impact of IT risk scenarios are estimated.
Risk Evaluation	In this step, levels of risk are compared according to risk evaluation criteria and risk acceptance criteria. The output is a prioritised list of risk elements and the incident scenarios that lead to the identified risk elements.	• Addresses this process step intrinsically as a part of 'risk aggregation'. It evaluates the risk according to 'risk tolerance of the management with regard to the risk appetite of the board'. • Uses a risk map for ranking and graphically displaying risk by defined ranges for frequency and impact.
Risk Treatment	Risk treatment options include: • Risk modification • Risk retention • Risk avoidance • Risk sharing	The treatments of identified risk is described in section 2B. These are: • Risk avoidance • Risk reduction/mitigation • Risk sharing/transfer • Risk acceptance
Information Security Risk Acceptance	The input is a risk treatment plan and the residual risk assessment subject to the risk acceptance criteria. This step comprises the formal acceptance and recording of the suggested risk treatment plans and residual risk assessment by management, with justfication for those that do not meet the enterprise's criteria.	Covers risk acceptance process step in section 2B, subsection 5.2 Risk Response. If an enterprise adopts a risk acceptance stance, it should carefully consider who can accept the risk—even more so with IT risk. IT risk should be accepted only by business management (and business process owners), in collaboration with and supported by IT, and acceptance should be communicated to senior management and the board.
Information Security Risk Communication and Consultation	This is a transversal process—information about risk should be exchanged and shared between the decision maker and other stakeholders throughout all the steps of the risk management process.	• Principle 1: Meeting stakeholder needs. • The Information enabler includes specific information to be communicated between the stakeholders.
Information Security Risk Monitoring and Review	Risk and its influencing factors should be monitored and reviewed to identify any changes in the context of the organisation at an early stage and to maintain an overview of the complete risk picture.	Includes goals and metrics that can be used to measure performance and a maturity model to set a road map for improving risk management processes.

Source: ISACA, *COBIT 5 for Risk*, USA, 2013, figure 48

Figure 2.9—Factor Analysis of Information Risk (FAIR)

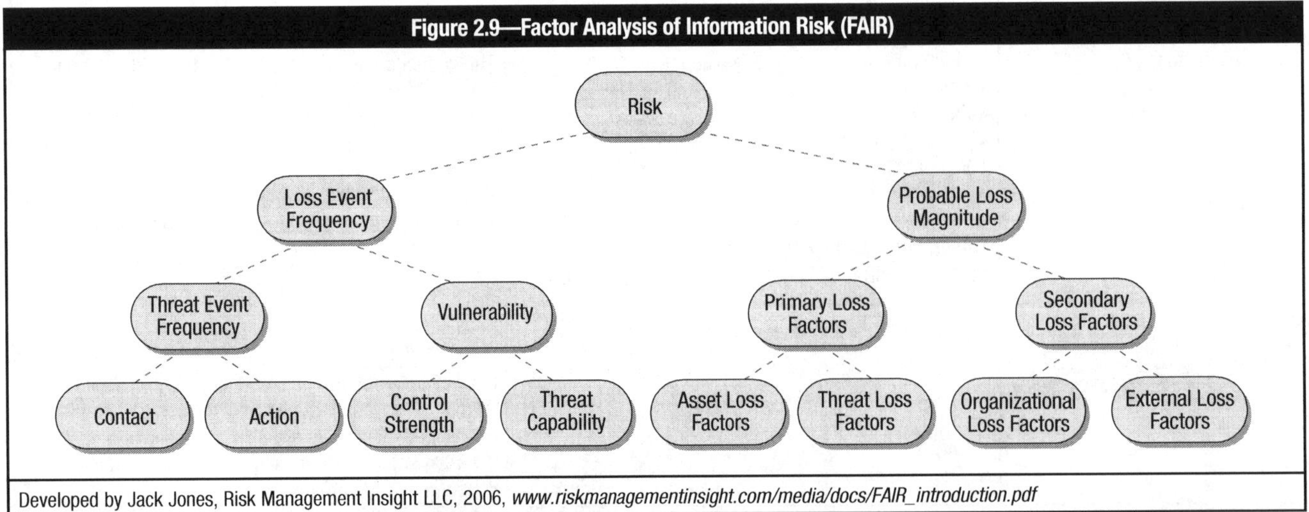

Developed by Jack Jones, Risk Management Insight LLC, 2006, *www.riskmanagementinsight.com/media/docs/FAIR_introduction.pdf*

- **A simulation model** that allows one to apply the taxonomy, measurement method and computational engine to build and analyze risk scenarios of virtually any size or complexity.

There are four primary components of risk taxonomy for which one wants to identify threat agent characteristics—those characteristics that affect:
- The frequency with which threat agents come into contact with an organization or assets
- The probability that threat agents will act against an organization or assets
- The probability of threat agent actions being successful in overcoming protective controls
- The probable nature (type and severity) of impact to assets

Probabilistic Risk Assessment

Probabilistic risk assessment (PRA) has emerged as an increasingly popular analysis tool, especially during the last decade. PRA is a systematic and comprehensive methodology to evaluate risk associated with every life cycle aspect of a complex engineered technological entity, from concept definition through design, construction and operation, and up to removal from service. Developed for use by the US National Aeronautics and Space Administration (NASA) and the nuclear power industry, the approach is complex and time-consuming, but may see increasing application in cases where high network security is required.

Basically, the approach looks to answer three questions:
- What can go wrong?
- How likely is it?
- What are the consequences?

The answers are based on a thorough deconstruction of all the elements at risk in a particular set of circumstances and then systematically identifying, modeling and quantifying scenarios that can lead to adverse consequences based on probability and statistical theory, reliability engineering, decision theory, and other information. Elements that go into the assessing a possible chain of events include hardware and software failure, human actions, control failures and other possible interactions that can lead to serious consequences.

2.7.9 IDENTIFICATION OF RISK

Risk identification is the process by which the type and nature of viable threats are determined and the organization's vulnerabilities that are subject to those threats are examined. The vulnerabilities that may be exploited by the identified threats constitute an identified risk. The process of risk identification is essential to effective risk management because only the risk identified can be assessed and treated appropriately.

It is necessary to identify all information assets including those that may exist with service providers or outsourcers, employees, contractors, and others elsewhere. Viable threats, both real and potential, must be determined. Viability reflects two factors: They exist or could reasonably be expected to materialize and they are subject to some form of control.

Risk identification is often accomplished through a knowledgeable group effort developing a variety of risk scenarios and what-ifs. This exercise assumes that all significant vulnerabilities of the organization are located and the type and nature of threat(s) that could exploit them are identified. Vulnerabilities take many forms. They may be commonly understood technical weaknesses, or they may exist obscured in particular business processes or unmonitored procedures. A lack of awareness by staff can also create vulnerabilities, such as exposure of information via conversations in elevators or restaurants that could be overheard. Outsourced services are a source of vulnerabilities that may be difficult to determine.

It is equally difficult to identify all viable threats. As determined and well-financed attackers dream up ever more sophisticated schemes to compromise organizations, determining the range of potential threats is a formidable challenge for even the most experienced defenders.

Each of the identified vulnerabilities must then be evaluated in terms of viable threats that might compromise them and result in an impact. This exercise will generate a list of identified risk for subsequent analysis to determine likelihood and the extent of potential impacts. Significant vulnerabilities not subject to an identified threat should be added to the analysis list as a possible threat may not have been uncovered or may develop in the future. A prudent approach may be to consider that what can go wrong eventually will and the process itself is subject to considerable error.

Despite these challenges, the odds of a successful attack can be significantly reduced by the diligent effort of a knowledgeable group identifying possible risk and remediation options on an ongoing basis. This is typically accomplished by brainstorming with various members of the various organizational units and considering that what can happen eventually will. The main areas to focus on for developing risk scenarios are shown in **figure 2.10**.

Technical methods, including the use of software, can be used in identifying and tracking risk as well as for providing reporting tools to record the analysis of risk. As with any process, the best tool to use is the one that is best suited to meet the unique needs of the organization. In applying risk identification and analysis methods, the information security manager should define resource requirements and establish a budget and timetable for these important tasks. Budgeting and project planning are covered in chapter 3 Information Security Program Development and Management.

Once an inventory of information assets is completed, a comprehensive list of sources of threats, vulnerabilities, exposures and risk is generated, as is a list of events that might have an impact on the ability of the organization to achieve its objectives as identified in the definition of scope and the framework. These events might prevent, degrade, delay or enhance the achievement of those objectives. In general, a risk can be related to or characterized by:
- **Its origin**—Threat agents such as hostile employees, employees not properly trained, competitors, governments
- **A certain activity, event or incident (i.e., threat)**—Unauthorized dissemination of confidential data, competitor deployment of a new marketing policy, new or revised data protection regulations, an extensive power failure

Figure 2.10—Risk Scenario Technique Main Focus Areas	
Focus/Issue	**Summary Guidance**
Maintain currency of risk scenarios and risk factors.	Risk factors and the enterprise change over time; hence, scenarios will change over time, over the course of a project or over the evolution of technology. For example, it is essential that the risk function develop a review schedule and the CIO works with the business lines to review and update scenarios for relevance and importance. Frequency of this exercise depends on the overall risk profile of the enterprise and should be done at least on an annual basis, or when important changes occur.
Use generic risk scenarios as a starting point and build more detail where and when required.	One technique of keeping the number of scenarios manageable is to propagate a standard set of generic scenarios through the enterprise and develop more detailed and relevant scenarios when required and warranted by the risk profile only at lower (entity) levels. The assumptions made when grouping or generalising should be well understood by all and adequately documented because they may hide certain scenarios or be confusing when looking at risk response. For example, if 'insider threat' is not well defined within a scenario, it may not be clear whether this threat includes privileged and non-privileged insiders. The differences between these aspects of a scenario can be critical when one is trying to understand the frequency and impact of events, as well as mitigation opportunities.
Number of scenarios should be representative and reflect business reality and complexity.	Risk management helps to deal with the enormous complexity of today's IT environments by prioritising potential action according to its value in reducing risk. Risk management is about reducing complexity, not generating it; hence, another plea for working with a manageable number of risk scenarios. However, the retained number of scenarios still needs to accurately reflect business reality and complexity.
Risk taxonomy should reflect business reality and complexity.	There should be a sufficient number of risk scenario scales reflecting the complexity of the enterprise and the extent of exposures to which the enterprise is subject. Potential scales might be a 'low, medium, high' ranking or a numeric scale that scores risk importance from 0 to 5. Scales should be aligned throughout the enterprise to ensure consistent scoring.
Use generic risk scenario structure to simplify risk reporting	Similarly, for risk reporting purposes, entities should not report on all specific and detailed scenarios, but could do so by using the generic risk structure. For example, an entity may have taken generic scenario 15 (project quality), translated it into five scenarios for its major projects, subsequently conducted a risk analysis for each of the scenarios, then aggregated or summarised the results and reported back using the generic scenario header 'project quality'.
Ensure adequate people and skills requirements for developing relevant risk scenarios.	Developing a manageable and relevant set of risk scenarios requires: • Expertise and experience, to not overlook relevant scenarios and not be drawn into highly unrealistic[6] or irrelevant scenarios. While the avoidance of scenarios that are unrealistic or irrelevant is important in properly utilising limited resources, some attention should be paid to situations that are highly infrequent and unpredictable, but which could have a cataclysmic impact on the enterprise. • A thorough understanding of the environment. This includes the IT environment (e.g., infrastructure, applications, dependencies between applications, infrastructure components), the overall business environment, and an understanding of how and which IT environments support the business environment to understand the business impact. • The intervention and common views of all parties involved—senior management, which has the decision power; business management, which has the best view on business impact; IT, which has the understanding of what can go wrong with IT; and risk management, which can moderate and structure the debate amongst the other parties. • The process of developing scenarios usually benefits from a brainstorming/workshop approach, where a high-level assessment is usually required to reduce the number of scenarios to a manageable, but relevant and representative, number.
Use the risk scenario building process to obtain buy-in.	Scenario analysis is not just an analytical exercise involving 'risk analysts'. A significant additional benefit of scenario analysis is achieving organisational buy-in from enterprise entities and business lines, risk management, IT, finance, compliance and other parties. Gaining this buy-in is the reason why scenario analysis should be a carefully facilitated process.

Figure 2.10—Risk Scenario Technique Main Focus Areas *(cont.)*	
Focus/Issue	Summary Guidance
Involve first line of defence in the scenario building process.	In addition to co-ordinating with management, it is recommended that selected members of the staff who are familiar with the detailed operations be included in discussions, where appropriate. Staff whose daily work is in the detailed operations are often more familiar with vulnerabilities in technology and processes that can be exploited.
Do not focus only on rare and extreme scenarios.	When developing scenarios, one should not focus only on worst-case events because they rarely materialise, whereas less-severe incidents happen more often.
Deduce complex scenarios from simple scenarios by showing impact and dependencies.	Simple scenarios, once developed, should be further fine-tuned into more complex scenarios, showing cascading and/or coincidental impacts and reflecting dependencies. For example: • A scenario of having a major hardware failure can be combined with the scenario of failed DRP. • A scenario of major software failure can trigger database corruption and, in combination with poor data management backups, can lead to serious consequences, or at least consequences of a different magnitude than a software failure alone. • A scenario of a major external event can lead to a scenario of internal apathy.
Consider systemic and contagious risk.	Attention should be paid to systemic and/or contagious risk scenarios: • Systemic—Something happens with an important business partner, affecting a large group of enterprises within an area or industry. An example would be a nationwide air traffic control system that goes down for an extended period of time, e.g., six hours, affecting air traffic on a very large scale. • Contagious—Events that happen at several of the enterprise's business partners within a very short time frame. An example would be a clearinghouse that can be fully prepared for any sort of emergency by having very sophisticated disaster recovery measures in place, but when a catastrophe happens, finds that no transactions are sent by its providers and hence is temporarily out of business.
Use scenario building to increase awareness for risk detection.	Scenario development also helps to address the issue of detectability, moving away from a situation where an enterprise 'does not know what it does not know'. The collaborative approach for scenario development assists in identifying risk to which the enterprise, until then, would not have realised it was subject to (and hence would never have thought of putting in place any countermeasures). After the full set of risk items is identified during scenario generation, risk analysis assesses frequency and impact of the scenarios. Questions to be asked include: • Will the enterprise ever detect that the risk scenario has materialised? • Will the enterprise notice something has gone wrong so it can react appropriately? Generating scenarios and creatively thinking of what can go wrong will automatically raise and, hopefully, cause response to, the question of detectability. Detectability of scenarios includes two steps: visibility and recognition. The enterprise must be in a position that it can observe anything going wrong, and it needs the capability to recognise an observed event as something wrong.

- **Its consequences, results or impact**—Service unavailability, loss or increase of market share/profits, increase in regulation, increase or decrease in competitiveness, penalties
- **A specific reason for its occurrence**—System design error, human intervention, prediction or failure to predict competitor activity
- **Protective mechanisms, exposure and controls (together with an estimate of effectiveness)**—Access control and detection systems, policies, security training, market research and surveillance of market
- **Time and place of occurrence**—A flood in the computer room during extreme environmental conditions

High-quality information and thorough knowledge of the organization and its internal and external environments are very important in identifying risk. Historical information about the organization or similar organizations may also prove very useful since the information can lead to reasonable predictions about current and evolving issues that have not yet been faced by the organization.

Identifying what may happen is rarely sufficient. The fact that there are many ways an event can occur makes it important to study all reasonably possible and significant causes and scenarios. Methods and tools used to identify risk and its occurrence include checklists, judgments based on experience and records, flowcharts, brainstorming, systems analysis, scenario analysis and systems engineering techniques.

In selecting a risk identification methodology, the following techniques should be considered:

- Team-based brainstorming, where workshops can prove effective in building commitment and making use of different experiences
- Structured techniques such as flowcharting, system design review, systems analysis, hazard and operability studies, and operational modeling
- What-if and scenario analysis for less clearly defined situations, such as the identification of strategic risk and processes with a more general structure, as shown in **figure 2.11**
- Threats identified internally and externally mapped to identified and suspected vulnerabilities

Figure 2.11—Risk Scenario Approaches

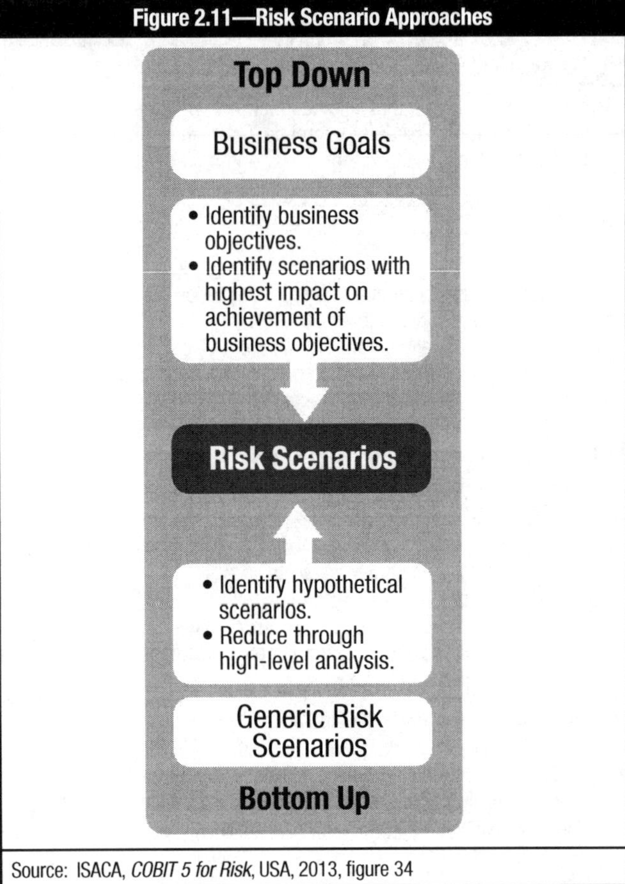

Source: ISACA, *COBIT 5 for Risk*, USA, 2013, figure 34

The development of risk scenarios is based on describing a potential risk event and documenting the factors and areas that may be affected by the risk event. Risk events may include system failure, loss of key personnel, theft, network outages, power failures, natural disasters or any other situation that could affect business operations and mission. Each risk scenario should be related to a business objective or impact. The key to developing effective scenarios is to focus on real and relevant potential risk events. Examples are developing a risk scenario based on a radical change in the market for an organization's products, a change in government or leadership, or a supply chain failure. **Figure 2.12** shows an example of the many inputs that are required to develop risk scenarios.

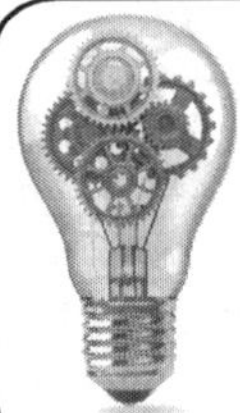

In Practice: Examine a risk scenario used by your organization for the elements described in this section. If none are available, walk through the risk scenario development process described in this section to evaluate a possible risk event in your organization.

2.7.10 THREATS

Threats to information resources and the likelihood that they will exploit an existing vulnerability must be assessed. In this context, threats are any circumstances or events with the potential to cause harm to an information resource by exploiting vulnerabilities in the system.

Threats can be external or internal, intentional or unintentional. They may be caused by natural events or political, economic or competitive factors. Threats always exist and are typically beyond the direct control of the risk practitioner or asset owner. Not all conceivable threats need to be considered by every organization. For example, an organization that operates in a region with a seismic rating of zero does not have to document exposure to volcanoes or earthquakes. However, it is important to identify the various types of threats that do apply and may compromise systems or otherwise affect the organization.

Threats may be divided into multiple categories, including:
- Physical
- Natural events
- Loss of essential services
- Disturbance due to radiation
- Compromise of information
- Technical failures
- Unauthorized actions
- Compromise of functions

The information security manager should document all the threats that may apply to the systems and business processes under review. This requires using the resources noted previously but should also include examining the cause of past failures, audit reports, media reports, information from national computer emergency response teams (CERTs), data from security vendors and communication with internal groups.

Threats may be the result of accidental actions, intentional/deliberate actions or natural events. Threats may originate from either internal or external sources, and actual attacks may leverage a combination of both internal and external sources. Where threats may be directed by individual agents, the information security manager should know and remain aware that such threat agents tend to be imaginative, creative and determined and will explore new methods and avenues of attack. To counter threat agents, the information security manager must also be determined and creative and seek to discover as many threats as possible. An unidentified threat is one for which the organization is more likely to be unprepared and vulnerable than a threat that is well documented.

Sources for information regarding threats are:
- Assessments
- Audits
- Business continuity plans
- Finance
- Government publications
- Human resources
- Insurance companies
- Management
- Media
- Product vendors
- Security companies
- Service providers
- Threat monitoring agencies
- Users

Figure 2.12—Risk Scenario Structure

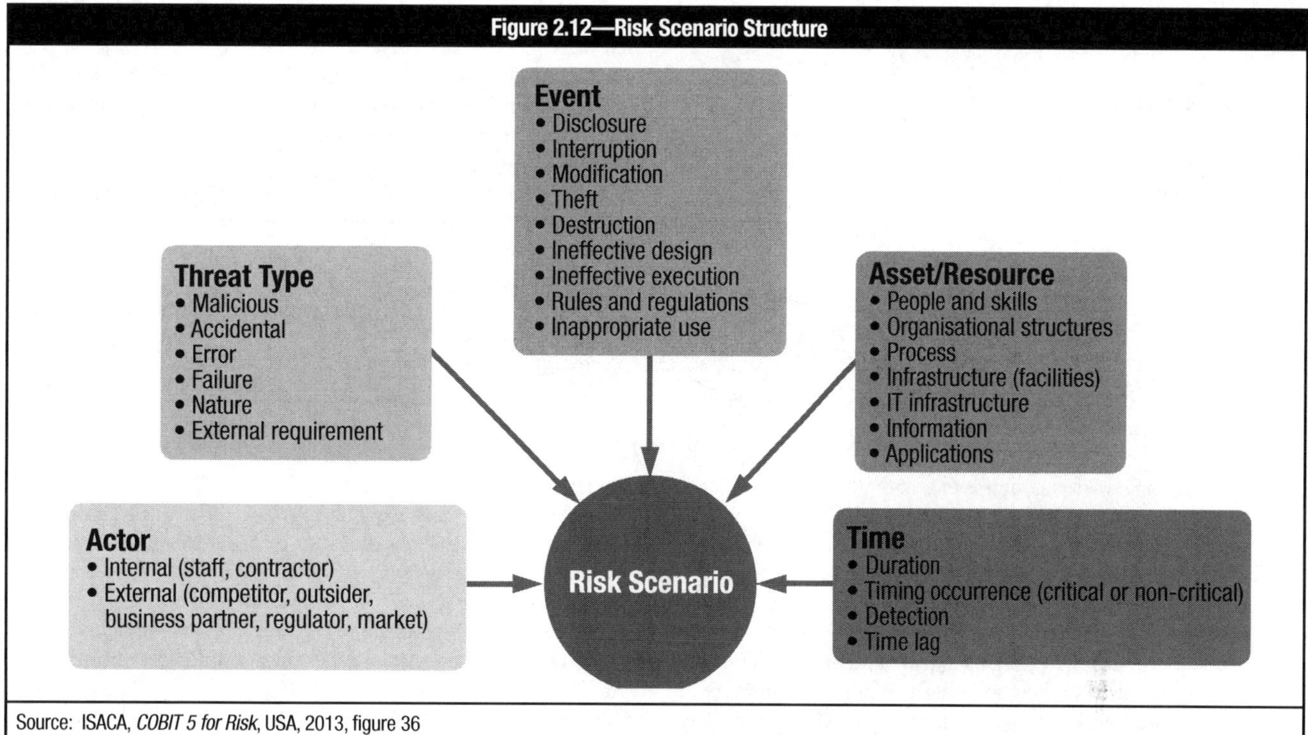

Source: ISACA, *COBIT 5 for Risk*, USA, 2013, figure 36

Internal Threats

Although many organizations label their employees and other staff as their best asset, employees may be unhappy if they are inadequately trained, treated poorly or not given enough time to do their jobs properly. Disillusionment and resentment can lead to a higher risk of errors, negligence and more conscious actions such as theft. Key personnel may be drawn to another company and leave serious gaps in the knowledge and skills needed to operate systems effectively.

Employees are the cause of a significant number of business impacts, which can be intentional and unintentional. A disgruntled employee may intentionally compromise systems or release data that expose the organization to legal or reputational risk. Employees may be convinced, bribed or threatened to disclose trade secrets for ideological or economic reasons. Many employees have a level of access to systems and data that far exceeds their actual job requirements, which can be exploited in an attack. The solution to the employee problem, therefore, lies at least in part in the application of need-to-know and least privilege, but this is an imperfect solution. Any system has trusted insiders, and one of them choosing to violate trust is difficult to either predict or prevent.

The typical malicious insider is a current or former employee, contractor or other business partner who has or had authorized access to an organization's network, system or data and intentionally intercepted (exfiltrated), interrupted, modified or fabricated data on the organization's information systems. The first step in addressing personnel threats is to start with the hiring process and review the qualifications and attitude of prospective employees. Employment candidates may submit incorrect information on job applications and claimed education, certification or experience they do not actually possess. A review of references and performance of background checks may be worthwhile to the extent permitted by law. At the time of hiring, the employee should be required to sign a nondisclosure agreement and be advised of the ethics and policies of the organization.

Throughout employment, employees should be reminded of organizational policies and their responsibilities through awareness sessions and regular management reviews. One of the best employee-based controls is to interact with employees to understand any frustrations, complaints or issues they may be facing and then seek to resolve those issues. During times of strike, layoffs, mergers, relocation and reorganization, an employee is more likely to be a risk. An employee who has been recently demoted or bypassed for a promotion may also pose a risk.

At the end of employment, an employee should return all organizational assets, including identification badges, equipment (e.g., laptops, mobile phones, access cards) and uniforms so that he/she cannot use those to gain unauthorized access in the future. In addition, systems, network and facility access should be removed immediately prior to the employee's departure to minimize the potential for crimes of opportunity.

External Threats

In a networked environment, where data are stored offsite or hosted by cloud service providers, threats to information systems from outside the organization can originate from anywhere and may take a number of forms, including, but not limited to:
- Criminal acts
- Data corruption
- Disease (epidemic)
- Espionage
- Facility flaws (freezing pipe/pipe burst)
- Fire
- Flooding

- Hardware flaws
- Industrial accidents
- Lost assets
- Mechanical failures
- Power surge/utility failure
- Sabotage
- Seismic activity
- Severe storms
- Software errors
- Supply chain interruption
- Terrorism
- Theft

Natural events such as a flood, storm, earthquake or tornado are unpredictable and may be extremely damaging. The use of governmental data and weather monitoring services may identify the threats associated with natural events and allow the risk practitioner to take necessary steps to be prepared.

An external personnel threat includes a hacker, a thief or an advanced persistent threat (APT) that is skilled and determined to break into systems for military or economic purposes. The term APT refers to advanced, highly skilled attackers that are determined (persistent) in their attempts to exploit systems and networks. The increased skills available to the hacking community and the effectiveness of the tools they possess make the risk of compromise much more significant. APTs may be sponsored by governments, organized crime or competitors.

Most breaches are the result of targets of opportunity, not determined attacks. As seen in the annual reports from Verizon and other organizations, many organizations are breached because they were discovered to be easy targets and threat agents took advantage of their vulnerabilities.

Advanced Persistent Threat

The information security manager must be aware that APTs pose a significant risk to organizations of virtually all types globally and ensure that appropriate steps are in place to detect and identify this threat.

APT is defined in NIST publication 800-39 as follows:

> *An APT is an adversary that possesses sophisticated levels of expertise and significant resources which allow it to create opportunities to achieve its objectives using multiple attack vectors (e.g., cyber, physical and deception). These objectives typically include establishing and extending footholds within the IT infrastructure of the targeted organizations for purposes of exfiltrating information, undermining or impeding critical aspects of a mission, program or organization; or positioning itself to carry out these objectives in the future. The advanced persistent threat: (i) pursues its objectives repeatedly over an extended period of time; (ii) adapts to defenders' efforts to resist it; and (iii) is determined to maintain the level of interaction needed to execute its objectives.*

Typical APT attacks have generally exhibited the following life cycle:

- **Initial compromise**—Attackers use social engineering and spear phishing via email, using zero-day viruses. They may plant malware on a web site that the victimized employees are likely to visit.
- **Establish foothold**—Attackers may plant remote administration software in the victim's network and/or create network back doors and tunnels that allow stealth access to network infrastructure.
- **Escalate privileges**—APTs use exploits and password cracking to acquire administrator privileges over the victim's computer and possibly expand it to Windows domain administrator accounts.
- **Internal reconnaissance**—Attackers collect information on surrounding infrastructure, trust relationships and Windows domain structure.
- **Move laterally**—They expand control to other workstations, servers and infrastructure elements and perform data harvesting on them.
- **Maintain presence**—APTs ensure continued control over access channels and credentials acquired in previous steps.
- **Complete mission**—Attackers exfiltrate stolen data from the victim's network.

Figure 2.13 shows the typical sources of APT. Due to the required sophistication, expertise, time and costs involved, these efforts are normally mounted only against high-value targets.

Figure 2.13—Typical Sources of APT

Threat	What They Seek	Business Impact
Intelligence agencies	Political, defense or commercial trade secrets	Loss of trade secrets or commercial, competitive advantage
Criminal groups	Money transfers, extortion opportunities, personal identity information or any secrets for potential onward sale	Financial loss, large-scale customer data breach or loss of trade secrets
Terrorist groups	Production of widespread terror through death, destruction and disruption	Loss of production and services, stock market irregularities, and potential risk to human life
Activist groups	Confidential information or disruption of services	Major data breach or loss of service
Armed forces	Intelligence or positioning to support future attacks on critical national infrastructure	Serious damage to facilities in the event of a military conflict

Source: ISACA, *Advanced Persistent Threats: How to Manage the Risk to Your Business*, USA, 2013, figure 2, page 29

Emerging Threats

Indications of emerging threats may include unusual activity on a system, repeated alarms, slow system or network performance, or new or excessive activity in logs. In many cases, compromised organizations have evidence of emergent threats in their logs well in advance of the actual compromise, but the evidence is not noticed or acted on. Lack of effective monitoring, when combined with a threat, can lead to a breach.

Most technologies are built with an emphasis on function and purpose without due consideration for the security implications. As a result, new technology tends to be a source of new vulnerabilities and may even itself be a threat agent within an information system. The risk practitioner must be alert to the emergence of new technologies and prepare for their introduction into the organization, particularly if these technologies promise cost savings or competitive advantage.

Bring your own device (BYOD) is an example of a revolution in how organizations view technology assets. It is an opportunity whose risk is self-evident yet it has tempted a wide variety of organizations by promising to greatly reduce the cost of initial procurement of IT assets and the rate at which they need to be refreshed. A threat strategy that emphasizes rejection of new technology is unlikely to remain in place long beyond the point at which it gains executive sponsorship.

2.7.11 VULNERABILITIES

The term "vulnerability," often referred to as a "weakness," is often used as if it is a binary condition. Something "is vulnerable" or it "is not vulnerable." In most cases, assets are vulnerable to varying degrees (i.e., a particular control condition might represent a high degree of vulnerability, while another control condition represents a lower degree of vulnerability). The extent of exposure must also be considered as it affects the probability that a vulnerability is compromised. These distinctions become critical in the process of prioritizing risk management efforts, when determining the level of risk within a scenario, and also when explaining conclusions and recommendations to management.

The *National Institute of Standards and Technology (NIST) Special Publication 800-30 Revision 1 Guide to Conducting Risk Assessments* provides a list of vulnerabilities to consider as well as "predisposing conditions" that may lead to the rapid or unpredictable emergence of new vulnerabilities. Many vulnerabilities are conditions that exist in systems and must be identified so they can be addressed. The purpose of vulnerability identification is to find the problems before they are found by an adversary and exploited, which is why an organization should conduct regular vulnerability assessments and penetration tests to identify, validate and classify its vulnerabilities. An assessment of vulnerabilities must consider process and procedural weaknesses as well as logical ones. Where vulnerabilities exist, there is a potential for risk.

Note: Other sources of vulnerabilities include:
- National Vunerability Database at *nvd.nist.gov*
- Common Weakness Enumeration *nvd.nist.gov*
- Harmonized Threat and Risk Assessement at *www.cse-cst.gc.ca*
- Contingency Planning and Management at *contingencyplanning.com.*
- Open Web Application Security Project (OWASP) at *www.owasp.org*

Estimating the degree of vulnerability can be accomplished through various forms of testing (when time permits and when the stakes are high) or through subject matter expert estimates. As with other valuations, estimates can be quantitative or qualitative. Whatever the approach, it is important to communicate the imprecise nature of these estimates so management is not misled. Effective approaches for reflecting uncertainty in values include using ranges or distributions to indicate both unlikely maximums and values that are more probable.

Determining the ultimate relevancy of a weak control also requires an understanding of the other controls in play that may mitigate the overall exposure. It would be inaccurate and a disservice to portray a control as a severe problem when, in fact, the combination of controls is relatively robust.

Many IT system weaknesses are identified using automated scanning equipment, and these can serve as leading indicators of potential compromise. Process and performance vulnerabilities are more difficult to ascertain and may require careful review and analysis to uncover. To be effective, the assessment must consider process, procedural and physical vulnerabilities in addition to technology weaknesses.

Consider the organization that does not have a formal information security training and awareness program. One set of vulnerabilities in this instance would stem from a lack of user awareness of security policies, standards and guidelines. Absent such awareness training, it has been shown that an organization is considerably more likely to suffer compromise from social engineering attacks such as phishing. Vulnerabilities can occur from numerous causes, ranging from a lack of processes for configuration control to poor personnel morale, from a vast variety of technical issues to countless human factors.

It may be useful to categorize and consider vulnerabilities in the following categories because the approach will vary for different areas, including some of the more common areas:
- Network vulnerabilities
- Physical access
- Applications and web-facing services
- Utilities
- Supply chain
- Processes
- Equipment
- Cloud computing
- Internet of things

Audits, security reviews, vulnerability scans and penetration tests are among the approaches that are usually helpful in identifying vulnerabilities. Some typical examples of vulnerabilities include:

- Defective software
- Improperly configured hardware/software
- Inadequate compliance enforcement
- Poor network design
- Uncontrolled or defective processes
- Inadequate governance or management
- Insufficient staff
- Lack of knowledge to support users or operations
- Lack of security functionality
- Lack of proper maintenance
- Poor choice of passwords
- Transmission of unprotected communications
- Lack of redundancy
- Poor management communications

2.7.12 RISK, LIKELIHOOD AND IMPACT

The traditional assessment of risk is expressed in the equation, threats × vulnerabilities = risk. The result is that if there are more threats against more or greater vulnerability, there is more risk. Another current, more generalized definition of risk, also subject to considerable controversy, from ISO 73:2009 is "the combination of the probability of an event and its consequence." The consequence could be a compromise or, in this context, it could be a positive event as well. For most, "positive risk" is generally considered an opportunity rather than a risk, and the ISO definition does not appear to clarify the matter. Currently, most practitioners consider the notion of risk in a negative sense, and the term is used from that perspective in this manual.

Increasingly, the concept of consequences is included in the risk equation, resulting in threats × vulnerabilities × consequences = risk. This is consistent with the current definition of risk being "the probability of an event and its consequences." Probability, in this case, is derived from the likelihood that a threat will exploit a vulnerability. The relevance is in the fact that if there are no consequences, the risk is not important and can be considered nonexistent. From the practitioner's viewpoint, it must be considered that entirely different processes are used to determine the probability of an event and the potential impact or consequences.

The likelihood, or probability, is a measure of the frequency that an event may occur. When identifying risk, likelihood is used to calculate the level of risk based on the number of events, combined with the impact that may occur in a given time period, usually a year. The likelihood or frequency combined with the magnitude (of impact) is used in determining ALE. The greater the frequency, the greater the likelihood, and, consequently, the greater the risk.

Determining likelihood requires consideration of a variety of factors including:

- **Volatility**—The probability of a risk may vary depending on the volatility of the situation. When conditions vary a great deal, there may be times when the risk is greater than at other times, increasing unpredictability and, therefore, requiring a higher estimation of risk.
- **Velocity**—In this context, velocity is a consideration of the extent of prior warning of an event and the amount of time between an event occurrence and the subsequent impact.
- **Proximity**—This is a term used to indicate the time between an event and the impact (i.e., the greater the velocity, the closer the proximity).
- **Interdependency**—It is important not to consider risk in isolation, but rather in the organizational context and in relationship to other assets and functions. The materialization of several types of risk might affect the organization differently depending in part on whether the risk events occur concurrently or sequentially.
- **Motivation**—The extent of an attacker's motivation will affect the chances of success. The nature of the motivation affects, to some extent, the type of assets at risk (i.e., politically motivated nation states will typically target different assets than criminals seeking financial gain, thereby affecting the risk to particular assets).
- **Skill**—The level of skill of potential attackers will typically determine the type and value of assets attacked. (i.e., high-value assets are likely to attract more skilled attackers).
- **Visibility**—High-visibility targets are more likely to be attacked.

Whatever definition for risk is used, the information security manager must understand the business risk profile (in the negative sense) of the organization in terms of adversely affecting the ability to achieve its objectives.

No model provides a complete picture of all possible risk, but logically categorizing the risk areas of an organization (as illustrated in **figure 2.14**) facilitates focusing on key risk management strategies and decisions. It also enables the organization to develop and implement risk treatment approaches that are relevant to the business and cost-effective.

A high-level categorization of risk is an inherent part of business and, because it is impractical and costly to eliminate all risk, every organization has a level of risk it will accept. The level of acceptable risk is a decision that must be made by senior management, although the information security manager may need to assist management in developing a concrete set of criteria. Approaches to assist in determining the level of acceptable risk are discussed in detail in section 1.8.4 Risk Objectives.

The essential concept is that the cost of protection should be proportional to the value of the asset and should not exceed the value of the asset being protected. With most remedial activities, there is a point of diminishing returns at which costs of additional controls rise faster than the benefits derived.

Many factors will affect what the organization ultimately considers acceptable based on considerations such as culture, the ability of the organization to absorb losses (sometimes referred to as risk capacity), the effects and costs of mitigation, the extent and kind of potential impacts, and possibly legal and/or contractual requirements. In terms of culture, the organization may be generally risk-averse and opt for tighter security and higher control costs, or the organization may be risk-aggressive and choose to take more chances.

Figure 2.14—Operational Risk Categories

Operational Risk Area	Description	Information or IT Mapping
Facilities and operating environment risk	Loss or damage to operational capabilities caused by problems with premises, facilities, services or equipment	Business continuity management for IT facilities
Health and safety risk	Threats to the personal health and safety of staff, customers and members of the public	Confidentiality of home addresses, travel schedules, etc.
Information risk	Unauthorized disclosure or modification to information, loss of availability of information, or inappropriate use of information	All aspects of information and IT security
Control frameworks risk	Inadequate design or performance of the existing risk management infrastructure	Business process analysis to identify critical information flows and control points
Legal and regulatory compliance risk	Failure to comply with the laws of the countries in which business operations are carried out; failure to comply with any regulatory, reporting and taxation standards; failure to comply with contracts; or failure of contracts to protect business interests	Compliance with data protection legislation, cryptographic control regulations, etc.; accuracy, timeliness and quality of information reported to regulators; and content management of all information sent to other parties
Corporate governance risk	Failure of directors to fulfill their personal statutory obligations in managing and controlling the company	Information security policy making, performance measurement and reporting
Reputation risk	The negative effects of public opinion, customer opinion and market reputation, and the damage caused to the brand by failure to manage public relations	Controlling the disclosure of confidential information; presenting a public image of a well-managed enterprise
Strategic risk	Failure to meet the long-term strategic goals of the business, including dependence on any estimated or planned outcomes that may be in the control of third parties	Managing the quality and granularity of information on which strategic business decisions are based (e.g., mergers, acquisitions, disposals)
Processing and behavioral risk	Problems with service or product delivery caused by failure of internal controls, information systems, employee integrity, errors and mistakes, or through weaknesses in operating procedures	All aspects of information systems security and the security-related behavior of employees in carrying out their tasks
Technology risk	Failure to plan, manage and monitor the performance of technology-related projects, products, services, processes, staff and delivery channels	Failure of information and communications technology systems and the need for business continuity management
Project management risk	Failure to plan and manage the resources required for achieving tactical project goals, leading to budget overruns, time overruns or both, or leading to failure to complete the project; the technical failure of a project or the failure to manage the integration aspects with existing parts of the business and the impact that changes can have on business operations	Management of all information security-related projects
Criminal and illicit acts risk	Loss or damage caused by fraud, theft, willful neglect, gross negligence, vandalism, sabotage, extortion, etc.	Provision of security services and mechanisms to prevent all types of cybercrime
Human resources risk	Failure to recruit, develop or retain employees with the appropriate skills and knowledge or to manage employee relations	Need for policies protecting employees from sexual harassment, racial abuse, etc., through corporate e-mail systems, etc.
Supplier risk	Failure to evaluate adequately the capabilities of suppliers leading to breakdowns in the supply process or substandard delivery of supplied goods and services; failure to understand and manage the supply chain issues	Outsourced service delivery of IT or other business information processing activities

Figure 2.14—Operational Risk Categories *(cont.)*		
Operational Risk Area	**Description**	**Information or IT Mapping**
Management information risk	Inadequate, inaccurate, incomplete or untimely provision of information to support the management decision-making process	Managing the accuracy, integrity, currency, timeliness and quality of information used for management decision support
Ethics risk	Damage caused by unethical business practices, including those of associated business partners. Issues include racial and religious discrimination, exploitation of child labor, pollution, environmental issues, behavior to disadvantaged groups, etc.	Ethical collection, storage and use of information; management of information content on web sites, intranets, and in corporate e-mails and instant messaging systems
Geopolitical risk	Loss or damage in some countries, caused by political instability, poor quality of infrastructure in developing regions, or cultural differences and misunderstandings	Managing all aspects of information security and IT systems' security in regions where the enterprise has business operations but where there is special geopolitical risk
Cultural risk	Failure to deal with cultural issues affecting employees, customers or other stakeholders. These include language, religion, morality, dress codes, and other community customs and practices.	Management of information content on web sites, intranets, and in corporate e-mails and instant messaging systems
Climate and weather risk	Loss or damage caused by unusual climate conditions, including drought, heat, flood, cold, storm, winds, etc.	Business continuity management for IT facilities
Source: Copyright SABSA Institute, *www.sabsa.org.* Reproduced with permission.		

For most organizations the failure to adequately identify, analyze, evaluate and respond to risk results in risk management efforts not being properly allocated and directed at those situations with the greatest probability of compromise and the greatest impact.

It is fair to say that, for most organizations, the primary risk management efforts are typically directed at controlling unauthorized system access—which, on average, represents a relatively small proportion of losses—while the majority of losses from the other identified causes such as outsourced services, electronic backup, lost or stolen laptops, and paper records generally receive less attention.

For the prevalent sources of compromise the information security manager needs to consider how to allocate protection resources for the greatest benefit. Because many of the most significant threats can be addressed directly (e.g., employees and former employees, service providers, consultants, suppliers), it is prudent to consider the specific controls addressing the risk. Strong access controls, limiting access to need-to-know, network segmentation, effective termination procedures and good monitoring will all contribute to reducing an organization's risk.

It is important to recognize that the major risk to organizations does not come from technology and, to a large extent, cannot be solved by technology. Rather, the majority of risk is largely a management and personnel problem.

2.7.13 RISK REGISTER

During the process of identifying risk and its components, a risk register should be established. The register should serve as a central repository for all information security risks including specific threats, vulnerabilities, exposures and assets affected. It should include the asset owner, the risk owner and other stakeholders.

Content of the risk register should be filled out as the assessment process is underway. Once the identification, analysis, evaluation and risk response efforts have completed and other pertinent information entered into the register, it will serve aa a master reference point for all risk-management-related activities.

The risk register is part of an organization's risk profile. A risk profile is an essential element for effective information risk management. It will serve to provide a comprehensive overview of the overall (known) risk to which the organization is exposed as well as other pertinent information. There are a variety of approaches to accomplishing this requirement. The COBIT 5 approach is a detailed and effective process to achieve this objective (**figure 2.15**).

Figure 2.16 contains a sample template for a risk register entry, which is a part of the risk profile.

2.7.14 ANALYSIS OF RISK

Risk analysis is the phase where the level of the identified risk and its nature are assessed and understood and the potential consequences of compromise determined. This phase also includes determining the effectiveness of existing controls and the extent to which they mitigate the identified risk. Risk analysis involves:

- Thorough examination of the risk sources (threats and vulnerabilities) determined in the risk identification phase
- The extent to which information assets have exposure to potential threats and the effect on likelihood
- The potential negative consequences (impact) if the assets are successfully attacked
- The likelihood that those consequences may occur and the factors that affect them
- Assessment of any existing controls or processes that tend to minimize negative risk or enhance positive outcomes (these

Figure 2.15—Risk Profile

A risk profile is a description of the overall (identified) risk to which the enterprise is exposed. A risk profile consists of:

- **Risk register**
  - **Risk scenarios**
  - **Risk analysis**
- **Risk action plan**
- **Loss events (historical and current)**
- **Risk factors**
- **Independent assessment findings**

Life Cycle and Stakeholders	Life Cycle Stage	Internal Stakeholder	External Stakeholder	Description/Stake
	Information planning	ERM committee, board	External audit, regulator	• Internal stakeholders: Initiate and drive the implementation and appoint a CRO. Have adequate information on the exposure. • External stakeholders: To have comfort on the risk management capabilities
	Information design	Risk function, compliance, CIO, CISO, business process owners, internal audit		• CRO: To obtain information from the other roles in order to provide the overview for the governance bodies • CIO: To be able to develop an adequate information system • Other roles: To be able to provide relevant information and to ensure completeness/adequacy
	Information build/acquire	Risk function, internal audit		• CRO: Provides functional requirements and consults others. • Internal audit: Provides quality assurance services on the implementation.
	Information use/operate: store, share, use	Board, ERM committee, business executive, CIO, risk function, CISO, business process owners, compliance, internal audit	External audit, regulator	• Business process owners, business executives and CIO: To efficiently provide relevant information • Board and ERM committee: To receive relevant information and to enable decision making • Internal audit, external audit and regulator: Receive relevant information • CRO: Oversees the caption, processing and interpretation of information
	Information monitor	Board, ERM committee, risk function, internal audit	External audit	• CRO: Ongoing monitoring on adequacy, completeness and accuracy of information; semi-annual assessment of performance (MEA01) and controls (MEA02) to maintain the information • Internal audit: Annual validation of format and level of contents
	Information dispose	Risk function		• CRO: According to data retention policy, to ensure confidentiality of information and to reduce the amount of information

Source: ISACA, *COBIT® 5 For Risk*, USA, 2013, figure 61, page 147

controls may derive from a wider set of standards, controls or good practices selected according to an applicability statement [i.e., the application of controls identified in the risk register] and may also come from previous risk treatment activities)

The level of risk can be estimated in various ways, including using statistical analysis and calculations combining impact and likelihood. Impact may not be clear unless BIAs have been performed to provide an understanding of the cost to the organization as the result of the loss of a particular function or asset.

Any formulas and methods for combining impact and likelihood must be consistent with the criteria defined when establishing the risk management context. This is because an event may have multiple consequences and affect different objectives; therefore, consequences and likelihood need to be combined to calculate the level of risk. If no reliable or statistically reliable and relevant past data are available (e.g., incident data kept for an incident database), other estimates based on such things as impacts affecting other organizations may be made as long as they are appropriately communicated and approved by the decision makers.

Information used to estimate impact and likelihood usually comes from:

- Past experience or data and records (e.g., incident reporting)
- Reliable practices, international standards or guidelines
- Market research on other organizations and analysis
- Experiments and prototypes
- Economic, engineering or other models
- Specialist and expert advice

Risk analysis techniques include:

- Interviews with experts in the area of interest and questionnaires
- Use of existing models and simulations
- Statistical and other analysis

Figure 2.16—Risk Register

Part I—Summary Data						
Risk statement						
Risk owner						
Date of last risk assessment						
Due date for update of risk assessment						
Risk category	○ STRATEGIC (IT Benefit/Value Enablement)		○ PROJECT DELIVERY (IT Programme and Project Delivery)		○ OPERATIONAL (IT Operations and Service Delivery)	
Risk classification (copied from risk analysis results)	○ LOW	○ MEDIUM	○ HIGH	○ VERY HIGH		
Risk response	○ ACCEPT	○ TRANSFER	○ MITIGATE	○ AVOID		
Part II—Risk Description						
Title						
High-level scenario (from list of sample high-level scenarios)						
Part III—Risk Response						
Detailed scenario description—Scenario components	Actor					
	Threat Type					
	Event					
	Asset/Resource					
	Timing					
Other scenario information						
Part III—Risk Analysis Results						
Frequency of scenario (number of times per year)	0	1	2	3	4	5
	N ≤ 0,01 ○	0,01 < N ≤ 0,1 ○	0,1 < N ≤ 1 ○	1 < N ≤ 10 ○	10 < N ≤ 100 ○	100 < N ○
Comments on frequency						
Impact of scenario on business	0	1	2	3	4	5
1. Productivity	Revenue Loss Over One Year					
Impact rating	I ≤ 0,1% ○	0,1% < I ≤ 1% ○	1% < I ≤ 3% ○	3% < I ≤ 5% ○	5% < I ≤ 10% ○	10% < I ○
Detailed description of impact						
2. Cost of response	Expenses Associated With Managing the Loss Event					
Impact rating	I ≤ 10k$ ○	10K$ < I ≤ 100K$ ○	100K$ < I ≤ 1M$ ○	1M$% < I ≤ 10M$ ○	10M$ < I ≤ 100M$ ○	100M$ < I ○
Detailed description of impact						
3. Competitive advantage	Drop-in Customer Satisfaction Ratings					
Impact rating	I ≤ 0,5 ○	0,5 < I ≤ 1 ○	1 < I ≤ 1,5 ○	1,5 < I ≤ 2 ○	2 < I ≤2,5 ○	2,5 < I ○
Detailed description of impact						
4. Legal	Regulatory Compliance—Fines					
Impact rating	None ○	< 1M$ ○	< 10M$ ○	< 100M$ ○	< 1B$ ○	> 1B$ ○
Detailed description of impact						
Overall impact rating (average of four impact ratings)						
Overall rating of risk (obtained by combining frequency and impact ratings on risk map)	○ LOW	○ MEDIUM	○ HIGH	○ VERY HIGH		
Risk response for this risk	○ ACCEPT	○ TRANSFER	○ MITIGATE	○ AVOID		
Justification						

Figure 2.16—Risk Register *(cont.)*

	Response Action	Completed	Action Plan
Detailed description of response (NOT in case of ACCEPT)	1.	O	O
	2.	O	O
	3.	O	O
	4.	O	O
	5.	O	O
	6.	O	O
Overall status of risk action plan			
Major issues with risk action plan			
Overall status of completed responses			
Major issues with completed responses			
Part IV—Risk Indicators			
Key risk indicators for this risk	1. 2. 3. 4.		

Source: ISACA, *COBIT® 5 for Risk*, USA, 2013, figure 62, page 151

Risk analysis may vary in detail according to the risk, the purpose of the analysis, and the required protection level of the relevant information, data and resources. Analysis may be qualitative, semiquantitative, quantitative or a combination of these. In any case, the type of analysis performed should, as stated earlier, be consistent with the criteria developed as part of the definition of the risk management context and consensus gained on the approach used.

Qualitative Analysis

In qualitative analysis, the magnitude and likelihood of potential consequences are presented and described in detail. The scales used can be formed or adjusted to suit the circumstances, and different descriptions may be used for different risk. Qualitative analysis may be used:

- As an initial assessment to identify risk that will be the subject of further, detailed analysis
- Where nontangible aspects of risk are to be considered (e.g., reputation, culture, image)
- Where there is a lack of adequate information and numerical data or resources necessary for a statistically acceptable quantitative approach

A common approach to qualitative analysis can be accomplished by using a 5 by 5 matrix, as shown in **figure 2.17**.

Figure 2.17—Semiquantitative Impact Matrix

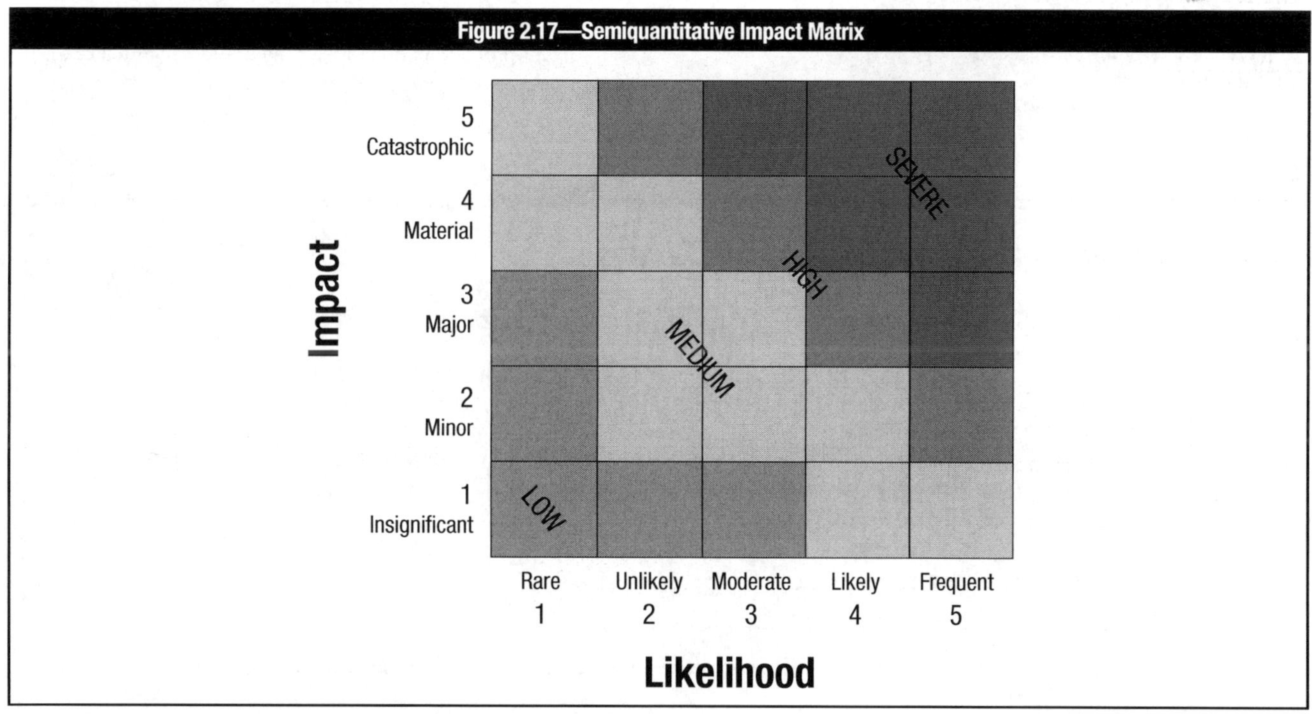

Semiquantitative Analysis

In semiquantitative analysis, the objective is to assign values to the scales used in the qualitative assessment. These values are usually indicative and not real, which is the prerequisite of the quantitative approach. Therefore, as the value allocated to each scale is not an accurate representation of the actual magnitude of impact or likelihood, the numbers used must only be combined using a formula that recognizes the limitations or assumptions made in the description of the scales used. It also should be mentioned that the use of semiquantitative analysis may lead to various inconsistencies due to the fact that the numbers chosen may not properly reflect analogies among risks, particularly when either consequences or likelihood are extreme.

For this approach to be useful, the values selected should be generally indicative and sufficient for the prioritization of one risk above another risk. For any process to operate successfully there should be a common understanding of these values and of the terms employed. The definitions that follow can be substituted with ones already in use within the organization, and it may be possible to use a subset from the ERM framework where one exists.

Typical values for impact are:
- Insignificant (value = 1): No meaningful impact, or of limited consequence
- Minor (value = 2): Impact on a small part of the business only, or less than US $1 million impact
- Major (value = 3): Impact on the organization's brand, or more than US $1 million impact
- Material (value = 4): Impact more than US $200 million and requiring external reporting
- Catastrophic (value = 5): Failure or significant downsizing of the organization

Typical values for likelihood are:
- Rare (value = 1)
- Unlikely (value = 2): Not seen within the last five years
- Moderate (value = 3): Seen within the last five years but not within the last year
- Likely (value = 4): Seen within the last year
- Frequent (value = 5): Happens on a regular basis

These values should be sufficient to allow risk prioritization according to a semiquantitative approach.
Typically, the risk probability and consequences would be calculated as:
- Risk = impact × likelihood

An example would be Risk = 4 (material) × 3 (moderate) = 12.

Quantitative Analysis

In quantitative analysis, numerical values are assigned to both impact and likelihood. These values are derived from a variety of sources. The quality of the entire analysis depends on the accuracy of the assigned values and the validity of the statistical models used. Impact can be determined by evaluating and processing the various results of an event or by extrapolation from experimental studies or past data. Consequences may be expressed in various terms of:
- Monetary
- Technical
- Operational
- Human impact criteria

As is made clear from the preceding analysis, the specification of the risk level is not unique. Impact and likelihood may be expressed or combined differently, according to the type of risk and the scope and objective of the risk management process.

Annual Loss Expectancy

Quantitative risk assessments attempt to arrive at a numerical value, usually expressed in financial terms. The most common form is either single loss expectancy (SLE) or ALE. SLE is the product of the asset value (AV) multiplied by the exposure factor (EF): SLE = AV × EF.

EF is the probability that an event will occur and its likely magnitude, and equals the percentage of asset loss caused by the identified threat. The result is that the greater the value, coupled with a greater probability and greater magnitude, the greater the potential risk of loss.

Knowledge Check: Semiquantitative Analysis

You, as an information security manager, have been asked to determine the relative value for some identified risk within an organization. Using the information provided in this section, determine the relative value of the following:
1. Reputational risk if a product line fails: The product development team has indicated that the market is ready for this particular product, but the infrastructure needed to launch the product is new to the organization and has been rushed into production to meet the desired launch date.
2. Noncompliance with new local regulation: Local government has passed a new law mandating businesses operating within the jurisdiction to update HVAC systems to more energy-efficient models. The cost of upgrading the existing system would be US $500,000, whereas the annual fine for noncompliance would be $10,000.
3. Email quarantine system is outdated: The company's email quarantine system is outdated, and messages are not being filtered as successfully as they had been in the past.

Answers on page 130.

ALE adds the annualized rate of occurrence (ARO) to the equation with the result that multiple occurrences will result in greater potential losses. ALE is usually expressed as:
ALE = SLE × ARO.

ALE is the annual expected financial loss to an asset, resulting from one specific threat.

ARO is the number of times a threat on a single asset is estimated to occur. The higher the risk associated with the threat, the higher the ARO. For example, if insurance data suggest that a serious fire is likely to occur once in 25 years, then the annualized rate of occurrence is 1/25 = 0.04.

EF represents the percentage of loss that a realized threat could have on a specific asset when the specific threat matches up with a specific vulnerability. Said another way, EF is the proportion of an asset's value that is likely to be destroyed by a particular risk, expressed as a percentage.

Value at Risk

Another approach required in certain financial sectors is value at risk (VAR), which can also have general risk management utility and benefit. This approach has been studied by various researchers, suggesting the suitability of the approach for information security management.

VAR is a computation based on historical data of the probability distribution of loss for a given period of time at a certainty factor typically of 95 percent or 99 percent. The probability distribution is arrived at using Monte Carlo simulations typically run through thousands of iterations with random variables based on historical information.

Operationally Critical Threat Asset and Vulnerability Evaluation® (OCTAVE®)

OCTAVE is an approach to risk assessment and ranking that is used to assist an organization in understanding, assessing and addressing its information security risk from the perspective of the organization. The OCTAVE methodology is process-driven and used to identify, prioritize and manage information security risk. It is intended to help organizations:

- Develop qualitative risk evaluation criteria based on operational risk tolerances
- Identify assets that are critical to the mission of the organization
- Identify vulnerabilities and threats to the critical assets
- Determine and evaluate potential consequences to the organization if threats are realized
- Initiate corrective actions to mitigate risk and create practice-based protection strategy

OCTAVE focuses on critical assets and the risk to those assets using a comprehensive, systematic, context-driven and self-directed evaluation approach. It can help an organization to maintain a proactive security posture and apply the organizational point of view to information security risk management activities.

Some characteristics of OCTAVE are that it:

- Identifies critical information assets
- Focuses risk analysis activities on these critical assets
- Considers the relationships among critical assets, the threats to these assets and the vulnerabilities (both organizational and technological) that can expose assets to threats
- Evaluates risk in operational context (i.e., how the critical assets are used to conduct the organization's business and how they are at risk due to security threats and vulnerabilities)
- Creates practice-based protection strategy for organizational improvement as well as risk mitigation plans to reduce the risk to the organization's critical assets

The OCTAVE process is based on three phases:

- **Phase 1: Build asset-based threat profiles (organizational evaluation)**—The analysis team determines critical assets and what is currently being done to protect them. The security requirements for each critical asset are then identified. Finally, the organizational vulnerabilities with the existing practices and the threat profile for each critical asset are established.
- **Phase 2: Identify infrastructure vulnerabilities (technological evaluation)**—The analysis team identifies network access paths and the classes of IT components related to each critical asset. The team then determines the extent to which each class of component is resistant to network attacks and establishes the technological vulnerabilities that expose the critical assets.
- **Phase 3: Develop security strategy and mitigation plans (strategy and plan development)**—The analysis team establishes risk to the organization's critical assets based on analysis of the information gathered and decides what to do about the risk. The team creates a protection strategy for the organization and mitigation plans to address identified risk. The team also determines the next steps required for implementation and gains senior management's approval on the outcome of the entire process.

Other Risk Analysis Methods

A number of other analysis options exist that may prove useful although they are typically more sophisticated than most organizations employ. Some of the more common approaches include:

- **Bayesian analysis**—A Bayesian analysis is a method of statistical inference that uses prior distribution data to determine the probability of a result. This technique's effectiveness and accuracy rely on the accuracy of the prior distribution data.
- **Bow tie analysis**—A bow tie analysis provides a diagram to communicate risk assessment results by displaying links among possible causes, controls and consequences. The cause of the event is depicted in the middle of the diagram (the "knot" of the bow tie) and triggers, controls, mitigation strategies and consequences branch off of the "knot."
- **Delphi method**—The Delphi method uses expert opinion, which is often received using two or more rounds of questionnaires. After each round of questioning, the results are summarized and communicated to the experts by a facilitator. This collaborative technique is often used to build a consensus among the experts.
- **Event tree analysis**—An event tree analysis is a forward-looking, bottom-up model that uses inductive reasoning to assess the probability of different events resulting in possible outcomes.
- **Fault tree analysis**—A fault tree analysis starts with an event and examines possible means for the event to occur (top down)

and displays these results in a logical tree diagram. This diagram can be used to generate ways to reduce or eliminate potential causes of the event.
- **Markov analysis**—A Markov analysis is used to analyze systems that can exist in multiple states. The Markov model assumes that future events are independent of past events.
- **Monte-Carlo analysis**—According to IEC 31010:2009:

> *Monte Carlo simulation is used to establish the aggregate variation in a system resulting from variations in the system, for a number of inputs, where each input has a defined distribution and the inputs are related to the output via defined relationships. The analysis can be used for a specific model where the interactions of the various inputs can be mathematically defined. The inputs can be based upon a variety of distribution types according to the nature of the uncertainty they are intended to represent. For risk assessment, triangular distributions or beta distributions are commonly used.*

2.7.15 EVALUATION OF RISK

During the risk evaluation phase, decisions have to be made concerning risk treatment and the treatment priorities based on the foregoing analysis, with allowances made for the likely margin of error, which may be considerable if reliable data are not available.

If the risk meets the acceptable risk criteria, the treatment option is likely to be acceptance. If the risk exceeds the acceptable level and is not within the tolerance variance, then treatment will most likely be some form of mitigation. The mitigation options include modification or addition of controls or business process reengineering that would render a process less risky. A system redesign can serve to reduce technical risk (e.g., eliminating a single point of system failure), or risk transfer (or sharing) might be the most cost-effective option. If there are no cost-effective options for mitigating excessive risk, it may be decided that the activity is not worth the risk or management may decide to accept the risk if the benefits are sufficiently high.

Typically risk transfer is selected for risk that has low likelihood and high impact. If risk is mitigated through the use of controls, control risk must also be considered (i.e., control risk is the probability the control will fail or be inadequate).

In some cases, the risk evaluation may lead to a decision to undertake further analysis if the results are ambiguous or are considered inaccurate or misleading.

The criteria used by the risk management team must also take into account the organization objectives, the stakeholder views, and, of course, the scope and objective of the risk management process itself as well as its probable margins of error. The decisions made are usually based on the level of risk but may also be related to thresholds specified in terms of:
- Consequences (e.g., impacts)
- The likelihood of events
- The cumulative (aggregated) impact of a series of events that could occur simultaneously
- The effect of cascading risk (domino effect) in closely coupled systems
- The cost of treatment
- The ability of the organization to absorb losses

2.7.16 RISK RANKING

The risk practitioner uses the results of risk assessment to place risk in an order that can be used to direct the risk response effort. Risk ranking is derived from a combination of all the components of risk including the recognition of the threats and the characteristics and capabilities of a threat source, the severity of a vulnerability, the likelihood of attack success when considering effectiveness of controls, control risk, and the impact to the organization of a successful attack. Taken together, these indicate the level of risk associated with a threat.

2.7.17 RISK OWNERSHIP AND ACCOUNTABILITY

Risk requires ownership and accountability. After a risk has been identified, analyzed and evaluated, a manager or senior official in the organization must be identified as its owner. A risk owner is accountable for accepting risk based on the organizational risk appetite and should be someone with the budget, authority and mandate to select the appropriate risk response based on analyses and guidance provided by the information security manager. This accountability extends to approving controls when mitigation is the chosen risk response. Types of controls are examined in more detail in chapter 3. However, the concept is to create a direct link, so that all risk is addressed through appropriate treatment and all controls are justified by the risk that mandates their existence.

The owner of a risk also owns any controls associated with that risk and is accountable for ensuring monitoring of their effectiveness. In some areas where there are regulations or laws that apply to risk, the risk owner may have to prepare standard reports on the status of risk, any incidents that may have occurred, the level of risk currently faced by the organization and the tested effectiveness of controls.

2.7.18 RISK TREATMENT (RESPONSE) OPTIONS

Faced with risk, organizations have four strategic choices:
- Avoid the risk by terminating the activity that gives rise to it.
- Transfer risk to another party (note that risk transfer usually results from a transfer of impact).
- Mitigate risk with appropriate control measures or mechanisms.
- Accept the risk.

Another alternative is that an organization may choose to ignore the risk, which can be dangerous. The distinction with accepting risk is understanding the probability and the consequences and finding them acceptable under the circumstances. Ignoring a risk may lead to a serious underestimation of both the likelihood and possible magnitude, with severe or disastrous consequences. Accordingly, this is an inadvisable course of action. The only time it may be prudent to ignore a risk is when the likelihood, exposure or impact is so small that the risk is not considered material to the organization or the impact is so great and rare that there is no possibility of addressing it (e.g., a comet strike or nuclear war).

Terminate the Activity

There are often ways activities might be modified or processes reengineered that can serve to mitigate or manage risk to acceptable levels. Analysis of the activity could also lead to the conclusion that it is not worth the risk. In this circumstance, it should be noted that even though the organization has determined to terminate the continuation of the product or service, the liability remains as long as the product or service is being used.

Transfer the Risk

An example of risk transference is the decision by an organization to purchase insurance to address areas of risk. When an organization buys insurance, some of the risk is transferred to the insurance company in exchange for premium payments that reflect the insurance company's assessment of the degree of risk it is assuming. It should be recognized that the risk is actually not transferred; rather, impact to the organization is reduced to the extent that insurance covers some or all of the costs associated with a compromise.

Risk can also be transferred by outsourcing IT functionality to a third party, provided indemnification clauses are in the contracts. However, in transferring operational risk, third-party agreements and contracts must specifically address the liability and responsibilities of both parties in specific indemnification clauses.

Indemnity agreements that can be part of an outsourced service agreement provide a level of protection against harmful incidents. **While some of the possible financial impacts associated with the risk can be transferred, the legal responsibility for the consequences of compromise generally cannot be transferred.**

Risk is typically transferred to insurance companies when the probability of an incident is low, but the impact is high. An example would be earthquake or flood insurance. For the information security manager, this means that a well-managed risk program must interface with other organizational assurance providers such as the insurance department to understand what risk is covered and to what extent. Obviously, if some identified risk is adequately managed by other entities, duplication of effort is generally not warranted.

Mitigate the Risk

Risk can be mitigated in a variety of ways such as implementing or improving security controls, instituting countermeasures, or modifying or eliminating risky processes. These controls can be preventive and directly address the risk or it may be possible to reduce exposure thereby reducing risk. In some cases, risk can be reduced by appropriate countermeasures reducing or eliminating a threat.

Potential impact may be reduced through compensating or corrective controls including contractual, procedural or technical processes.

Accept the Risk

There are a variety of circumstances where a defined risk may be accepted. One condition is if the cost of mitigating it is too high in proportion to the benefit or value of the asset. In other cases, it may simply not be feasible to effectively mitigate a risk or the potential impact may be low. It must be considered that not all impacts can be readily reduced to strictly financial terms and, in many cases, will not be the only consideration.

Elements such as customer trust and confidence, legal liability or breach of regulatory requirements may need to be considered as well. In any case, the information risk management procedure should enable accurate and appropriate documentation of the risk for a business manager to make the decision to accept the risk based on sufficient knowledge and understanding. Acceptance of risk should be regularly reviewed to ensure that the rationale for the initial acceptance is still valid within the current business context.

Risk Acceptance Framework

A risk acceptance framework can be a useful tool used to set the criteria for the acceptance of risk and the level at which management acceptance is executed.

A typical risk acceptance framework is shown in **figure 2.18**.

Figure 2.18—Risk Acceptance Framework

Risk Level	Level Required for Acceptance
Low	Risk acceptance possible by local management
Medium	Risk acceptance possible by chief information officer (CIO)
High	Risk acceptance possible by CIO, director or chief information security officer (CISO), depending on potential impact
Severe	Risk acceptance only at board level, depending on potential impact. Risk reduction is mandatory through rigorous controls or monitoring. Management notification process is required.

2.7.19 RESIDUAL RISK

A risk prior to mitigation is called inherent risk. The risk that remains after countermeasures and controls have been implemented is **residual risk**. Risk is never eliminated; residual risk always remains. It should be noted that reducing one risk inevitably introduces another risk, hopefully of a lesser nature. For example, requiring dual control to access sensitive information introduces the risk that two individuals collude to provide unauthorized access.

The objective is to ensure that residual risk is equal to the organization's criteria for acceptable risk and risk tolerance. Risk tolerance is defined as the allowable deviation from acceptable risk and is usually described as a percentage or range (e.g., plus or minus 10 percent). Acceptable residual risk should also be the outcome of meeting the defined control objectives and be equivalent to the defined security baselines for the organization.

Residual risk reported through a subsequent risk assessment can be used by management to identify those areas in which more control is required to further mitigate risk. Acceptable levels of risk are established as a part of an information security strategy, as described in chapter 1. If a strategy is not developed, management must determine the acceptable risk levels, usually in terms of allowable potential impacts. Residual risk in excess of this level

should be further treated with the option of additional mitigation through implementing more stringent controls. Risk below this level should be evaluated to determine whether an excessive level of countermeasures or control is being applied, and whether cost savings can be achieved by removing or modifying them. Final acceptance of residual risk takes into account:
- Regulatory compliance
- Organizational policy
- Sensitivity and criticality of relevant assets
- Acceptable levels of potential impacts
- Uncertainty inherent in the risk assessment approach
- Cost and effectiveness of implementation

In judging the appropriateness of controls or countermeasures, the full life cycle cost must be considered, including the cost of implementing and operating specific measures or mechanisms balanced against the risk and potential impacts being addressed.

2.7.20 IMPACT

Ultimately, all risk management activities are designed to reduce impacts to acceptable levels to create or preserve value to the organization. The result of any vulnerability exploited by a threat that causes a loss is an impact. Threats and vulnerabilities that do not cause an impact are usually irrelevant and generally not considered a risk.

In commercial organizations, impact is generally quantified as a direct financial loss in the short term or an ultimate (indirect) financial loss in the long term. Examples of such losses can include:
- Direct loss of money (cash or credit)
- Criminal or civil liability
- Loss of reputation/goodwill/image
- Reduction of share value
- Conflict of interests to staff or customers or shareholders
- Breach of confidence/privacy
- Loss of business opportunity/competition
- Loss of market share
- Reduction in operational efficiency/performance
- Interruption of business activity
- Noncompliance with laws and regulations resulting in penalties

As with risk calculations, impact calculations can be done either qualitatively or quantitatively. Some impacts lend themselves to quantitative representation such as the range of possible financial impacts. Others, such as loss of reputation or market share, may be more difficult and may be adequately presented by qualitative statements. Impacts are determined by performing a business impact assessment and subsequent analysis, which generally includes gathering industry statistics providing a semiquantitative analysis approach. This analysis will determine the criticality and sensitivity of information assets. It will provide the basis for setting access control authorizations and for business continuity planning (BCP), and will include RTOs, RPOs, maximum tolerable outages (MTOs), SDOs and AIW. A BIA serves to prioritize risk management and, coupled with asset valuations, it provides the basis for the levels and types of protection required as well as a basis for the development of a business case for controls.

2.7.21 CONTROLS

Controls are any technology, process, practice, policy, standard or procedure that serves to regulate an activity to mitigate or reduce risk. It can be of an administrative, technical, management or legal nature. Controls are discussed in detail in section 3.12 Controls and Countermeasures.

Because it is normal to find a number of different controls in different parts of a typical process, it is important to understand the entire risk mitigation process from end to end. While layering of controls is a prudent approach, using an excessive number of controls addressing the same risk is wasteful and usually reduces productivity. It is also important to ensure that the various controls are not subject to the same risk, which defeats the purpose of layering them. For risk assessments to be effective and reasonably accurate, it is necessary to ensure that they are conducted from the beginning of processes through to the end. This will facilitate understanding if upstream controls minimize or eliminate some risk that may preclude the need for subsequent controls. It will also help determine whether there is unnecessary control redundancy or duplication.

2.7.22 LEGAL AND REGULATORY REQUIREMENTS

Legal and regulatory requirements must be considered in terms of risk and impact. This is necessary for senior management to determine the appropriate level of compliance and priority. Regulations must first be evaluated to determine to what extent the organization is subject to the regulation and is already compliant. If it is found that the organization is noncompliant, then the regulations must be evaluated to determine the level of risk they pose to the organization. The organization must take into consideration the level of enforcement and its relative position in relation to its peers because enforcement actions are usually initiated against those that are least compliant. The potential impact of full compliance, partial compliance and noncompliance, in both direct financial and reputational impacts, must be evaluated as well. These evaluations provide the basis for senior management to determine the nature and extent of compliance activities appropriate for the organization. The information security manager must be aware that senior management may decide that risking sanctions is less costly than achieving compliance, or that because enforcement is limited, or even nonexistent, compliance is not warranted. This is a management decision that should be based on risk and impact.

2.7.23 COSTS AND BENEFITS

When controls or countermeasures are planned, an organization should consider the costs and benefits. If the costs of specific controls or countermeasures (control overhead) exceed the benefits of mitigating a given risk, the organization may choose to accept the risk rather than incur the cost of mitigation. This follows the general principle that the cost of a control should never exceed the expected benefit. This is the principle of proportionality described in generally accepted security systems principles (GASSP) or its successor, generally accepted information security principles (GAISP).

Cost-benefit analysis helps provide a monetary impact view of risk and helps determine the cost of protecting what is important. However, cost-benefit analysis is also about making smart choices based on potential risk mitigation costs vs. potential losses (risk exposure). Both of these concepts tie directly back to good governance practices.

Unfortunately, most information security crime and loss metrics are not as established as traditional robbery and theft statistics. The annual Computer Security Institute (CSI) Computer Crime and Security Survey has been one measurement involving losses in the information security realm, but some practitioners suggest that its loss figures are understated, while others believe they are overstated.

Rather than debate the validity of these measures, it may be more useful to look at a few metrics that most organizations can quantify with some accuracy. Three common measurements of potential losses are employee productivity impacts, revenue losses and direct cost loss events. Virus and worm incidents are the ones most frequently cited when discussing impact on productivity. Another is the result of the compromise of PII.

Revenue losses can also be determined in a similar manner. If a business has an e-commerce web site that is producing US $1 million of revenue each day, then a denial of service (DoS) attack that lasts half a day creates a $500,000 loss. It is debatable whether this type of attack would merely force customers to delay their purchases or whether they would simply go to a competitor. However, any public perception of an organization being the victim of a hacking attack, whether sensitive information was compromised or not, will often result in the loss of customer trust.

While productivity and revenue losses are considered direct losses, indirect losses may include the number of additional hours employees have to work to respond and recover from an incident. Another direct cost may be additional protection efforts resulting from a compromise.

When considering costs, the total cost of ownership (TCO) must be considered for the full life cycle of the control or countermeasure. This can include such elements as:
- Acquisition costs
- Deployment and implementation costs
- Recurring maintenance costs
- Testing and assessment costs
- Compliance monitoring and enforcement
- Inconvenience to users
- Reduced throughput of controlled processes
- Training in new procedures or technologies as applicable
- End-of-life decommissioning

2.7.24 EVENTS AFFECTING SECURITY BASELINES

Baseline security is defined as the minimum security level across the enterprise. It should be noted that baselines may be different for assets of different classification levels. Obviously, the higher the classification, the higher and more restrictive the baseline should be set. As an example, the higher-classification assets might require two-factor authentication for access whereas lower classifications can be accessed with ID and passwords.

A number of factors may change the risk or impact equation, necessitating that baseline security be changed. Baseline security is determined by the collective ability of controls to protect the organization's information assets. Baseline security is essentially controlled by the least restrictive aspect of collective standards and is the minimum level of security across the organization. Baseline security levels should also be reflected in control objectives.

Knowledge Check: Cost-benefit Analysis

A malware application infects 10,000 employees in a 40,000-person organization. Each infected system costs each affected employee one hour of productivity. The average hourly wage of the employees in the organization is US $30.

Anti-malware scanning software would cost the organization $20,000 per 10,000 devices as part of the organization's desktop management system. An employee training session on email security awareness would require each employee to complete a 30-minute module on the company's intranet.

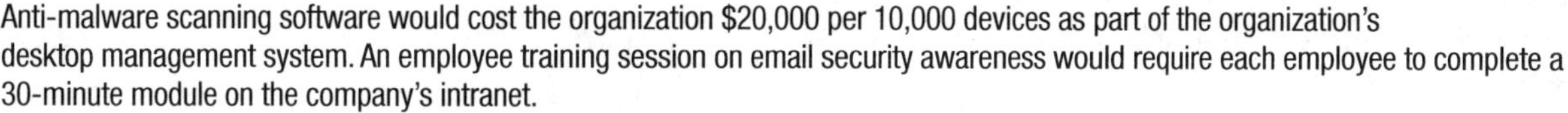

Questions:

A. Determine the financial cost in employee-hours lost in this situation.
B. Determine the costs of each of the risk remediation options.
C. This is the third time this year that a significant malware infection has occurred in the organization. Based on this additional knowledge, what would be the appropriate risk remediation approach for this organization to take?
  A. Anti-malware scanning software
  B. Employee awareness training
  C. Both options

Answers on page 130.

Any security incident can be attributed to either a control failure or the lack of a control. Any significant incident will require a risk reassessment and analysis of the root cause of the failure, which may, in turn, require increasing or modifying baseline security by changing the relevant standards and/or controls.

Information security managers need to monitor and assess events that affect security baselines and, thus, might affect the organization's security program. Based on this assessment, the information security manager must determine if the organization's security plans and test plans require modification to address changing risk.

Security baselines may be modified for various reasons. As an example, a vendor may identify that a parameter in its software or hardware that must be changed to achieve or maintain the desired protection levels. Another reason could be an outside event that requires increased baselines. For example, if there is a protest or other civil unrest near the organization's facility, the baseline for physical security may need to be increased for a period of time until that threat passes. Other drivers could include new or changing laws or regulations, identifying unacceptable risk levels as a result of new or changing threats, degradation of controls over time, and changes to systems creating new vulnerabilities.

2.8 INFORMATION ASSET CLASSIFICATION

Information asset classification is required to determine the relative sensitivity and criticality of information assets, sometimes referred to collectively as business value. Criticality is determined by the impact on the organization as a result of the loss of an asset (i.e., how important the asset is to the business). Sensitivity is based on the potential damage to the organization as a result of unauthorized disclosure. This will provide the basis for protection efforts, business continuity planning and user access control. For larger organizations, this can be a daunting task because there are likely to be peta- or terabytes of electronic data, warehouses of documents, and thousands of individuals and devices. Yet, without determining the business value, sensitivity and criticality (and, increasingly, legal and regulatory requirements) of information resources, it is not possible to develop an effective risk management program that provides appropriate protection proportional to business value or sensitivity and criticality.

In cases where comprehensive classification is not possible due to resource constraints or other reasons, a less effective option is a business dependency assessment. This can be used to provide a basis for allocating proportional protective activities. This approach is based on the information resources that critical business functions utilize.

The first step in the classification process is to ensure the information asset inventory is complete and the location of each asset is identified. Because this is also a necessary step in the risk identification phase of a risk assessment, it is an essential activity. In many organizations, this may prove difficult because there is often no comprehensive inventory of information-related assets. This may be especially true in larger organizations with multiple independent business units lacking a strong centralized security function. The identification process must include determining the location of the data, the data owners, data users and data custodians. The security manager must also determine what data are housed by external service providers. These service providers can include media vaulting and archival firms, mailing list processors, firms that process mail containing company information, firms that act as couriers or transporters of information, and third-party service providers. Service providers may also include data centers providing hosting functions, payroll services or health insurance administration.

The information security manager working with the business units should ascertain the appropriate information classification based on business value or levels of sensitivity and criticality of information assets, and ensure that all business and IT stakeholders have the opportunity to review and approve the established guidelines for access control levels. The number of levels should be kept to a minimum. Classifications should be simple, such as designations by differing degrees for sensitivity and criticality. End-user managers, in coordination with the security administrator, can use these classifications in their risk and impact assessment process and to determine access levels of users. The information security manager must also be aware that in a risk-averse organization or one with a blame culture there will be an incentive to overclassify, and it will be necessary to consider controls to limit this tendency.

A major benefit of information asset classification is the reduced risk of underprotection and the cost of overprotection of information assets by tying security to business objectives. Although it is a complicated undertaking to implement data classification, the long-term benefits to the organization are substantial. The alternative is to provide all information assets with the same level of protection, which is unlikely to be cost-effective. If the organization is risk averse and requires a high level of security, providing the same high level of protection to all assets can be very costly.

There are a number of questions that should be asked in any information asset classification model, including but not limited to:

- How many classification levels are suitable for the organization?
- How will information be located?
- What process is used to determine classification?
- How will classified information be identified?
- How will it be marked?
- How will it be handled?
- How will it be transported?
- How will confidential information be stored and archived?
- What is the life cycle of the information (create, update, retrieve, archive, dispose)?
- What are the processes associated with the various stages in the information asset life cycle?
- How will it be retained according to policy or law?
- How will it be safely destroyed at the end of the retention period?
- Who has ownership of information?
- Who has access rights?
- Who has authority for determining access to the data?
- What approvals are needed for access?

An important part of information classification is not just applying a classification label to each piece of information, but identifying security measures that can consistently be applied to each level. As the level of sensitivity or criticality increases, security measures should increase in rigor so that, at the highest level, security mechanisms are the most restrictive or provide the greatest level of protection. It must also be remembered that sensitivity and criticality require different security mechanisms and processes. For any piece of information, an owner may have to make both a sensitivity and a criticality decision.

2.8.1 METHODS TO DETERMINE CRITICALITY OF ASSETS AND IMPACT OF ADVERSE EVENTS

A number of methods exist to determine the sensitivity and criticality of information resources and the impact of adverse events. A BIA is the usual process to identify the impact of adverse events. Methods outlined within COBIT, NIST and the OCTAVE framework are representative of the resources the information security manager may utilize in this effort.

It is a generally accepted practice to focus on the impact that a loss of information assets has on the organization rather than on a specific adverse event. Because numerous adverse events could occur, it is not practical or cost-effective to list them all.

The first step to determine information asset importance is to break the corporate or organizational structure into business units or departments (**figure 2.19**). Under the corporate or top-level organizational structure, each of the business units should be rated by its importance or value to the business.

In **figure 2.19**, Business Unit B is given a number one rating since it is the most important. The importance usually equates to revenue, but the value may equate to critical functions being performed. There can be as many units or departments as exist, but it may be more manageable to stay at a higher level.

The rating should be done by the senior management team, based on its understanding of the organization. This is the foundation for establishing the risk management structure. The relative importance or value of the business units or departments will flow down into critical functions, assets and resources.

The next step is identifying the critical organizational functions. The focus for each business unit or department is to define what tasks are important to the unit in achieving its goals. There can be a two-level structure within the critical function layer to represent complex operations. The critical business functions are also numerically rated against each other to help with prioritization of subsequent risk remediation efforts.

Once critical functions have been identified, as shown in **figure 2.20**, the basic structure of the organization has been mapped. It is important to recognize that the structure has been focused only on operational elements, not technologies,

Figure 2.19—Top Layer of Business Risk Structure

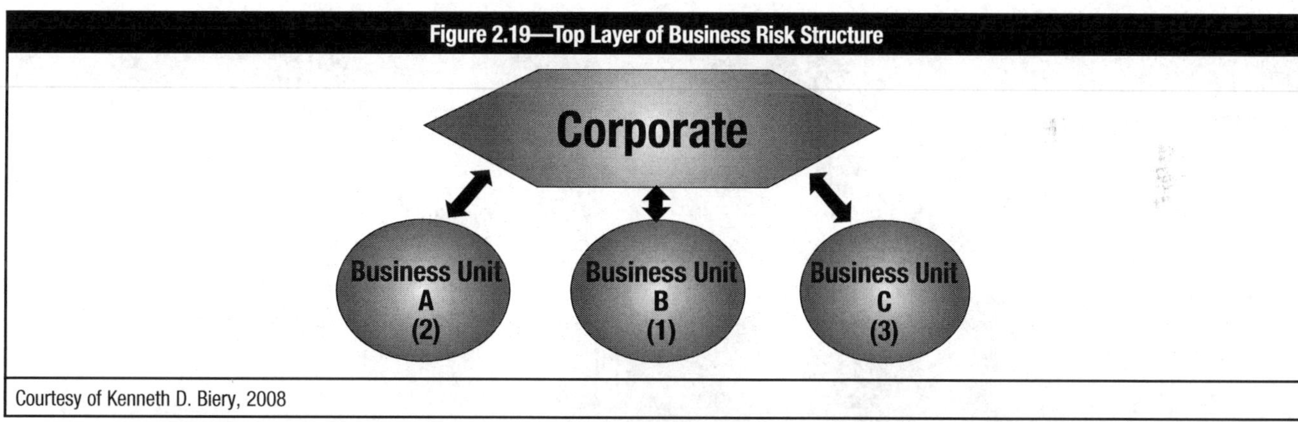

Courtesy of Kenneth D. Biery, 2008

Figure 2.20—Critical Function Layer of Business Risk Structure

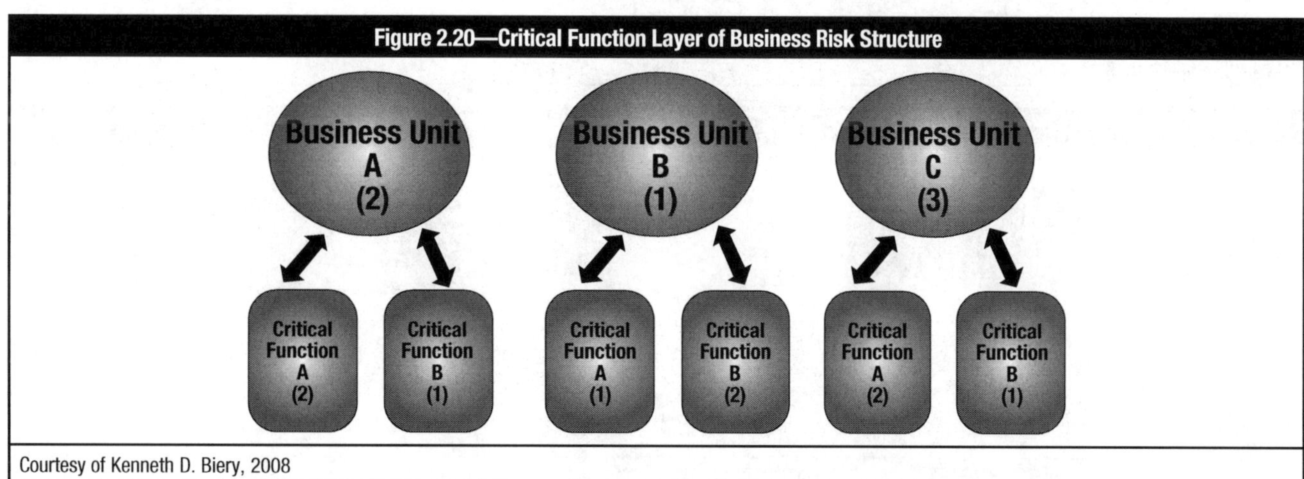

Courtesy of Kenneth D. Biery, 2008

applications or data. This progressive drill-down structure looks similar to a BIA that is performed when conducting business continuity planning. The structure will provide a management-level view of risk and where it resides in the organization.

In the structure shown in **figure 2.21**, assets and resources are the containers of risk. Because there are vulnerabilities identified with the assets that can be exploited by threats, there is risk. Assets, like business units and critical functions, are numerically rated as well. Because assets are associated with the critical business function they support, they will be rated against each other from the most important to the least important in their group.

The risk represented in **figure 2.22** is the composite of vulnerabilities that a threat can exploit to cause a negative impact to an asset. With the approach represented in the preceding figures, an organization can see where risk originates and how it can potentially impact business operations. The roll-up and drill-down nature of this approach is useful for management throughout the organization. For example, business owners may want to see their critical functions' level of risk to be able to set a prioritization schedule for fixing vulnerabilities or prioritizing protection efforts.

The ability to determine how identified risk can impact business operations is demonstrated in **figure 2.23**, which is the combined elements discussed above. It shows how risk exposure can potentially impact some of the company's most valuable assets. The structure shown in the diagram helps both management and the security team align and prioritize their efforts accordingly.

2.8.2 IMPACT ASSESSMENT AND ANALYSIS

A BIA is performed to determine the impact of losing the availability of any resource to an organization; it establishes the escalation of that loss over time, identifies the minimum resources needed to recover, and prioritizes the recovery of processes and supporting systems. The BIA is often thought of in the context of business continuity and disaster recovery. Other methods aside from the BIA may be used to determine potential impact. Impact is the bottom line of risk, and the range of severity in terms of the organization must be determined to provide the information needed and guide risk management activities. Obviously, even high probability with little or no impact is not of concern.

It is common for these assessments to determine only a worst-case outcome, which represents a minority of events. As a result of this "impact inflation," management often discounts these assessments as unrealistic and excessively pessimistic.

A more effective approach involves performing a reasonably small set of scenario analyses with key organization stakeholders, where a range of potential outcomes is determined. This range of outcomes is then used to define a quantitative distribution of impact magnitudes, including minimum, maximum and most likely, including values as well as a confidence level. These values can then be used as inputs to quantitative analysis methods (e.g., Monte Carlo simulations to determine probability distribution) that more accurately describe for management the real impact potential.

Figure 2.21—Aligning Assets to the Critical Function Layer

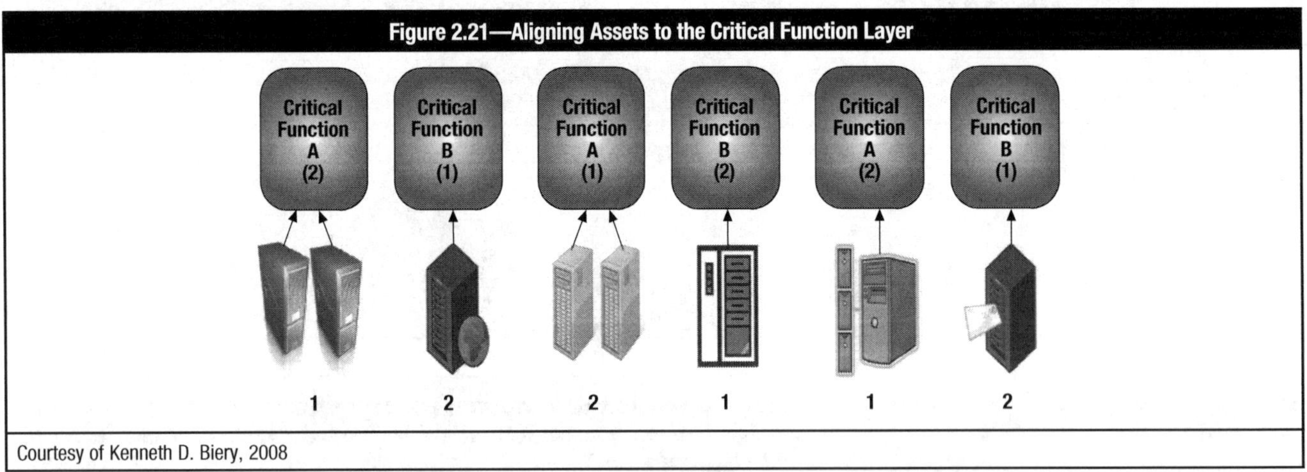

Courtesy of Kenneth D. Biery, 2008

Figure 2.22—Asset Vulnerabilities

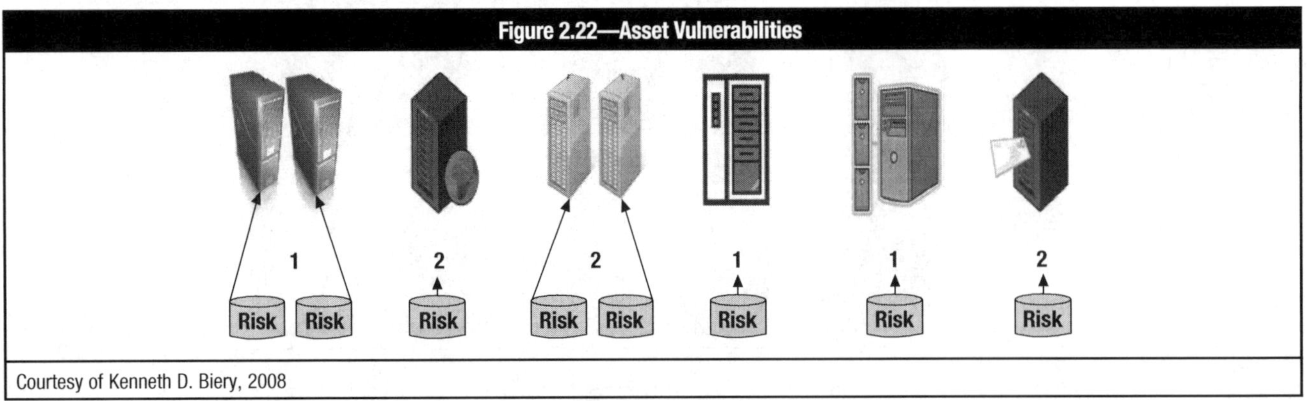

Courtesy of Kenneth D. Biery, 2008

Figure 2.23—Combined Risk Impact Structure

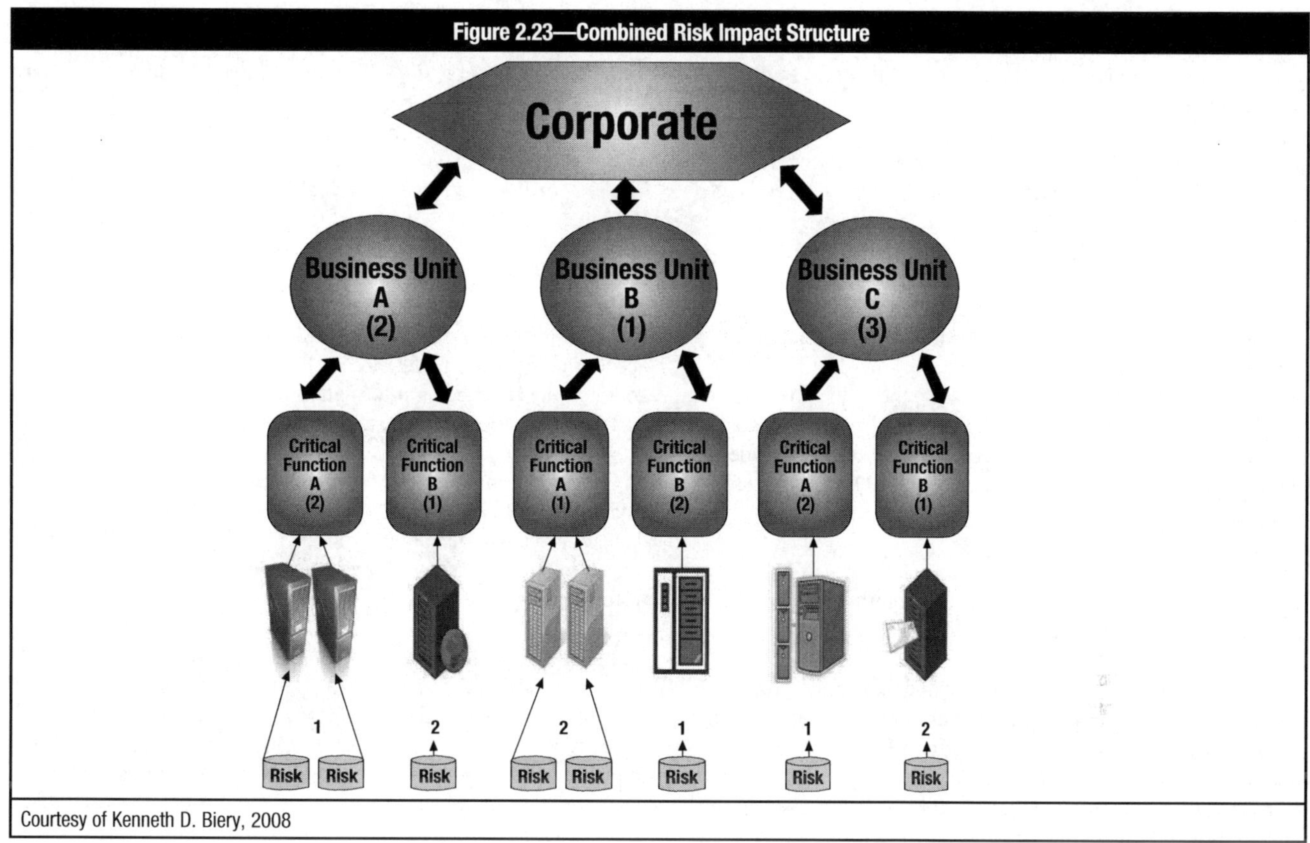

Courtesy of Kenneth D. Biery, 2008

The other advantage this approach provides is that it more closely resembles the type of impact data that management receives from other business risk domains (e.g., investment, marketing, credit). This alignment improves management's ability to make an apples-to-apples comparison and well-informed risk decisions.

Before beginning the impact analysis of a specific set of assets, it is necessary to obtain the following information:

- System mission (e.g., objectives of the processes performed by the IT system or personnel)
- System (manual or technical) and data criticality (e.g., the system's value or importance to an organization)
- System, personnel and data criticality (the impacts associated with unintended disclosure)

This information can be obtained from performing a BIA or from organizational documentation such as a mission impact analysis report or asset criticality assessment report, if they exist. A mission impact assessment and analysis or BIA prioritizes the impact levels associated with the compromise of an organization's information assets based on a qualitative or quantitative assessment of the criticality of those assets. An asset criticality or resource dependency assessment identifies and prioritizes the critical organization information assets (e.g., hardware, software, systems, services and related technology assets) that support the organization's critical missions.

If this documentation does not exist or such assessments for the organization's information assets have not been performed, the system and data sensitivity can be determined based on the level of protection required to maintain the availability, integrity and confidentiality of the system and data.

The adverse impact of a security event can be described in terms of loss or degradation or any combination of integrity, availability and confidentiality:

- **Loss of integrity**—System and data integrity refers to the requirement that information be protected from improper modification. Integrity is lost if unauthorized or erroneous changes are made to the data or IT system by either intentional or accidental acts. If the loss of system or data integrity is not corrected, continued use of the contaminated system or corrupted data could result in broader corruption, fraud or misinformed decisions. Also, violation of integrity may be the first step in a successful attack against system availability or confidentiality.
- **Loss of availability**—If a mission-critical IT system or process is unavailable to its end users, the organization's mission may be affected. Loss of system functionality and operational effectiveness, for example, may result in loss of productive time, thus impeding the end users' performance of their functions in supporting the organization's mission.
- **Loss of confidentiality**—In this context, confidentiality refers to the protection of information from unauthorized disclosure. The impact of unauthorized disclosure of confidential information can range from the jeopardizing of national security to the disclosure of private consumer data. Unauthorized, unanticipated or unintentional disclosure of private consumer or other regulated data can result in loss of public confidence, loss of customer base and legal action against the organization.

Some tangible impacts can be measured quantitatively in lost revenue, the cost of repairing the system or the level of effort required to correct problems caused by a compromise. Other impacts (e.g., loss of public confidence, loss of credibility, damage to an organization's interest) cannot be measured in specific units but can be described qualitatively in terms of high, medium and low impacts.

In conducting an impact analysis, consideration should be given to the advantages and disadvantages of quantitative vs. qualitative assessments. The main advantage of the qualitative impact analysis is that it prioritizes the risk and identifies areas for immediate improvement in addressing the vulnerabilities. The disadvantage of the qualitative analysis is that it does not provide specific quantifiable measurements of the magnitude of the impacts, therefore making a cost-benefit analysis of any recommended controls difficult.

The major advantage of a quantitative impact analysis is that it provides a measurement of the impacts' magnitude, which can be used in the cost-benefit analysis of recommended controls.

The disadvantage is that, depending on the numerical ranges used to express the measurement, the meaning of the quantitative impact analysis may be unclear, requiring the result to be interpreted in a qualitative manner. Additional factors must also be considered to determine the magnitude of impact, such as the range of possible errors in estimation or computations.

2.9 OPERATIONAL RISK MANAGEMENT

Operational risk is defined as the risk of loss resulting from inadequate or failed processes, people and systems, or from external events. Interruption of business operations is a major concern and preventing it the primary focus of risk management. In most cases, incident management is adequate to manage the materialized risk and minimize significant disruption to operations. In some cases, incidents will escalate to disasters and fall under the purview of business continuity and disaster recovery. In either event, the backstop to limit risk and ensure it is managed is the understanding and capability to address these issues sufficiently to ensure the survival of the organization. The following section covers the main elements that must be addressed by the information security manager and other organizational units to manage risk effectively under most foreseeable circumstances.

2.9.1 RECOVERY TIME OBJECTIVES

The information security manager must understand RTOs and how they apply to the organization's information resources as part of the overall evaluation of risk. The organization's business needs will dictate the RTO, which is usually defined as the amount of time to recover an acceptable level of normal operations. The acceptable level is defined by the SDO. The information asset's functional criticality, recovery priorities and interdependencies offset by costs are variables that will determine the RTO.

Determining RTOs can depend upon a number of factors including the cyclical need (e.g., time of day, week, month or year) of the information and organization, interdependencies among the information and the organization's requirements, and the cost of available options. The organization's requirements can be based on customer needs, contractual obligations or SLAs, and, possibly, regulatory requirements.

The information security manager should consider that the RTO may vary with the timing of the month or year. Financial information may not be as critical at the beginning of the month when the new fiscal month is being opened. This same information is likely to be highly critical at the end of the month when the monthly financial reports are being prepared and the accounting period is being closed. The timing of business cycles and their dependence on information need to be considered as part of information classification.

RTOs are determined by performing a BIA in coordination with developing a BCP. The interconnection of systems and their dependencies will generally require a BIA for most or all systems related to critical business functions as this will affect the order of restoration.

The BIA is generally conducted by interviewing information owners to obtain their perspective on the cost associated with an interruption in service for a business system or process. Often there are two perspectives for RTOs. One is the perspective of the individuals whose job it is to use the information, and the other is the view of senior management who must consider costs and may need to arbitrate among business units competing for resources. An information asset that a divisional supervisor may believe is critical may not be critical in the eyes of the vice president of operations, who is able to include the overall organizational risk in the evaluation of the RTO.

The information security manager should understand that both perspectives are important and work toward an RTO that considers both. The result will factor into the BCP, the scope of the services to be restored and priority order for the recovery of systems. In the end, the final decision is that of senior management. Senior management is in the best position to arbitrate the needs and requirements of the different components of the business—such as the regulatory requirements to which the organization is subject—and determine which processes are the most critical to the continued survival of the business as well as determine acceptable costs.

2.9.2 RTO AND ITS RELATION TO BUSINESS CONTINUITY PLANNING AND CONTINGENCY PLANNING OBJECTIVES AND PROCESSES

Knowledge of the RTO for information systems and their associated data is needed for an organization to develop and implement an effective BCP program. Once the RTOs are known, the organization can identify and develop contingency strategies that will meet the RTOs of the information resources. The RTOs will drive the order of priority for restoration of services and, in certain cases, the selection of specific recovery technologies in situations where the RTO is short.

One critical factor when developing contingency processes is cost. System owners invariably prefer shorter RTOs, but the tradeoffs in cost may not be warranted. Near-instantaneous recovery can be achieved, where needed, using technologies such as mirroring of information systems, so that, in the event of a disruption, the information systems are always available quickly. In general, the cost of recovery is less if the RTO for a given resource is longer.

There is a breakeven point of the time period to determine the RTO, where the impact of the disruption begins to be greater than the cost of recovery. The length of this time period depends on the nature of the business disruption and the assets involved. Qualitative as well as quantitative issues must be taken into consideration because loss of customer confidence, even if it cannot be estimated, can have a long-term negative impact on the organization. Most organizations can reduce their RTOs, but there is an associated cost.

2.9.3 RECOVERY POINT OBJECTIVES

The RPO is determined based on the acceptable data loss in case of disruption of operations. It indicates the most recent point in time to which it is acceptable to recover the data, which generally is the latest backup. RPO effectively quantifies the permissible amount of data loss in case of interruption. Depending on the volume of data, it may be advisable to reduce the time between backups to prevent a situation where recovery becomes impossible because of the volume of data to be restored. It may also be the case that the time required to restore a large volume of data makes it impossible to achieve the RTO.

While this is typically the scope of business continuity and disaster recovery planning, it is an important consideration when developing a risk management strategy.

2.9.4 SERVICE DELIVERY OBJECTIVES

SDOs are defined as the minimal level of service that must be restored after an event to meet business requirements until normal operations can be resumed. SDOs will be affected by both RTOs and RPOs and must also be considered in any risk management strategy and implementation. Higher levels of service will generally require greater resources as well as more current RPOs.

2.9.5 MAXIMUM TOLERABLE OUTAGE

MTO refers to the maximum time an organization can operate in alternate (or recovery) mode. Various factors may affect the MTO, such as availability of fuel to operate emergency generators, accessibility of a recovery site that might be located remotely and limited operational capacity of the recovery site. This variable will affect the RTO, which, in turn, affects the RPO.

From a risk management perspective, the relationship among the MTO, RTO and RPO must be considered to minimize risk of inadequate recovery to the organization.

2.9.6 ALLOWABLE INTERRUPTION WINDOW

AIW is the amount of time the normal operations can be down before the organization faces major financial difficulties that threaten its existence. The MTO should in any event be as long as the AIW to minimize the risk to the organization in the event of a disaster.

2.10 THIRD-PARTY SERVICE PROVIDERS

A typical organization uses many information resources in support of its business processes. These resources can originate within the organization or be provided by entities external to the organization. Most organizations will use a combination of the two. The information security manager needs to be aware of the location and access permissions for all information resources because they all require protection regardless of who is processing them.

The information security manager has a number of considerations to address when outsourcing, including:
- Ensuring that the organization has appropriate controls and processes in place to facilitate outsourcing
- Ensuring that there are appropriate information risk management clauses in the outsourcing contract
- Ensuring that a risk assessment is performed for the process to be outsourced
- Ensuring that an appropriate level of due diligence is performed prior to contract signature
- Managing the information risk for outsourced services on a day-to-day basis
- Ensuring that material changes to the relationship are flagged and new risk assessments are performed as required
- Ensuring that proper processes are followed when relationships are ended

Considerations when outsourcing services include the following:
- Outsourcing or planning to outsource business-critical functions generally increases information risk.
- The complexity of managing information risk is increased in outsourcing arrangements by the separation of responsibility for control specification and control implementation.
- The separation of responsibility for control specification and control implementation is bridged by the outsourcing contract. This underlines the contract's importance as the primary method through which the organization can manage its information risk.
- Where the outsourced business function operates within a regulated industry, the outsourcing contract needs to explicitly address regulatory requirements.
- The complexity of information risk assessment is increased in outsourcing arrangements because there are three different areas of information risk to assess: the business function, the outsourcing provider and outsourcing itself.
- The style of the overall contract and the amount of innovation contributed by the provider have a major impact on the way in which information risk management requirements are specified.
- The relationship between the organization and the outsourcing provider often contributes more to effective information risk management in an outsourcing arrangement than the contract.
- Because few businesses remain static, information risk management within the outsourcing arrangement must evolve so that it continues to be relevant to the organization's needs.
- The exit strategy for the outsourcing arrangement is at least as important as the initial transition. It should be developed at the planning stage and included in the contract to facilitate the continued availability of the outsourced business function. The exit strategy is far too important to leave until the outsourcing arrangement comes to its conclusion.

The information risk management requirements for outsourced business functions are different from those for in-house functions and, in many instances, are greater. After the information risk has been analyzed and the controls have been identified, these controls need to be defined within the contract for the provider to implement. Outsourcing results in a disconnection between setting the controls and their implementation. **Figure 2.24** provides a simplistic view of this separation of the responsibilities of the organization vs. those of the provider.

The information security manager should be aware that, although the organization can outsource information risk management to a third party, it generally cannot outsource responsibility.

The disconnection between control definition and control implementation makes managing the risk associated with outsourcing business functions complex and renders the outsourcing contract essential in managing information risk. The challenge for the information security manager is in how to define and implement information risk management controls in different outsourced business functions throughout the organization.

The business problem is the need to define and implement information risk management measures to protect the information within business functions that have been handed to a third party to operate on a day-to-day basis.

Information risk recommendations for outsourcing initiatives include ensuring the following:

- Timely involvement of information risk management professionals to ensure the risk assessment and controls definition
- Negotiation of key information risk management controls within the contract
- Mechanisms to negotiate small changes to information risk management controls as they may be costly
- Avoidance of difficult and complicated changes to information risk management controls
- Mechanisms to get information on whether information risk is being managed effectively
- Mechanisms to compensate for the lack of trust in the provider's staff

2.10.1 OUTSOURCING CHALLENGES

Outsourced information resources may present an information security manager with other challenges, including external organizations that may be reluctant to share technical details on the nature and extent of their information protection mechanisms. This makes it critical to ensure that adequate specified levels of protection are included in SLAs and other outsourcing contracts. One common approach is to specify requirements for specific audits such as SOC 2 (*Report on Controls at a Service Organization Relevant to Security, Availability, Processing Integrity, Confidentiality or Privacy*, developed by the AICPA) or ISO 27001:2013 certification. It is also important to analyze SOC 2 or other audit reports upon receipt for third-party auditor comments and, if present, comments about customer control effectiveness and policy compliance. Note that SOC 2 reports are often not sufficient on their own because the criteria have been defined by the organization in question. For high-risk relationships, it is generally preferable to rely on periodic compliance assessments conducted directly by the sourcing organization or a contracted third party.

From a risk management perspective, it is also important that incident management and response, BCP/DRP, and testing include all important outsourced services and functions. This includes implementing a well-defined and tested incident detection, escalation and response plan in concert with the outsourcing entities.

For regulated organizations such as financial institutions, a time frame for notification of regulatory agencies regarding suspicious events involving regulated information is often in in place. Contracts with third parties must ensure that processes are established to support such notifications.

Third-party vendor financial viability is another consideration. Because outsourcing contracts are often awarded to low-cost bidders, the risk of the outsourcing organization to continue to operate according to the contract and to honor any indemnity agreements may be a function of their financial capabilities. Financial information can be obtained from a variety of sources including credit reports, US Securities and Exchange Commission filings of publicly traded firms, and annual reports. If such certifications are not available, information must be obtained from providers—information sufficient to determine how external entities are securing information assets.

Figure 2.24—Disconnect of Responsibilities With Outsourced Providers

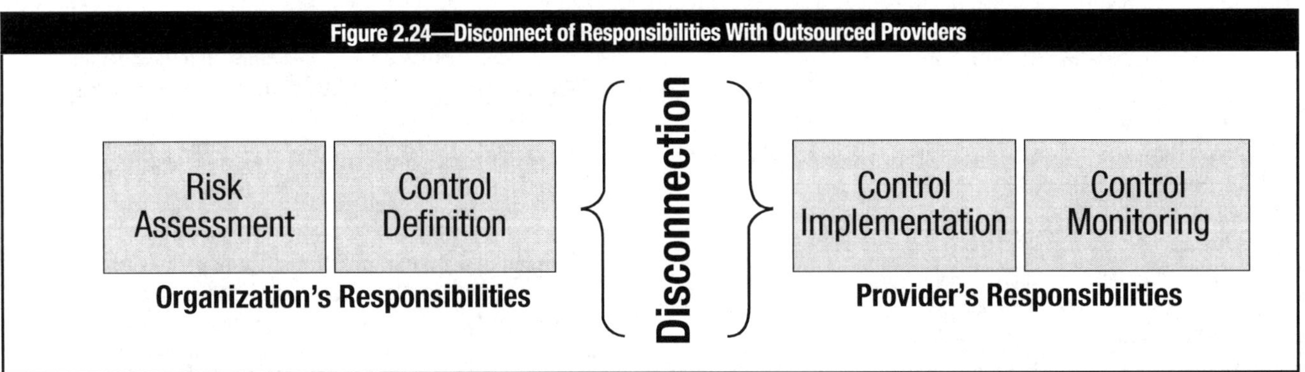

Some portion of risk associated with outsourced information services can be transferred by incorporating indemnity clauses in SLAs. Key clauses that should be part of a third-party contract must include, but are not restricted to:

- Right to source code in event of default of provider (e.g., source code escrow)
- Requirement that the vendor remain timely with compliance to industry and regulatory requirements
- Right to audit the vendor's books of accounts and premises
- Right to review the vendor's processes
- Insistence on standard operating procedures (SOPs)
- Right to assess the skill sets of the vendor resources
- Advance information if the resources deployed are to be changed

An SLA should also be developed and formally agreed that explains timeliness of response (system and human), scanning expectations and other key issues.

2.11 RISK MANAGEMENT INTEGRATION WITH LIFE CYCLE PROCESSES

Ensuring that risk identification, analysis, evaluation and mitigation activities are integrated into life cycle processes is an important task for information security management. Most organizations have change management procedures that can provide the information security manager with an approach to implementing risk management processes on an ongoing basis. Because changes to any information resource are likely to introduce new vulnerabilities and change the overall risk equation, it is important that the information security manager is aware of proposed modifications.

Change management is a tenet of well-managed organizations. As distributed computing has become the norm and changes made more easily in a dispersed environment by people with limited knowledge, organizations often experience a lack of standardization in their hardware and software environments. Realizing this, management in most organizations has instituted more robust change management procedures and, as a consequence, has begun to achieve better control over the enterprise information resources. This is, of course, a moving target, and organizations with remote operations are, in some cases, still finding effective change management an elusive goal.

Change management processes may already be established for other business areas and activities. The benefit of this is that many organizations now have change management procedures that span the entire organization. The information security manager must be aware of these change management activities and ensure that security is well entrenched so changes are not made without considering the implications to the overall security of the organization's information assets. One method of helping to ensure this is for information security management to participate as a member of the change management committee and ensure that all significant changes are subject to review and approval by security and meet policy and standards requirements.

While the normal focus in change management addresses hardware and software changes (testing, sign-offs, etc.) and security impact, the change management process should extend well beyond the system owners and IT population. The change management process must include facilities management with respect to data center infrastructure and any other area that may impact overall information security (e.g., physical access control of sensitive or critical areas).

The impact of change management must address system and facilities maintenance windows with facilities personnel (often outsourced) and business continuity management. Quite often changes are not documented on a timely basis within these areas. Facilities may not have current single-line drawings and blueprints. Correspondingly, computer infrastructure/configuration management may not have the changes properly documented or updated on a timely basis. Business continuity may also fall behind on relevant updates when those updates occur in cycles. Emergency response and business continuity may also suffer communications lapses when current and changing processes are not reviewed within facility infrastructure areas.

Facilities personnel often have access to environmental monitoring and control systems (building management system [BMS]/supervisory control and data acquisition [SCADA] systems) for heating, ventilating and air conditioning (HVAC), water and electricity, or even physical access control systems. These are often programmed for remote computer access, an area that often escapes information security oversight. Physical security and control systems have been designed within the context of their being deployed within a closed/controlled environment.

Closed-circuit camera systems once configured to transmit data over the backbone network are often integrated with the IT infrastructure, posing the risk of compromise to these systems from IT facilities and the network.

IT systems can also be compromised due to vulnerabilities in the physical security system. Physical security and control systems, because of their importance in the protection of facilities and people, may be part of the organization's critical infrastructure. The protection of these systems, their system code and data may need to be integrated into security classification schemes. Facilities service contracts in support of physical and control systems often have emergency-response clauses. However, those contracts should be reviewed because they may lack adequate SLAs or fail to provide sufficient details and accountability for an adequate emergency response. A high-availability application may suffer unforeseen business impacts as a result of facilities failures.

Internal or external parties managing the facilities are sometimes overlooked from a risk management perspective. These are also the vulnerable areas because human beings pose the greatest threat to information security. Human beings are subject to negligence in compliance with established processes, ignorance of known threats, collusion or emotional behavior. It is important to include this factor in assessing the risk for critical information assets. Examples of internal parties are the operators who are granted, by job roles, direct physical access to the systems and facilities. External parties that may pose a risk include the servicing agents (e.g., cleaning or maintenance crew). Possible controls are background screening, employment terms, annual confirmation (via signature) of compliance to code of conducts, or contractual terms in the case of external parties.

By integrating risk identification, analysis and mitigation activities into change management (life cycle processes), the information security manager can ensure that critical information resources are adequately protected. This is a proactive approach, enabling the information security manager to better plan and implement security policies and procedures in alignment with the business goals and objectives of the organization. It also permits information security controls to be interjected into an activity that holds the greatest potential for degrading existing controls.

2.11.1 RISK MANAGEMENT FOR IT SYSTEM DEVELOPMENT LIFE CYCLE

According to NIST Special Publication 800-30, minimizing negative impact on an organization and the need for a sound basis for decision making are the fundamental reasons organizations implement a risk management process for their IT systems. Effective risk management must be fully integrated into the system development life cycle (SDLC). An IT system's SDLC has five phases: initiation, development or acquisition, implementation, operation or maintenance, and disposal. In some cases, an IT system may occupy several of these phases at the same time. However, the risk management methodology is the same, regardless of the SDLC phase for which the assessment is being conducted. Risk management is an iterative process that can be performed during each major phase of the SDLC. **Figure 2.25** describes the characteristics of each SDLC phase and indicates how risk management can be performed in support of each phase.

2.11.2 LIFE CYCLE-BASED RISK MANAGEMENT PRINCIPLES AND PRACTICES

Because risk management is a continuous process, the information security manager should view risk management itself as having a life cycle. This life cycle includes assessment, treatment and monitoring phases, as illustrated in **figure 2.26**. Employing a life-cycle-based risk management approach and integration with change management improves costs in that a full risk assessment does not have to be performed periodically. Instead, updates may be made to the risk assessment and risk management processes on an incremental basis.

Figure 2.25—Characteristics of the SDLC Phases

SDLC Phase	Phase Characteristics	Support from Risk Management Activities
Phase 1—Initiation	The need for an IT system is expressed and the purpose and scope of the IT system is documented.	Identified risk is used to support the development of the system requirements, including security requirements and a security concept of operations (strategy).
Phase 2—Development or Acquisition	The IT system is designed, purchased, programmed, developed, or otherwise constructed.	Risk identified during this phase can be used to support the security analyses of the IT system that may lead to architecture and design trade-offs during system development.
Phase 3—Implementation	The system security features should be configured, enabled, tested and verified.	The risk management process supports implementation against its requirements and within its modeled operational environment. Decisions regarding risk identified must be made prior to system operation.
Phase 4—Operation or Maintenance	The system performs its functions. Typically the system will undergo periodic updates or changes to hardware and software; the system may also be altered in less obvious ways due to changes to organizational processes, policies and procedures.	Risk management activities are performed for periodic system reauthorization (or reaccreditation or whenever major changes are made to an IT system in its operational, production environment (e.g., new systems interfaces).
Phase 5—Disposal	This phase may involve the disposition of information, hardware and software. Activities may include moving, archiving, discarding or destroying information and sanitizing the hardware and software.	Risk management activities are performed for system components that will be disposed of or replaced to ensure that the hardware and software are properly disposed of, that residual data are appropriately handled, and that system migration is conducted in a secure and systematic manner.

Figure 2.26—The IT Risk Management Life Cycle

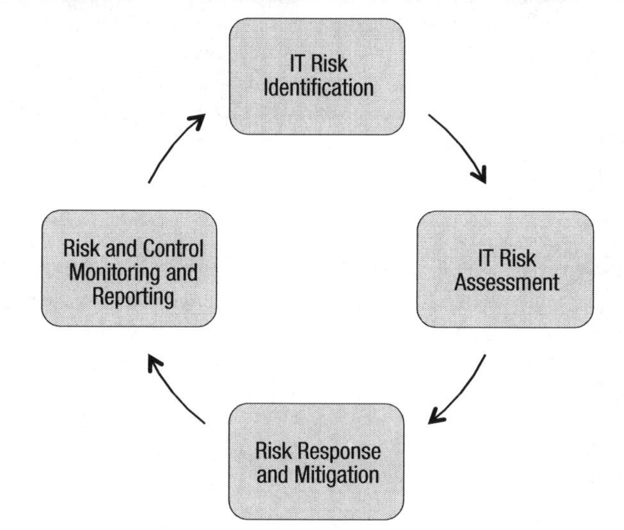

Source: ISACA, *CRISC Review Manual 6th Edition*, USA, 2015

2.12 SECURITY CONTROL BASELINES

A baseline is defined as "an initial set of critical observations or data used for comparison or a control." In order to formulate a baseline of security controls, some measurement of the effectiveness and efficiency of controls is necessary. A baseline is usually not based on a single test of controls, but on the overall capacity of controls to collectively mitigate risk to acceptable levels.

To establish control baselines, security managers can refer to many of the published standards that may be implemented within the organization. Based on these standards, a test of the control is performed multiple times to establish an evaluation of the effectiveness and efficiency of the control required by the standard.

Implementing baselines for security processes sets the minimum security requirements throughout the organization so they are consistent with acceptable risk levels. Different baselines must be set for different security classifications, with more stringent security controls required for the higher classification levels of the more critical or sensitive assets. If the organization has not implemented a classification scheme, it will be difficult for the information security manager to develop a rational basis for setting baselines without the risk of overprotecting some resources and underprotecting others.

Setting security baselines for an organization's operational enterprise has a number of benefits. It standardizes the minimum amount of security measures that must be employed throughout the organization; this results in positive benefits for risk management. It also provides a convenient point of reference to measure changes to security and identify corresponding effects on risk.

Working in conjunction with the organization's enterprise architecture group, an information security technology baseline of controls can be developed that is appropriate for the organization's operating environment.

There is a wealth of information available from NIST, COBIT, ISO and security vendors regarding standards for information security controls. However, the information security manager has to keep in mind that every organization has its own needs and priorities. While NIST, ISO, COBIT and vendors can provide starting points of support for developing controls, specific analysis should always be performed.

Controls suitable for the organization must be developed based on a variety of factors such as culture, structure, and risk appetite and tolerance. It is also important to keep in mind that, in addition to technology in the definition of a risk analysis program, people and processes must be considered. The information security manager also needs to develop procedural and physical security baselines. This will often present more of a challenge because these are areas typically outside the areas of security department control. Appropriate standards as described in chapter 1, with approval of the steering committee, can be the most effective approach to address this issue. Internal audit and regular security reviews can help provide assurance of compliance.

There is a general consensus among many vendors, security organizations, information security professionals and systems auditors about security configuration specifications that represent a prudent level of due care. These cooperative efforts continue to define consensus-based, good practice security configurations for various systems and platforms. The information security manager should examine these specifications and, where appropriate, they should be tailored and incorporated into organizational security baselines.

While industry standard baselines are important for the information security manager to be aware of, he/she must assess the level of security that should be employed in the organization in various security domains. The commingling of different technologies can often introduce new risk and change a secure system or platform into one that has new vulnerabilities. A tailored risk assessment that recognizes these interactions and dependencies will enable the information security manager to determine whether security processes and procedures above the accepted baselines are necessary to provide adequate security commensurate with the organization's defined levels of acceptable risk. Some organizations and industries may require higher baselines. Regulatory requirements for certain industries and regions may set a higher standard. Another issue may be that some of the organization's information is classified as highly critical or sensitive, and it must have control mechanisms that provide a higher level of security commensurate with its classification.

Standards must be developed or modified to set the lower boundaries of protection for each security domain. Standards provide the basis for measurement and testing approaches for evaluating whether security baselines are being met by existing controls. The objective is for residual risk across the security domain to be at a consistent acceptable level. This should be validated by periodic risk assessments to address the ever-changing nature of threats, vulnerabilities and technologies.

Example: To establish a baseline for antivirus processes, the information security manager will require periodic (weekly, monthly) reports of infected systems, virus alerts, virus incidents reported to the help desk, definition file updates and other pertinent information. Based on this information, it is possible to evaluate the effectiveness of the control, and based on additional information—such as man-hours required, software costs and latent risk—to evaluate the efficiency of the antivirus process as an IT control. This demonstrates how metrics can be used to monitor baselines and provide the basis for adjusting them as necessary to manage risk.

Selecting relevant metrics among the numerous available choices can be achieved by using a set of criteria. Effective metrics for risk management, including evaluating controls, can be selected based on the best ranking of the following characteristics:

- **Specific**—Based on a clearly understood goal; clear and concise
- **Measurable**—Able to be measured; quantifiable (objective), not subjective
- **Attainable**—Realistic; based on important goals and values
- **Relevant**—Directly related to a specific activity or goal
- **Timely**—Grounded in a specific time frame
- **Meaningful**—Understood by the recipients
- **Accurate**—A reasonable degree of accuracy
- **Cost-effective**—Not too expensive to acquire or maintain
- **Repeatable**—Able to be acquired reliably over time
- **Predictive**—Indicative of outcomes
- **Actionable**—Clear to the recipient what action must be taken

2.13 RISK MONITORING AND COMMUNICATION

Implementing an effective risk management program requires monitoring and communication. Monitoring the effectiveness of controls is an ongoing effort required to manage risk. Communication channels must be established both for reporting and disseminating information relevant to managing risk as well as providing the information security manager with information about risk-related activities throughout the enterprise, which includes reporting significant changes in risk and training and awareness.

2.13.1 RISK MONITORING

An important component of the risk management life cycle is continuously monitoring, evaluating, assessing and reporting risk. The results and status of this ongoing analysis need to be documented and reported to senior management on a regular basis. To facilitate such reporting, visual aids such as graphs or charts and summarized overviews can be useful. **Figure 2.27** shows a risk report format from COBIT 5 that can serve as a standard approach for communicating risk status.

Senior management will typically have little interest in technical details and is likely to want an overview of the current status and indicators of any immediate or impending threat that requires attention.

Red-amber-green reports, often referred to as security dashboards, heat charts or stoplight charts, are often used to show an overall assessment of the security posture. Depending on the recipients, other forms of representing security status, such as bar graphs or spider charts, are often more effective at conveying trends. Whatever the form of reporting, the information security manager is responsible for managing this reporting process to ensure that it takes place and the results are analyzed adequately and acted on appropriately in a timely manner. This responsibility includes identifying the types of events that will trigger reporting required by regulatory agencies and/or law enforcement and advising management of this requirement.

2.13.2 KEY RISK INDICATORS

One approach seeing increasing use is to report and monitor risk through the use of KRIs. KRIs can be defined as measures that, in some manner, indicate when an enterprise is subject to risk that exceeds a defined risk level. Typically, these indicators are trends in factors known to increase risk and are generally developed based on experience. They can be as diverse as increasing absenteeism or turnover in key employees or rising levels of security events or incidents.

KRIs can provide early warnings on possible issues or areas that pose particular risk.

A variety of risk indicators can be developed for various parts of an organization as a means of ongoing monitoring. A KRI is differentiated by being highly relevant and possessing a high probability of predicting or indicating a significant change in risk. Selection of KRIs, in addition to experience, can be based on sources such as industry benchmarks, external threat reporting services, or any other factor that can be monitored that indicates changes in risk to the organization.

KRIs are specific to each enterprise, and their selection depends on a number of parameters in the internal and external environment—such as the size and complexity of the organization, whether it operates in a highly regulated market, and its business strategy. Identifying useful risk indicators includes the following considerations:

- Including the different stakeholders in the enterprise. Risk indicators should not focus solely on the operational or the strategic side of risk. Rather, they should be identified for all stakeholders. Involving the right stakeholders in the selection of risk indicators will also ensure greater buy-in and ownership.
- Balancing the selection of risk indicators covering performance indicators that indicate risk after an event has occurred, lead indicators that indicate what capabilities are in place to prevent events from occurring, and trends based on analyzing indicators over time or correlating indicators to gain insight
- Ensuring that the selected indicators drill down to the root cause of events rather than just focusing on symptoms

Additionally, it is important to determine which measures are likely to serve as effective KRIs. These are differentiated by being highly relevant and possessing a high probability of predicting or indicating important risk. The criteria for selecting effective KRIs include:

- **Impact**—Indicators for risk with high potential impact are more likely to be KRIs.
- **Effort to implement, measure and report**—For different indicators of equivalent sensitivity to changing risk, the ones that are easier to measure are preferred.

Figure 2.27—Risk Report

A risk report includes information on current risk management capabilities and actual status and trends with regard to risk. This report will be based on the risk profile and will be tailored to the requirements of recipients.

	Life Cycle Stage	Internal Stakeholder	External Stakeholder	Description/Stake
Life Cycle and Stakeholders	**Information planning**	Risk function		Risk management capabilities, actual status and trends are communicated timely and accurately to the right people, based on their needs and requirements.
	Information design	Risk function		Risk management capabilities, actual status and trends are communicated timely and accurately to the right people, based on their needs and requirements.
	Information build/acquire	Risk function, business process owners/CIO		Ensure that the risk report includes the latest information on risk management capabilities, latest status and trends with regard to risk of the entire organisation.
	Information use/operate: store, share, use	Board, executive management, CIO, risk function, business process owners/CIO, compliance, internal audit	External audit, regulator	Effective utilisation and availability of information item by all involved stakeholders.
	Information monitor	Board, risk committee, audit committee, risk function, business process owners/CIO	External audit	Verifies that information remains actual and alerts on changes.
	Information dispose	Risk function		Ensures information is disposed of in a timely, secure and appropriate manner.

		Quality Subdimension and Goals	Description—The extent to which information is...	Relevance	Goal—The risk report...
Goals	Intrinsic	**Accuracy**	correct and reliable	High	should accurately define the risk management capabilities and actual status and trends with regard to risk, in such a way that it does not arouse confusion
		Objectivity	unbiased, unprejudiced and impartial	High	information is based on the enterprise's risk culture and confirmed by observations.
		Believability	regarded as true and credible	Medium	should be realistic and accurate
		Reputation	regarded as coming from a true and credible source	High	source information is collected from competent and recognised sources.

Source: ISACA, *COBIT® 5 for Risk*, USA, 2013, figure 64, page 155

- **Reliability**—The indicator must possess a high correlation with the risk and be a good predictor or outcome measure.
- **Sensitivity**—The indicator must be representative of the risk and capable of accurately indicating variances in the risk level.

Because the enterprise's internal and external environments are constantly changing, the risk environment is also highly dynamic, and the set of KRIs will more than likely change over time. Each KRI is related to the risk appetite and tolerance so trigger levels can be defined that will enable stakeholders to take appropriate action in a timely manner.

2.13.3 REPORTING SIGNIFICANT CHANGES IN RISK

As changes occur in an organization, the risk assessment must be updated to ensure its continued accuracy. Reporting these changes to the appropriate levels of management at the proper time is a primary responsibility of the information security manager. The information security manager should have periodic meetings with senior management to present a status on the organization's overall risk profile, including changes in risk level as well as the status of any open (untreated) risk.

In addition, the security program should include a process in which a significant security breach or security event will trigger a report to senior management and a reassessment of risk and applicable controls because all security events or incidents are the result of the failure of, or lack of, controls. The information security manager should have defined processes by which security events are evaluated based on impact to the organization. This evaluation may warrant a special report to upper management to inform them of the event, the impact and the steps being taken to mitigate the risk. Refer to chapter 4 for more detailed information on establishing and managing an incident response program.

2.14 TRAINING AND AWARENESS

People typically constitute the greatest risk to any organization, generally through accident, mistake, a lack of knowledge/information and, occasionally, malicious intent. Appropriate training and awareness campaigns can have a significant positive contribution on managing risk. Many controls are procedural and require some operational knowledge and compliance. Technical controls must be configured and operated correctly to provide the expected level of assurance. Ensuring users are educated in procedures and understand risk management processes is the responsibility of the information security manager, and appropriate training and awareness activities should be included in any risk management program.

The training and awareness program should be targeted to different staffing and security levels (e.g., senior management, middle management/IT staff and end users).

End-user information security training should include, among other things, sessions on:
- The importance of adhering to the security policies and procedures of the enterprise
- Responding to emergency situations
- Significance of logical access in an IT environment
- Privacy and confidentiality requirements
- Recognizing and reporting security incidents
- Recognizing and dealing with social engineering

2.15 DOCUMENTATION

Appropriate documentation that is readily available regarding risk management policies and standards, as well as other relevant risk-related matters, is required to effectively manage risk. Decisions concerning the nature and extent of documentation involve costs and related benefits. The risk management strategy, policy and program define the documentation needed. Specifically, at each stage of the risk management process, documentation should include:
- Objectives
- Audience
- Information resources
- Assumptions
- Decisions

A risk management policy document includes information such as:
- Objectives of the policy and rationale for managing risk
- Scope and charter of information risk management
- Links between the risk management policy and the organization's strategic and corporate business plans
- Extent and range of issues to which the policy applies
- Guidance on what is considered acceptable risk levels
- Risk management responsibilities
- Support expertise available to assist those responsible for managing risk
- Level of documentation required for various risk-management-related activities (e.g., change management)
- A plan for reviewing compliance with the risk management policy
- Incident and event severity levels
- Risk reporting and escalation procedures, format and frequency

In some circumstances, a compliance and due diligence statement may be required to ensure that managers formally acknowledge their responsibility to comply with risk management policies and procedures.

Typical documentation for risk management should include (see section 2.7.13 Risk Register):
- A risk register—For each risk identified, the register should record the:
  - Source of risk
  - Nature of risk
  - Risk owner
  - Risk ranking by severity
  - Selected treatment option
  - Existing controls
  - Recommended controls not implemented and the reasons why they should be implemented
- Consequences and likelihood of compromise, including:
  - Income loss
  - Unexpected expense
  - Legal risk (compliance and contractual)
  - Interdependent processes
  - Loss of public reputation or public confidence
- Initial risk rating
- Vulnerability to external/internal factors
- An inventory of information assets, including IT and telecommunication assets, that lists:
  - Description of the asset
  - Technical specifications
  - Number/quantity
  - Location
  - Special licensing requirements, if any
- A risk mitigation and action plan, providing:
  - Who has responsibility for implementing the plan
  - Resources to be utilized
  - Budget allocation
  - Timetable for implementation
  - Details of mechanism/control measures
  - Policy compliance requirements
- Monitoring and audit documents, which include:
  - Outcomes of audits/reviews and other monitoring procedures
  - Follow-up of review recommendations and implementation status

Finally, it is essential that all documentation be subject to an effective version control process as well as a standard approach to marking and handling. Documentation should be conspicuously labeled with classification level, revision date and number, effective dates and document owner.

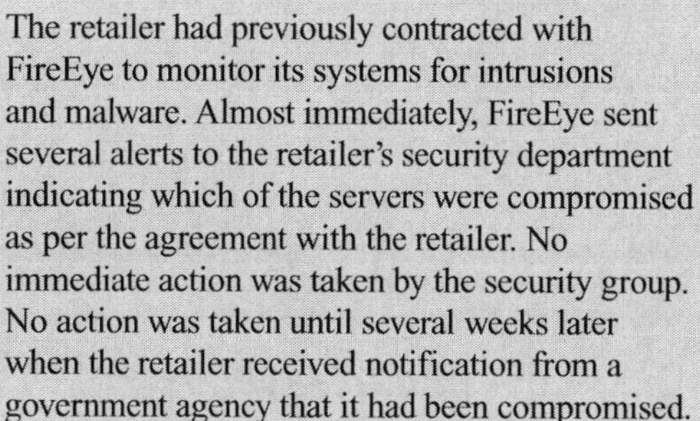

2.16 CASE STUDY

A major retailer was compromised, resulting in the loss of about 45 million credit and debit card records. Estimates of the ultimate total costs to the retailer were up to US $18 billion, including costs incurred by banks having to replace the cards and covering the resultant fraudulent charges. It also included the possible costs of lawsuits, governmental sanctions and fines levied by the major credit card organization for failure to be PCI-compliant. The retailer also suffered a 46 percent decrease in holiday sales when the breach was made public.

Upon analysis of the incident, it was discovered that the breach was facilitated initially by a phishing email to one of the retailer's service providers—a small HVAC company. The retailer had a service provider and vendor portal to facilitate purchasing and billing. Once the hackers gained access to the service company, they used the retailer's portal to gain access to the retailer's network. Although the portal was in a segmented portion of the network, the intruders managed to bridge the segment and gain access to the internal network. From there, they gained access to the point-of-sale (POS) system where they installed malware on the POS terminals that captured credit and debit card transactions because the POS traffic was in cleartext to avoid the system overhead associated with encryption.

The unencrypted customer information intercepted was stored on the compromised retailer's server and then subsequently exfiltrated to a series of Internet service providers before finally being sent to overseas in an effort to hide the trail. To mask the exfiltration, the data were sent only during normal business hours. Analysis discovered that the exfiltration did not start until several weeks after the malware had been installed.

The retailer had previously contracted with FireEye to monitor its systems for intrusions and malware. Almost immediately, FireEye sent several alerts to the retailer's security department indicating which of the servers were compromised as per the agreement with the retailer. No immediate action was taken by the security group. No action was taken until several weeks later when the retailer received notification from a government agency that it had been compromised.

1. **From a risk management standpoint, who in the organization should bear primary responsibility for this event?**
   A. Board of directors
   B. CEO
   C. Information security manager
   D. IT manager

2. **What could have been done to ensure the most effective action was taken when the company's security department was notified by FireEye that servers had been compromised?**
   A. Investigate and analyze the potential impact of the compromised servers.
   B. Isolate the POS devices.
   C. Initiate a tested incident management plan.
   D. Locate and remove the malware on the POS devices.

3. **What could the information security manager suggest from a risk management standpoint to prevent a similar incident from occurring in the future?**

Answers on page 130.

CHAPTER 2 ANSWER KEY

KNOWLEDGE CHECK: SEMIQUANTITATIVE ANALYSIS (PAGE 110)

1. If this product fails or if its launch is hampered by technical problems, a major negative impact to the organization's reputation could result. Using the scale on page 110, the value assigned would be Major, or 3. The likelihood of this product failing is unlikely if all goes smoothly, but the failure of the infrastructure is likely, due to the rush to market. Considering these two factors, a fair estimate would be Moderate, or value 3. Risk would then equal 9.

2. The cost for a violation annually would be US $10,000, so the impact would be Minor, or value 2. The likelihood of a violation is Likely, as it will be incurred annually, or value 4. Risk would then equal 8.

3. The impact of the email quarantine system could be considered Minor, as it would impact only the business's email system, or value 2. Likelihood of spam or other malicious mail content getting through the system is Frequent, as it is commonly happening, or value 5. Therefore, risk would equal 10.

KNOWLEDGE CHECK: COST-BENEFIT ANALYSIS (PAGE 115)

1. In this situation, 10,000 employees losing one hour of productivity at US $30 per hour would equal a cost in productivity of $300,000 (10,000 × 1 × 30 = 300,000).

2. The cost of the anti-malware scanning software would equal US $80,000 ($20,000 per 10,000 devices times 4 for 40,000 devices). The training session would cost the organization 20,000 hours of employee productivity time, or $600,000 (40,000 × 0.5 × 30 = 600,000).

3. Anti-malware software would be the least expensive option, but it would not help employees to be proactive protecting their systems from malicious software. Both options would help to eliminate potential future attacks in the event that malicious software goes undetected by the software.

CASE STUDY (PAGE 129)

Question 1:

A. It is the responsibility of the board of directors to ensure that executive management adequately addresses significant risks to the organization. These and other systemic failures in various departments were the result of lacks of oversight, resources, authority or competency, or perhaps inadequate integration or communication. Whichever the case or combination of causes, the responsibility for the string of failures that led to this catastrophic event inevitably is a failure of governance and resides with the board of directors.

B. The CEO would be responsible for aspects of the information security program and the circumstances leading up to the event; however, this incident was primarily caused by failures in governance.

C. The information security manager would be responsible for aspects of the information security program and the circumstances leading up to the event; however, this incident was primarily caused by failures in governance.

D. The IT manager is responsible for certain aspects of the retailer's IT program; however, this incident was primarily caused by failures in governance.

Question 2:

A. Isolating the affected servers and implementing encryption would have prevented the POS transactions, resulting in significant revenue loss unless backup servers in a recovery facility could have been brought on line to handle the traffic.

B. Once the POS devices were compromised, there were few good options available. Taking them offline would have had a massive impact on revenues. Because the credit and debit information was being stored for several weeks before being exfiltrated, isolating those servers while investigating and analyzing them would have thwarted the impact of the attack.

C. If a tested incident management plan existed, it is likely the risk and impact scenario that occurred would have been considered and tested and can be assumed that some level of preventive action would have been initiated.

D. Locating the malware on the POS devices would have been feasible, but removing it from tens of thousands of devices could pose a major problem unless it could have been automated.

Question 3:
A more robust training and awareness regarding potential information security risk to inform all parties, from vendors to senior management, would better position the organization to recognize and respond to potential risk events. Awareness training has a positive impact on risk management.

The information security manager may also want to suggest better third-party management, including ensuring that the outsourcers adhere to the information security policies of the organization and that they are granted the appropriate level of access to systems.

Compliance with PCI DSS would be another important suggestion to help prevent further incidents as it would set requirements for security to which the organization should adhere.

Page intentionally left blank

Chapter 3:

Information Security Program Development and Management

Section One: Overview

Section Two: Content

erview

areas of knowledge needed to ... ... mation security program.

DOMAIN DEFINITION

Develop and maintain an information security program that identifies, manages and protects the organization's assets while aligning to information security strategy and business goals, thereby supporting an effective security posture.

LEARNING OBJECTIVES

The objective of this domain is to ensure that the CISM candidate has the knowledge necessary to:
• Understand the broad requirements and activities needed to create, manage and maintain an information security program to implement an information security strategy.
• Define and utilize the resources required to achieve the IT goals consistent with organizational objectives.
• Understand the people, processes and technology necessary to execute the information security strategy.

CISM EXAM REFERENCE

This domain represents 27 percent of the CISM examination (approximately 41 questions).

TASK AND KNOWLEDGE STATEMENTS

TASK STATEMENTS

There are 10 tasks within this domain that a CISM must know how to perform:

T3.1 Establish and/or maintain the information security program in alignment with the information security strategy.

T3.2 Align the information security program with the operational objectives of other business functions (e.g., human resources [HR], accounting, procurement and IT) to ensure that the information security program adds value to and protects the business.

T3.3 Identify, acquire and manage requirements for internal and external resources to execute the information security program.

T3.4 Establish and maintain information security processes and resources (including people and technologies) to execute the information security program in alignment with the organization's business goals.

T3.5 Establish, communicate and maintain organizational information security standards, guidelines, procedures and other documentation to guide and enforce compliance with information security policies.

T3.6 Establish, promote and maintain a program for information security awareness and training to foster an effective security culture.

T3.7 Integrate information security requirements into organizational processes (e.g., change control, mergers and acquisitions, system development, business continuity, disaster recovery) to maintain the organization's security strategy.

T3.8 Integrate information security requirements into contracts and activities of third parties (e.g., joint ventures, outsourced providers, business partners, customers) and monitor adherence to established requirements in order to maintain the organization's security strategy.

T3.9 Establish, monitor and analyze program management and operational metrics to evaluate the effectiveness and efficiency of the information security program.

T3.10 Compile and present reports to key stakeholders on the activities, trends and overall effectiveness of the information security program and the underlying business processes in order to communicate security performance.

KNOWLEDGE STATEMENTS

The CISM candidate must have a good understanding of each of the domains delineated by the knowledge statements. These statements are the basis for the exam.

There are 16 knowledge statements within the Information Security Program Development and Management domain:

K3.1 Knowledge of methods to align information security program requirements with those of other business functions

K3.2 Knowledge of methods to identify, acquire, manage and define requirements for internal and external resources

K3.3 Knowledge of current and emerging information security technologies and underlying concepts

K3.4 Knowledge of methods to design and implement information security controls

K3.5 Knowledge of information security processes and resources (including people and technologies) in alignment with the organization's business goals and methods to apply them

K3.6 Knowledge of methods to develop and document information security standards, procedures and guidelines

K3.7 Knowledge of internationally recognized regulations, standards, frameworks and best practices related to information security program development and management

K3.8 Knowledge of methods to implement and communicate information security policies, standards, procedures and guidelines

K3.9 Knowledge of training, certifications and skill set development for information security personnel

K3.10 Knowledge of methods to establish and maintain effective information security awareness and training programs

K3.11 Knowledge of methods to integrate information security requirements into organizational processes (e.g., access management, change management, audit processes)

K3.12 Knowledge of methods to incorporate information security requirements into contracts, agreements and third-party management processes

K3.13 Knowledge of methods to monitor and review contracts and agreements with third parties and associated change processes as required

K3.14 Knowledge of methods to design, implement and report operational information security metrics

K3.15 Knowledge of methods for testing the effectiveness and efficiency of information security controls

K3.16 Knowledge of techniques to communicate information security program status to key stakeholders

RELATIONSHIP OF TASK TO KNOWLEDGE STATEMENTS

The task statements are what the CISM candidate is expected to know how to perform. The knowledge statements delineate each of the areas in which the CISM candidate must have a good understanding in order to perform the tasks. The task and knowledge statements are mapped, insofar as it is possible to do so. Note that although there is often overlap, each task statement will generally map to several knowledge statements.

Task and Knowledge Statements Mapping

Task Statement	Knowledge Statements
T3.1 Establish and/or maintain the information security program in alignment with the information security strategy.	K3.1 Knowledge of methods to align information security program requirements with those of other business functions K3.2 Knowledge of methods to identify, acquire, manage and define requirements for internal and external resources K3.5 Knowledge of information security processes and resources (including people and technologies) in alignment with the organization's business goals and methods to apply them K3.7 Knowledge of internationally recognized regulations, standards, frameworks and best practices related to information security program development and management K3.11 Knowledge of methods to integrate information security requirements into organizational processes (e.g., access management, change management, audit processes)
T3.2 Align the information security program with the operational objectives of other business functions (e.g., human resources [HR], accounting, procurement and IT) to ensure that the information security program adds value to and protects the business.	K3.1 Knowledge of methods to align information security program requirements with those of other business functions K3.5 Knowledge of information security processes and resources (including people and technologies) in alignment with the organization's business goals and methods to apply them K3.11 Knowledge of methods to integrate information security requirements into organizational processes (e.g., access management, change management, audit processes) K3.12 Knowledge of methods to incorporate information security requirements into contracts, agreements and third-party management processes
T3.3 Identify, acquire and manage requirements for internal and external resources to execute the information security program.	K3.2 Knowledge of methods to identify, acquire, manage and define requirements for internal and external resources K3.3 Knowledge of current and emerging information security technologies and underlying concepts K3.5 Knowledge of information security processes and resources (including people and technologies) in alignment with the organization's business goals and methods to apply them K3.11 Knowledge of methods to integrate information security requirements into organizational processes (e.g., access management, change management, audit processes) K3.12 Knowledge of methods to incorporate information security requirements into contracts, agreements and third-party management processes K3.13 Knowledge of methods to monitor and review contracts and agreements with third parties and associated change processes as required K3.14 Knowledge of methods to design, implement and report operational information security metrics
T3.4 Establish and maintain information security processes and resources (including people and technologies) to execute the information security program in alignment with the organization's business goals.	K3.2 Knowledge of methods to identify, acquire, manage and define requirements for internal and external resources K3.3 Knowledge of current and emerging information security technologies and underlying concepts K3.5 Knowledge of information security processes and resources (including people and technologies) in alignment with the organization's business goals and methods to apply them

Task and Knowledge Statements Mapping *(cont.)*			
Task Statement		**Knowledge Statements**	
T3.5	Establish, communicate and maintain organizational information security standards, guidelines, procedures and other documentation to guide and enforce compliance with information security policies.	K3.5 K3.6 K3.8 K3.10	Knowledge of information security processes and resources (including people and technologies) in alignment with the organization's business goals and methods to apply them Knowledge of methods to develop and document information security standards, procedures and guidelines Knowledge of methods to implement and communicate information security policies, standards, procedures and guidelines Knowledge of methods to establish and maintain effective information security awareness and training programs
T3.6	Establish, promote and maintain a program for information security awareness and training to foster an effective security culture.	K3.8 K3.9 K3.16	Knowledge of methods to implement and communicate information security policies, standards, procedures and guidelines Knowledge of training, certifications and skill set development for information security personnel Knowledge of techniques to communicate information security program status to key stakeholders
T3.7	Integrate information security requirements into organizational processes (e.g., change control, mergers and acquisitions, system development, business continuity, disaster recovery) to maintain the organization's security strategy.	K3.4 K3.5 K3.7 K3.10 K3.11	Knowledge of methods to design and implement information security controls Knowledge of information security processes and resources (including people and technologies) in alignment with the organization's business goals and methods to apply them Knowledge of internationally recognized regulations, standards, frameworks and best practices related to information security program development and management Knowledge of methods to establish and maintain effective information security awareness and training programs Knowledge of methods to integrate information security requirements into organizational processes (e.g., access management, change management, audit processes)
T3.8	Integrate information security requirements into contracts and activities of third parties (e.g., joint ventures, outsourced providers, business partners, customers) and monitor adherence to established requirements in order to maintain the organization's security strategy.	K3.4 K3.7 K3.12	Knowledge of methods to design and implement information security controls Knowledge of internationally recognized regulations, standards, frameworks and best practices related to information security program development and management Knowledge of methods to incorporate information security requirements into contracts, agreements and third-party management processes
T3.9	Establish, monitor and analyze program management and operational metrics to evaluate the effectiveness and efficiency of the information security program.	K3.14 K3.15	Knowledge of methods to design, implement and report operational information security metrics Knowledge of methods for testing the effectiveness and efficiency of information security controls
T3.10	Compile and present reports to key stakeholders on the activities, trends and overall effectiveness of the information security program and the underlying business processes in order to communicate security performance.	K3.14 K3.15 K3.16	Knowledge of methods to design, implement and report operational information security metrics Knowledge of methods for testing the effectiveness and efficiency of information security controls Knowledge of techniques to communicate information security program status to key stakeholders

TASK STATEMENT REFERENCE GUIDE

The following section contains the task statements a CISM candidate is expected to know how to accomplish mapped to the areas in the review manual with information that support the execution of the task. The references in the manual focus on the knowledge the information security manager must know to accomplish the tasks and successfully negotiate the exam.

TASK STATEMENT REFERENCE GUIDE			
Task Statement		Reference in Manual	
T3.1	Establish and/or maintain the information security program in alignment with the information security strategy.	3.1.2 3.6 3.8	Strategic Alignment Information Security Framework Components Information Security Infrastructure and Architecture
T3.2	Align the information security program with the operational objectives of other business functions (e.g., human resources [HR], accounting, procurement and IT) to ensure that the information security program adds value to and protects the business.	3.1 3.10 3.11	Information Security Program Management Overview Security Program Management and Administrative Activities Security Program Services and Operational Activities
T3.3	Identify, acquire and manage requirements for internal and external resources to execute the information security program.	3.10 3.11	Security Program Management and Administrative Activities Security Program Services and Operational Activities
T3.4	Establish and maintain information security processes and resources (including people and technologies) to execute the information security program in alignment with the organization's business goals.	3.10 3.11	Security Program Management and Administrative Activities Security Program Services and Operational Activities
T3.5	Establish, communicate and maintain organizational information security standards, guidelines, procedures and other documentation to guide and enforce compliance with information security policies.	3.10 3.13	Security Program Management and Administrative Activities Security Program Metrics and Monitoring
T3.6	Establish, promote and maintain a program for information security awareness and training to foster an effective security culture.	3.10.2 3.11	Security Awareness Training and Education Security Program Services and Operational Activities
T3.7	Integrate information security requirements into organizational processes (e.g., change control, mergers and acquisitions, system development, business continuity, disaster recovery) to maintain the organization's security strategy.	3.9 3.11 3.12	Architecture Implementation Security Program Services and Operational Activities Controls and Countermeasures
T3.8	Integrate information security requirements into contracts and activities of third parties (e.g., joint ventures, outsourced providers, business partners, customers) and monitor adherence to established requirements in order to maintain the organization's security strategy.	3.10.11 3.11.9 3.11.10 3.12	Vendor Management Outsourcing and Service Providers Cloud Computing Controls and Countermeasures
T3.9	Establish, monitor and analyze program management and operational metrics to evaluate the effectiveness and efficiency of the information security program.	3.12 3.13	Controls and Countermeasures Security Program Metrics and Monitoring
T3.10	Compile and present reports to key stakeholders on the activities, trends and overall effectiveness of the information security program and the underlying business processes in order to communicate security performance.	3.10.5 3.13	Documentation Security Program Metrics and Monitoring

SUGGESTED RESOURCES FOR FURTHER STUDY

Brotby, W. Krag; Gary Hinson; *PRAGMATIC Security Metrics: Applying Metametrics to Information Security*, Auerbach Publications, USA, 2013

Cloud Security Alliance (CSA), *cloudsecurityalliance.org*

Gadia, Sailesh; "Cloud Computing Risk Assessment: A Case Study," *ISACA Journal*, vol. 4, USA, 2011

International Federation of Accountants (IFAC), "Managing Security of Information: Guidelines," 2006, *www.ifac.org*

ISACA, *Cloud Computing: Business Benefits With Security, Governance and Assurance Perspectives,* USA, 2009, *www.isaca.org/Knowledge-Center/Research/ResearchDeliverables/Pages/Cloud-Computing-Business-Benefits-With-Security-Governance-and-Assurance-Perspective.aspx*

ISACA, COBIT 5, USA, 2012, *www.isaca.org/cobit*

ISACA, COBIT 5 for Information Security, USA, 2012, *www.isaca.org/cobit*

ISACA, COBIT 5: Enabling Processes, USA, 2012, *www.isaca.org/cobit*

National Institute of Standards and Technology (NIST) Special Publications, *csrc.nist.gov/publications/PubsSPs.html*

SANS Institute, *www.sans.org*

Senft, Sandra; Frederick Gallegos; Aleksandra Davis; *Information Technology Control and Audit, Fourth Edition*, CRC Press, USA, 2013

Wulgaert, Tim, and ISACA; *Security Awareness: Best Practices to Secure Your Enterprise*, ISACA, USA, 2005

Wysocki, Robert K.; *Effective Project Management: Traditional, Agile, Extreme, 6th Edition*, Wiley Publishing Inc., USA, 2011

Note: Publications in bold are stocked in the ISACA Bookstore.

SELF-ASSESSMENT QUESTIONS

CISM exam questions are developed with the intent of measuring and testing practical knowledge in information security management. All questions are multiple choice and are designed for one best answer. Every CISM question has a stem (question) and four options (answer choices). The candidate is asked to choose the correct or best answer from the options. The stem may be in the form of a question or incomplete statement. In some instances, a scenario or a description problem may also be included. These questions normally include a description of a situation and require the candidate to answer two or more questions based on the information provided. Many times a CISM examination question will require the candidate to choose the most likely or best answer.

In every case, the candidate is required to read the question carefully, eliminate known incorrect answers and then make the best choice possible. Knowing the format in which questions are asked, and how to study to gain knowledge of what will be tested, will go a long way toward answering them correctly.

3-1 When implementing an intrusion detection system (IDS), the information security manager should recommend that it be placed:

A. outside the firewall.
B. on the firewall server.
C. on a screened subnet.
D. on the external router.

3-2 Which of the following is the **BEST** metric for evaluating the effectiveness of security awareness training? The number of:

A. password resets
B. reported incidents
C. incidents resolved
D. access rule violations

3-3 Security monitoring mechanisms should **PRIMARILY**:

A. focus on business-critical information.
B. assist owners to manage control risk.
C. focus on detecting network intrusions.
D. record all security violations.

3-4 When contracting with an outsourcer to provide security administration, the **MOST** important contractual element is the:

A. right-to-terminate clause.
B. limitations of liability.
C. service level agreement.
D. financial penalties clause.

3-5 Which of the following is **MOST** effective in preventing security weaknesses in operating systems?

A. Patch management
B. Change management
C. Security baselines
D. Configuration management

3-6 Which of the following is the **MOST** effective solution for preventing internal users from modifying sensitive and classified information?

A. Baseline security standards
B. System access violation logs
C. Role-based access controls
D. Background investigations

3-7 Which of the following is the **MOST** important consideration when implementing an IDS?

A. Tuning
B. Patching
C. Encryption
D. Packet filtering

3-8 Which of the following practices is **BEST** used to remove system access for contractors and other temporary users when it is no longer required?

A. Log all account usage and send regular reports to their managers.
B. Establish predetermined automatic expiration dates.
C. Require managers to email the security department when the user leaves.
D. Ensure that each individual has signed a security acknowledgment.

3-9 Which of the following is **MOST** important for a successful information security program?

A. Adequate training on emerging security technologies
B. Open communication with key process owners
C. Adequate policies, standards and procedures
D. Executive management commitment

3-10 An enterprise is implementing an information security program. During which phase should metrics be established to assess the effectiveness of the program over time?

A. Testing
B. Initiation
C. Design
D. Development

ANSWERS TO SELF-ASSESSMENT QUESTIONS

3-1 A. Placing an IDS on the Internet side of the firewall is not usually advised (except to assess the traffic hitting the firewall) because the system will generate alerts on all malicious traffic—even though 99 percent will be stopped by the firewall and never reach the internal network.

B. Because firewalls should be installed on hardened servers with minimal services enabled, it would be inappropriate to install an IDS on the same physical device.

C. An IDS should be placed on a screened subnet, which is a demilitarized zone.

D. If placing it on the external router were feasible, it would not be advised (except to assess the traffic hitting the firewall) because the system will generate alerts on all malicious traffic—even though 99 percent will be stopped by the firewall and never reach the internal network.

3-2 A. Password resets may or may not have anything to do with awareness levels.

B. Reported incidents will provide an indicator of the awareness level of staff. An increase in reported incidents could indicate that the staff is paying more attention to security.

C. The number of incidents resolved may not correlate to staff awareness.

D. Access rule violations may or may not have anything to do with awareness levels.

3-3 **A. Security monitoring must focus on business-critical information to remain effectively usable by and credible to business users.**

B. Control risk is the possibility that controls would not detect an incident or error condition and, therefore, is not a correct answer because monitoring would not directly assist in managing this risk.

C. Network intrusions are not the only focus of monitoring mechanisms.

D. Although ideally all security violations should be recorded, this is only one objective of security monitoring.

3-4 A. Right to terminate is usually a part of the service level agreement (SLA) that can be invoked on failure of the outsourcer to meet the terms of the SLA.

B. The element of limitations of liability is also typically covered as a part of the SLA.

C. SLAs provide metrics to which outsourcing firms can be held accountable and will typically cover the other choices as well. This is the most encompassing answer.

D. Financial penalties are covered by the SLA in the event of failure of performance as specified in the SLA.

3-5 **A. Patch management is a preventive control in that it corrects discovered weaknesses by applying a patch to the original program code that eliminates the weakness preventing exploitation.**

B. Change management controls the process of introducing changes to systems that may introduce new vulnerabilities.

C. Security baselines provide minimum recommended settings to provide a consistent minimum level of security across the organization.

D. Configuration management ensures that incorrect configuration does not result in increased risk and controls the updates to the production environment.

3-6 A. Baseline security standards may require access controls, but alone do not prevent unauthorized access.

B. Violation logs are detective and do not prevent unauthorized access.

C. Role-based access controls help ensure that users have access only to files and systems appropriate for their job role and is a preventive control.

D. Background checks are not a preventive control although may be predictive based on past events.

3-7 **A. If an IDS is not properly tuned, it will generate an unacceptable number of false positives and/or fail to sound an alarm when an actual attack is underway.**

B. Patching is more related to system hardening and typically a part of maintenance activities.

C. Encryption is not a significant consideration when implementing an IDS.

D. Packet filtering is not used in an IDS.

3-8 A. Logging account usage does nothing to remove access for temporary employees.

B. Predetermined expiration dates that automatically remove access are the most effective means of removing systems access for temporary users.

C. Reliance on managers to promptly send in termination notices cannot always be counted on and by itself does not ensure access termination.

D. Requiring each individual to sign a security acknowledgment would have little effect in this case.

3-9 A. Adequate training on new technologies is one of many requirements for an effective security program, but it is not as crucial as management support.

B. Open communication with process owners is an important element, but by itself it is not sufficient to ensure a successful program.

C. Adequate policies, standards and procedures are essential, but they will not be effective absent management support for adequate resources and enforcement.

D. Sufficient executive management support is the most important factor for the success of an information security program.

3-10 A. The testing phase is too late because the system has already been developed and is in production testing.
B. In the initiation phase, the basic security objective of the project is acknowledged, but it is premature to consider where security checkpoints should be located and the development of a test plan.
C. In the design phase, security checkpoints are defined and a test plan is developed.
D. Development is the coding phase and is too late to consider security test points and test plans.

Section Two: Content

3.0 INTRODUCTION

The purpose of the information security program is to execute the strategy and achieve the organizational objectives for acceptable levels of risk and business disruption. Generally, a road map is constructed based on the strategy, a set of high-level objectives and/or goals and desired outcomes with a plan to achieve them. The road map consists of the step-by-step detailed plans to achieve those goals, which are specific projects or initiatives. The plans must also include the ongoing activities required to manage, maintain and improve the cost-effectiveness of the program.

The information security program encompasses the entire organization and all the activities that serve to provide protection for the organization's information assets. This includes both the development and ongoing management of the diverse information-security-related activities, processes and projects. The program exists solely to support and further the business objectives of the organization. To be effective and add value to the organization, the program must be demonstrably aligned with and focused on assisting the enterprise in achieving its goals by enabling business activities and managing risks to limit disruptions to acceptable levels.

The diverse elements and activities that make up the information security program require a variety of internal and external resources to achieve its strategic objectives. Resources may already exist, need to be acquired or be outsourced to external entities. Program objectives can be met in many ways, and it is up to the information security manager to identify optimal resource options, initiate the organization's acquisition process, and manage the required implementation or integration of the resources into the information security program.

The task is to develop information security processes—such as asset classification, escalation, notification and monitoring—needed to provide an appropriate level of control in support of business activities and then to manage them to ensure they are maintained properly to ensure effectiveness over time.

Administrative controls such as standards must initially be developed for all policies and then maintained and modified as necessary to ensure ongoing policy compliance. The information security manager ensures that underlying procedures are developed by the operational entities that are consistent with the standards. Circumstances such as changing environments, new or escalated risk, and new technologies may require changes in standards and procedures to maintain ongoing compliance with policies.

An organization's personnel constitute a critical and first line of defense for information security. With most individuals in the organization busy focusing on their own specific tasks, information security and risk are typically not primary focuses. Ongoing efforts to maintain security-related awareness and being alert to potential risk across the enterprise are essential components of an effective information security program.

Most organizational activities pose some degree of risk to the organization. An information security strategy must be developed and implemented to address those risks in a cost-effective manner. To accomplish this objective, all of the organization's processes must be assessed and the appropriate security requirements applied to minimize disruptions and achieve the business objectives.

A major part of managing information security requires processes to ensure that contracts with external service providers and anyone that has access to the organization's information systems contain language adequate to provide security consistent with the objectives of the strategy. This requires these entities to be monitored on an ongoing basis and periodically inspected and/or audited to ensure they live up to the security requirements. There must also be clear and tested escalation processes if things go wrong and, in all cases, strong authentication, authorization and network segmentation must be used to ensure that external parties have access to only what they need to meet their responsibilities.

Effective management requires effective metrics and monitoring. Metrics at the operational, tactical and strategic levels must be developed, then monitored and analyzed on an ongoing basis. Other essential monitoring requirements include essential controls, key risk indicators (KRIs) to warn of changing risk, internal and external environments, and compliance with policies and standards.

Once an information security program is operational, reporting on how well it is performing over time is essential. Trends can be strong indicators of improvements over time or that more must be done to deal with emerging risk.

3.1 INFORMATION SECURITY PROGRAM MANAGEMENT OVERVIEW

An information security program encompasses all the activities and resources that collectively provide information security services to an organization. Primary program activities include design, development and integration of enterprisewide controls related to information security, as well as the ongoing administration and management of these controls. Controls can range from simple policies and processes to highly complex technology solutions. Depending on the size and nature of the organization, these activities can be executed by a single individual or a chief information security officer (CISO) managing a large staff with diverse skills.

In some instances, the information security manager may be required to initiate an information security program from its inception. More often, the information security manager manages, modifies and improves an existing program. In either case, it is important to have a solid understanding of the many aspects and requirements of effective program design, implementation and management.

Successful security program management is not significantly different from managing any other organizational activity. The primary difference is that, despite increased visibility, information security remains a somewhat ill-defined and frequently misunderstood discipline.

Many information security managers have a technical background. Many did not embark on a management career; rather, they were technologists that found themselves increasingly faced with management functions and responsibilities. Additionally, this move toward a broader management role has been driven by the increased desire of the business to understand why specific security controls are required and how the business benefits specifically from them. In short, senior management wants to understand the specific risk the information security program is addressing and why the controls it mandates are a sound investment and actually benefit the business. As a result, many information security professionals have had to expand their view beyond technology and develop a greater understanding of the business activities they are seeking to protect.

The dual trends of growing security organizations and increased pressure from the business to ensure that the security program is aligned with and supports the business objectives have combined to broaden the body of knowledge security managers must master. An extensive range of business knowledge and understanding must now be added to the base of information security knowledge all security professionals should possess.

Information Security Management Trends

In an increasing number of organizations, the information security manager is a member of senior leadership, designated as vice president of security, CISO or chief security officer (CSO), as detailed in section 1.3.5 Chief Information Security Officer. Also, a number of security functions may report to an independent corporate-level security organization. These functions include physical and information security, IT security, compliance, privacy, business continuity planning/disaster recovery (BCP/DR) and security architecture. In some large multinational organizations, many or all these functions are currently included under a senior corporate risk manager or chief risk officer (CRO).

The benefits of this trend are apparent. All these functions serve the basic need of ensuring the safety and preservation of the organization and are interdependent. Aggregating these assurance activities under a single corporate function is likely to gain favor as it becomes increasingly apparent that a disintegrated, stovepipe approach to security is inefficient, costly and ineffective.

However, in other companies, information security continues to be a fragmented, low-level IT effort. In many sectors, such as manufacturing, the food industry and retail, information security is seen as primarily a technical activity largely dealing with compliance issues as opposed to a vital strategic business activity.

Regardless of trends and optimal considerations for security, the reality in the majority of organizations is something less than optimal, but security must still be managed on an ongoing basis.

In addition to understanding security concepts and technologies, information security managers will increasingly need to gain expertise in a range of management functions such as budgeting, planning, business case development, recruiting and other personnel related functions.

Essential Elements of an Information Security Program

Three elements are essential to ensure successful security program design, implementation and ongoing management:

1. The program must be the execution of a well-developed information security strategy closely aligned with and supporting organizational objectives.
2. The program must be well designed with cooperation and support from management and stakeholders.
3. Effective metrics must be developed for program design and implementation phases as well as the subsequent ongoing security program management phases to provide the feedback necessary to guide program execution to achieve the defined outcomes.

Many frameworks recommend a risk assessment as the starting point for developing an information security strategy. While this may be adequate in some cases, it is not optimal and does not address the balance of important outcomes for information security, including strategic alignment, resource management, value delivery, assurance process integration and performance measurement.

The comprehensive approach to security strategy development, as presented in chapter 1, goes beyond just addressing risk by also defining overall objectives for information security. These objectives should be explicitly linked to organizational objectives. The outlined approach also describes methodologies for defining the desired state of security and provides the basis for developing a comprehensive and effective strategy to achieve those objectives. Beyond focusing solely on managing risk, this approach considers how information security should be linked to and actively support the organization's strategic objectives, ensure its preservation, and optimize security resources and activities.

Many organizations are still not ready to undertake the costs and efforts to implement information security governance. In these cases, the information security manager may need to "shortcut" objectives development. Use of a standard framework, such as COBIT or ISO/IEC 27001:2013, in conjunction with capability maturity model integration (CMMI) or a process assessment model (PAM), can help to achieve this goal. This approach will allow the information security manager to determine the current state of the information security program, set specific goals and determine a strategy to achieve them.

The information security manager must develop defined objectives for the information security program and gain management and stakeholder consensus. Without defined objectives for information security, it will be impossible to devise effective information security management metrics because there will be no point of reference to show progress and development is likely to be *ad hoc* and haphazard.

Regardless of how the objectives of information security are devised and a strategy for achieving them developed, the goals of an information security program are to implement the strategy and achieve the defined objectives. Once developed, the information security program should clearly represent the elements of the strategy.

Information security program development entails a variety of activities, projects and initiatives involving people, processes and technology over a protracted period of time. The information security manager should keep in mind that the objectives and expected benefits of the information security program are most useful when they are defined in business terms to help nontechnical stakeholders understand and endorse the program goals. This is also more likely to promote feedback and participation from business owners that, in turn, will make it more likely that the security program will be aligned with overall organizational objectives. The information security manager should also consider that programs and initiatives that do not have specific identifiable business or organizational benefits should be carefully examined to determine whether they are justified or resources can produce greater benefits elsewhere.

3.1.1 IMPORTANCE OF THE INFORMATION SECURITY PROGRAM

Achieving adequate levels of information security at a reasonable cost requires good planning, an effective strategy and capable management. Information security program management is an ongoing requirement that serves to protect information assets, satisfy regulatory obligations, and minimize potential legal and liability exposures. Properly designed, implemented and managed, the information security program provides critical support for many business functions that simply would not be feasible without it.

To be of any use, a strategy must be implemented and made operational. One or more processes must be devised to achieve the objectives of the strategy. The security program is the process by which the organization's security systems are designed, engineered, built, deployed, modified, managed and maintained until they are removed from service. This is a critical aspect of the information security manager's responsibilities, covers a broad area, and requires substantial expertise and broad technical and managerial skills.

A well-executed security program serves to effectively design, implement, manage and monitor the security program, transforming strategy into actuality. While providing the capabilities to meet security objectives, it also accommodates the inevitable changes in security requirements, taking advantage of security expertise, tools and techniques already available in the infrastructure. It also increases the likelihood that the efforts are well integrated, decreasing costs of maintenance and administration and providing a consistent level of security across the enterprise.

It should be clear that an effective security program requires a great deal of planning as well as expertise and resources. Effective planning can be significantly aided by developing an enterprise security architecture at the conceptual, logical, functional and physical levels. Well-defined models and frameworks that can assist in this process are detailed in section 3.9 Architecture Implementation.

3.1.2 OUTCOMES OF INFORMATION SECURITY PROGRAM MANAGEMENT

Effective information security program management should achieve the objectives defined in the security strategy discussed in chapter 1. As with other management activities, the goals must be defined in specific, objective and measurable terms. Appropriate metrics must then be established to determine whether the goals have been achieved and, if not, by how much they were missed and how performance might be improved.

Whether formal information security governance has been implemented or not, an acceptable level of the following six outcomes should be considered the basis for developing the objectives of an effective information security program:
- Strategic alignment
- Risk management
- Value delivery
- Resource management
- Performance measurement
- Assurance process integration

Developing a strategy and defining the attributes of the information security program are primarily conceptual and logical exercises. Development of the security program will require transforming these concepts and logical relationships into technologies and processes. From an architectural perspective, this will require developing the physical, functional and operational components that are necessary to achieve the defined objectives.

If this is a major initiative, a full security architecture should be developed to ensure that the goals and desired outcomes are realized. If enterprise architecture(s) already exists, then a security architecture should be incorporated into the existing enterprise architecture(s). As with any complex structure with numerous moving parts, architecture can serve to define logical, physical and operational component and process relationships. It can also clarify potential issues and provide traceability from concept to implementation and operation.

Strategic Alignment

Effective alignment of information security with business or mission objectives requires regular interaction with business owners and an understanding of their plans and objectives. It often depends on garnering input from and building consensus among the major operating units within the organization. In the realm of the information security manager, this consensus involves topics such as understanding:
- Organizational information risk
- Selection of appropriate control objectives and standards
- Gaining agreement on acceptable risk and risk tolerance
- Definitions of financial, operational and other constraints

Strategic alignment can be accomplished through a security steering committee, if business owners or their delegates are members and active participants. The security program can support the alignment of business objectives and information security by implementing processes that ensure that defined business objectives provide the ongoing input to guide security activities.

Both old and new issues requiring attention should be tracked and communicated regularly. Action items related to issue investigation, resolution and disposition should be monitored and reported as organizational security performance metrics. A regular information security strategy report should be delivered to executive management to provide visibility into successes and setbacks around strategic alignment. This information can range from progress on projects of interest to new risk or capabilities that may affect a particular line of business.

Day-to-day operational interactions are powerful in their ability to build rapport and cooperation throughout the organization. In addition to facilitating timely action on information security issues, active relationships fostered by the information security manager can also increase awareness and a sense of responsibility around information security.

If the organization has a strategic business planning unit, active participation in its activities may also provide insight into future business directions and ensure that security considerations are included in the planning process. This may provide opportunities to orient security activities to support those objectives and identify potential risk.

Efforts to align security with business objectives must include consideration of security solutions that are a good fit for current and planned business initiatives. Alignment must also take into account enterprise processes as well as cost, culture, governance, existing technology and structure of the organization.

Risk Management

Managing the risk to information assets is a primary responsibility of the information security manager and provides the foundation and rationale for virtually all information security activities. Risk analysis must be based on business requirements and an understanding of the organization's processes, culture and technology. To effectively manage risk, the information security manager must develop a comprehensive understanding of threats the organization faces, its vulnerabilities, its risk profile and the level of risk that management determines is acceptable. The potential impacts of threats that can materialize must be evaluated and used to establish treatment priorities. Risk must be managed to a level that is acceptable to the organization. However, the risk landscape is always changing, and new risk will inevitably arise during program development and administration. It is important that a continuous process of risk management be maintained during program development, implementation and evolution.

Value Delivery

Value delivery as an objective requires information security to deliver the required level of security effectively and efficiently. The execution of the security program can have a considerable effect on achieving this goal. Good planning and project management skills are needed to implement the strategy efficiently.

Security investments should be managed to optimize support of business objectives and deliver clear value to the organization. The information security manager should direct efforts toward achievement of a standard set of security practices and establishment of security baselines proportionate to risk. Protection efforts must be prioritized to allocate limited resources to areas of greatest need and benefit.

Continued delivery of value requires that security solutions be institutionalized as normal and expected practices based on standards. Solutions must comprehensively address logical, technical, operational and physical concerns based on an understanding of the end-to-end operating processes of the organization. Security management cannot remain static; it must strive to develop a culture of continuous improvement.

Resource Management

Developing and managing a security program require people, technology and processes. The information security manager must endeavor to utilize human, financial, technical and knowledge resources efficiently and effectively. An important aspect of resource management is accomplished by ensuring that knowledge is captured and made available to those who need it.

Security processes and practices must be documented and they must be consistent with standards and policies. Project planning, technology selection, and skill acquisition or development significantly factor into the effectiveness of resource management. Security architectures should be developed to define and use infrastructures to achieve security objectives efficiently. These efforts help promote recognition of the resource needs and shortfalls and provide the basis for good resource management.

Performance Measurement

If an information security strategy has been developed, it should identify a variety of important monitoring and metrics requirements. It is likely that during the evolution and management of a security program, additional opportunities to develop meaningful metrics or points of useful monitoring will become apparent. There may be opportunities to roll up groups of metrics to provide a more holistic picture for managing security. Considerable new work has been done by ISACA with the goal of developing processes to provide better security management metrics. This topic is covered in greater detail in section 3.13 Security Program Metrics and Monitoring.

The development and implementation of the security program itself will require a means of measuring progress and monitoring activities. An effective information security program results in processes designed to achieve governance objectives as well as measurable artifacts that demonstrate whether the objectives are met. Security processes should be designed with measurable control points that allow independent auditors to attest that the program is in place and effectively managed.

The information security manager must develop monitoring processes and associated metrics to provide continuous reporting on the effectiveness of information security processes and controls. Good metrics design and implementation require an understanding of the information needed by various constituencies to manage effectively. Metrics must be developed at multiple levels, including strategic, tactical and operational levels. The metrics used should be defined, agreed on by management and aligned with strategic objectives. Care must be taken to ensure that metrics provide useful information relevant to managing security activities to achieve defined objectives. These measurement processes help identify shortcomings and failures of security activities and provide feedback on progress made in resolving issues.

Assurance Process Integration

It is important for the information security manager to be aware of and understand all organizational assurance functions because they invariably have significance for information security. The information security manager should develop formal relationships with other assurance providers and endeavor to integrate those activities with information security activities. In the typical organization, this might include physical security, risk management, privacy office, quality assurance, audit, change management, insurance, HR, business continuity and disaster recovery. Others may have a role to play as well.

As dictated by the governance objectives discussed in chapter 1, an information security manager should seek to increase information assurance and the predictability of business operations through mitigation measures that reduce information-related risk to defined and agreed-upon levels of acceptability. As discussed in chapter 2, acceptable risk at an acceptable cost can be determined by developing recovery time objectives (RTOs), which will serve to balance the cost of restoration against acceptable outages. In other cases, acceptable risk may be determined in terms of reliability, integrity, performance levels, confidentiality, acceptable downtimes and financial impacts.

3.2 INFORMATION SECURITY PROGRAM OBJECTIVES

The objective of the information security program is to implement the strategy in the most cost-effective manner possible, while maximizing support of business functions and minimizing operational disruptions. Chapters 1 and 2 explain how governance and risk management objectives for a security program are defined and incorporated into an overall strategy. The success of these initial steps will determine the degree of clarity in understanding the information security program development objectives.

If the security strategy has been well developed, the primary task will be turning high-level strategy into logical and physical reality through a series of projects and initiatives. Even with a well-developed security strategy, elements will need to be modified or reconsidered during program design, development and ongoing administration. This can occur for a variety of reasons, such as changes in business requirements, underlying infrastructure, topology, technologies or risk level. It can also be the case that better solutions become available during the course of the program development or subsequently. Even unanticipated resistance by those affected by the changes introduced by the new or modified program can drive significant changes to design, implementation or operation of a security program.

Whether the strategy has been developed in significant detail or only to the conceptual level, program development will include a great deal of planning and design to achieve working project plans. It is likely that standard system development life cycle (SDLC) approaches will be useful, including feasibility, requirements and design phases. Developing these plans in a collaborative fashion is important to gain consensus and cooperation from various stakeholders and to minimize subsequent implementation and operational problems.

3.2.1 DEFINING OBJECTIVES

An information security manager is rarely faced with a situation where no information security activity is present in an organization. Therefore, only rarely will the information security program need to be built from scratch. More often, developing the information security program will be a process of comparing the existing program to what will be required to reach the desired information security state of the organization. This is achieved by performing a gap analysis (as discussed in section 3.7.3 Gap Analysis—Basis for an Action Plan). As noted in chapter 1, the IT organization should have identified this at a high level, but it may be necessary to further define these objectives down to a more concrete and practical level. This may require a substantial amount of effort, but it is a critical component for developing the security program.

It is essential to determine the forces that drive the business need for the information security program. Primary drivers for an information security program include:

- The ever-increasing requirements for regulatory compliance
- Higher frequency and cost related to security incidents
- Concerns over reputational damage
- Growing commercial demands of Payment Card Industry Data Security Standard (PCI DSS)
- Business processes or objectives that may increase organizational risk

Determining the drivers will help clarify objectives for the program and provide the basis for the development of relevant metrics.

Once the objectives have been clearly defined, the purpose of the security program development activities is to develop the processes and projects that close the gap between the current state and those objectives. Typically, much of the basic work will be to identify necessary controls, implement them, develop suitable metrics and then monitor control points in support of control objectives.

Whether or not there is an existing information security program, there are some basic building blocks that need to be in place to support control activity and know that it is effective. As has been previously stated, the first step is always to determine management objectives for information security, develop key goal indicators (KGIs) that reflect those objectives, and then develop ways to measure whether the program is heading in the right direction to meet those objectives.

3.3 INFORMATION SECURITY PROGRAM CONCEPTS

As an information security program is developed, the developers must keep in mind the fundamental purpose of the security program (i.e., to implement the security strategy and achieve the defined outcomes, as detailed in chapter 1).

In situations where security governance has not been implemented and/or a strategy has not been developed, it will still be necessary to define overall objectives for security activities. Off-the-shelf, ready-made objectives can include conforming to a particular set of standards or achieving a defined maturity level based on the CMMI model.

Whether set forth in a strategy or not, a security program implementation effort should also include a series of specific control objectives as defined in COBIT or the ISO 27000 family of standards. It is likely that a great deal of any security program will consist of designing, developing and implementing controls, whether technical, procedural or physical. As these controls are developed, monitoring and metrics must also be considered. Processes to measure control effectiveness and determine control failure will be essential. Some metrics development approaches and processes are detailed in section 3.13 Security Program Metrics and Monitoring.

Implementation will typically consist of a series of projects and initiatives. It generally involves project management skills, including resource utilization, budgeting, setting and meeting time lines and milestones, quality assurance, and user acceptance testing (UAT).

Many projects involve unusual or complex technical elements and may require detailed specification, design and engineering efforts. This often requires skills outside of the information security manager's capabilities and it is generally prudent to consider engaging the services of consultants or contractors with subject matter expertise.

3.3.1 CONCEPTS

Implementing and managing a security program will require the information security manager to understand and have a working knowledge of a number of management and process concepts including:
- Architectures
- Budgeting, costing and financial issues
- Business case development
- Business process reengineering
- Communications
- Compliance monitoring and enforcement
- Contingency planning
- Control design and development
- Control implementation and testing
- Control monitoring and metrics
- Control objectives
- Deployment and integration strategies
- Documentation
- Personnel issues
- Problem resolution
- Project management
- Quality assurance
- Requirements development
- Risk management
- SDLCs
- Specification development
- Training needs assessments and approaches
- Variance and noncompliance resolution

> **Note:** This is not an all-inclusive list; rather, it is merely representative of many of the major concepts with which the information security manager must be familiar.

3.3.2 TECHNOLOGY RESOURCES

An information security program involves a variety of technologies in addition to processes, policies and people. The information security manager must be qualified to make decisions with respect to technology, including the viability and applicability of available solutions in terms of the program's goals and objectives. It is also essential that the information security manager understand where a given technology fits into the basic prevention, detection, containment, reaction and recovery framework, and how it will serve to implement strategic elements.

Some of the technologies related directly to information security that the information security manager should be familiar with include:
- Antivirus systems
- Application security methodologies
- Authentication and authorization mechanisms (one-time passwords [OTPs], challenge-response, public key infrastructure [PKI] certificates, multifactor authentication, biometrics)
- Backup and archiving approaches such as redundant array of inexpensive disks (RAID)
- Cryptographic techniques (e.g., PKI, Advanced Encryption Standard [AES])
- Data integrity controls (e.g., backups, data snapshots, data replication, RAID, SAN real-time replication)
- Data leak prevention methodologies (removable media security, content filtering, etc.) and associated technologies
- Digital signatures
- Identity and access management systems
- Firewalls
- IDSs, including host-based intrusion detection systems (HIDSs), network intrusion detection systems (NIDSs)
- Intrusion prevention systems (IPSs)
- Log collection, analysis and correlation tools (i.e., security information and event management [SIEM])
- Mobile computing
- Mobile devices
- Remote access methodologies (virtual private network [VPNs], etc.)
- Security features inherent in networking devices (e.g., routers, switches)
- Smart cards
- Vulnerability scanning and penetration testing tools
- Web security techniques
- Wireless security methodologies

While many of these technologies are specifically related to security and most function as controls, the information security manager must recognize that virtually all deployed technologies will have security implications.

In addition to technologies that are security-related, the information security manager must be familiar with the broader aspects of information technology including, but not limited to:
- Bring your own device (BYOD)
- Cloud computing
- Databases
- Enterprise architectures (two- and three-tier client servers, messaging, etc.)
- Internet and network protocols (TCP/IP, UDP, etc.)
- Local area networks (LANs)

- Network routing concepts and protocols
- Operating systems
- Servers
- Storage area networks (SANs)
- Virtualization
- Web-related technologies and architectures
- Wide area networks (WANs)
- Internet of things (IoT)
- Application server
- Middleware

Note: These lists are not meant to be comprehensive; instead, they are representative of the areas and types of technologies the information security manager should be familiar with to successfully manage an information security program. Any CISM candidate who is unfamiliar with any of the terms listed should review the glossary at the end of this manual. Additionally, the Knowledge Center on the ISACA web site (www.isaca.org) contains a fully searchable store of information that addresses many technologies, trends and concepts important to information security managers.

3.4 SCOPE AND CHARTER OF AN INFORMATION SECURITY PROGRAM

Whether forming a new information security department or coming into an existing one, the information security manager will need to determine the scope, responsibilities and charter of the department. It is rare to find these elements clearly identified and documented, unlike general technical responsibilities for items such as firewalls, IDSs and virus detection, which are usually documented. The lack of defined responsibilities will make it difficult to determine what to manage or how well the security function is meeting objectives.

The information security manager coming into a new situation is advised to invest considerable effort in gaining an understanding from those he or she reports to regarding expectations, responsibilities, scope, authority, budgets, reporting requirements, etc. It will be useful to specifically document these elements and obtain agreement with management.

In terms of the chain of command, it is vital to understand where the information security function fits into the overall organizational structure. In many situations, there will be inherent structural conflicts that the manager should be aware of and carefully consider. It may also be prudent to discuss any potential conflicts of interest with management and understand how they will be handled.

The information security department largely serves as an internal regulatory function, and its ability to function effectively precludes reporting to those it is supposed to regulate. While there may be exceptions, information security managers who report in the technology chain of command or to other operational managers tend to be limited in their ability to provide effective information security across the enterprise.

If the department is established and has been functioning well, it is likely that many security department functions will already be accepted practice. Nevertheless, it will be useful to determine whether responsibilities are clearly defined and well documented. In many instances, a prior manager will have assumed a variety of functions and taken on responsibilities that were not formally defined and documented. Such managers may have been effective through being persuasive and influential rather than through defined and documented processes and structure. If the prior manager is available for orientation, it would be prudent to use whatever time is available to gain insight into the existing situation. If the prior manager is not available, it may take a thorough investigation to determine what the manager's responsibilities were and how tasks were accomplished.

The ability to be effective in a particular organization will be heavily impacted by culture and the information security manager's understanding of it. Security is often politically charged, and success may hinge more on developing the right relationships than on any particular expertise. To a varying extent, organizations do not operate in the manner defined by organizational charts, but by undocumented relationships and influence. Much as Google™ determines relative importance by the number of links to a particular site, the influence of particular individuals is affected by their relationships to other people within the organization.

It is also essential to gain a thorough understanding of the current state of security functions in the organization. This may include the assessment elements such as risk assessment and business impact analysis (BIA). The state of governance and strategy will need to be ascertained, as will the condition of policies and standards, compliance, etc. Reviews of recent audits, incidents and other pertinent reports will be useful as well. Once completed, the balance of the elements identified in chapter 1 on governance can be considered, and the information security manager will have a basis to move forward with implementing an information security management framework.

Figure 3.1 shows a summary of the steps in developing an information security program. While many of these tasks will fall to the information security manager, some may be under the purview of other departments.

The scope of an information security program is established by the development of a strategy as discussed in chapter 1, in combination with risk management responsibilities covered in chapter 2. The extent to which management supports the implementation of the strategy and risk management activities determines the charter.

Implementation of a security program impacts an organization's established way of doing things. Within an existing structure of people, processes and technology, the information security manager must strive to integrate changes to these established processes and policies. Inevitably, this will result in some degree of resistance to change, which the information security manager should plan to address.

In the absence of an adopted information security strategy and where no formal charter is documented, an information security manager can fall back to industry standards, such as a customized

Figure 3.1—Steps in Information Security Program Development

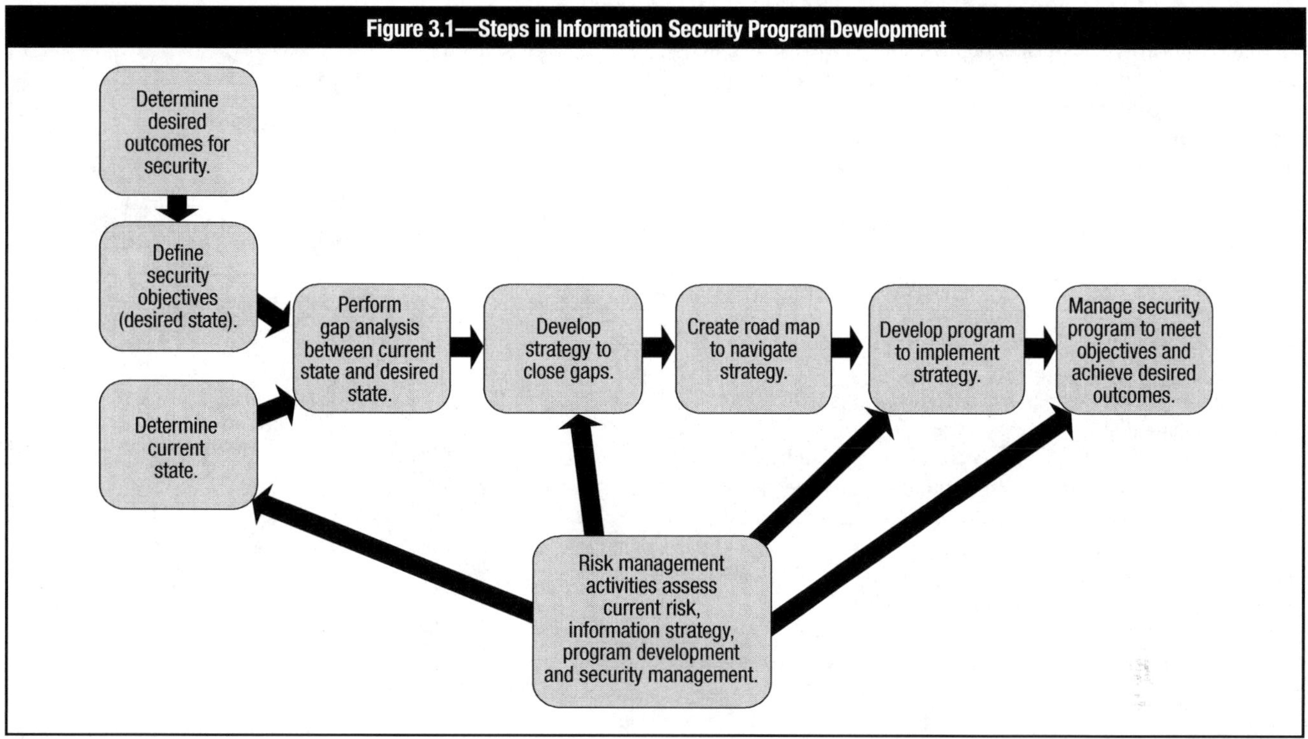

version of ISACA's description of a mature information security program, that might read:

> *Information security is a joint responsibility of business, information security and IT management and is integrated with corporate business objectives. Information security requirements are clearly defined, optimized and included in a verified security plan. Security functions are integrated with applications at the design stage and end users are increasingly accountable for managing security. Information security reporting provides early warning of changing and emerging risk, using automated active monitoring approaches for critical systems. Incidents are promptly addressed with formalized incident response procedures supported by automated tools. Periodic security assessments evaluate the effectiveness of implementation of the security plan. Information on new threats and vulnerabilities is systematically collected and analyzed, and adequate mitigating controls are promptly communicated and implemented. Intrusion testing, root cause analysis of security incidents and proactive identification of risk are the basis for continuous improvements. Security processes and technologies are integrated organizationwide.*

This approach, coupled with COBIT 5 or *ISO/IEC 27001:2013 Information technology—Security techniques—Information security management systems—Requirements*, can serve as the basis for a strategy and help define the scope and charter of the information security program.

3.5 THE INFORMATION SECURITY MANAGEMENT FRAMEWORK

The information security management framework is a conceptual representation of an information security management structure. It should define the technical, operational, managerial, administrative and educational components of the program; the organizational units and leadership responsible for each component; the control or management objective that each component should deliver; the interfaces and information flow among the components; and each component's tangible outputs. Although formats and detail levels vary, the framework should fundamentally describe the information security management components (e.g., roles, policies, standard operating procedures [SOPs], management procedures, security architectures) and their interactions in broad strokes. This is, in essence, an operational architecture as described in section 3.9 Architecture Implementation.

Other outcomes of an effective security management framework focus on shorter-term needs. For example, organizational decision makers require awareness of risk and mitigation options in support of corporate initiatives such as external hosting of information systems. Implementers of solutions often require the services of technical security subject matter expertise, which the information security manager should facilitate either through internal or external resources. Ensuring that initiatives and existing operations adhere to policies and standards is also an area that the information security manager and the security department are expected to manage.

The information security manager is typically expected to craft information security management options that deliver outcomes that are less direct, but no less important, to achieving security goals. These objectives include demonstrating, both directly and indirectly, that:

- The program adds tactical and strategic value to the organization.
- The program is being operated efficiently and with concern to cost issues.
- Management has a clear understanding of information security drivers, activities, benefits and needs.
- Information security knowledge and capabilities are growing as a result of the program.
- The program fosters cooperation and goodwill among organizational units.
- There is facilitation of information security stakeholders' understanding of their roles, responsibilities and expectations.
- The program includes provisions for the organization's continuity of business.

These soft goals revolve around directly and indirectly demonstrating to security steering committees, senior management and boards of directors that the information security program is delivering results and the information security manager is managing the program effectively.

There are a number of different frameworks that can be used by the information security manager for developing the information security program. Two of the most common internationally recognized approaches, COBIT and ISO/IEC 27001, are described in the following sections.

3.5.1 COBIT 5

COBIT 5 provides a comprehensive framework with a focus on helping enterprises create optimal value from IT by maintaining a balance between realizing benefits and optimizing risk levels and resource use. COBIT 5 enables IT and information to be governed and managed in a holistic manner for the enterprise, addressing the business and IT functional areas of responsibility, considering the information-related interests of internal and external stakeholders.

COBIT 5 is based on five key principles (shown in **figure 3.2**) for governance and management of enterprise IT, as discussed in section 1.8.3 The Desired State.

COBIT® 5 for Information Security leverages the comprehensive view of COBIT 5 while focusing on providing guidance for professionals involved in maintaining the confidentiality, availability and integrity of enterprise information. The framework provides tools to help understand, use, implement and direct core information-security-related activities and make more informed decisions. It enables information security professionals to effectively communicate with business and IT leaders and manage risk associated with information, including risk related to compliance, continuity, security and privacy.

Figure 3.2—COBIT 5 Principles

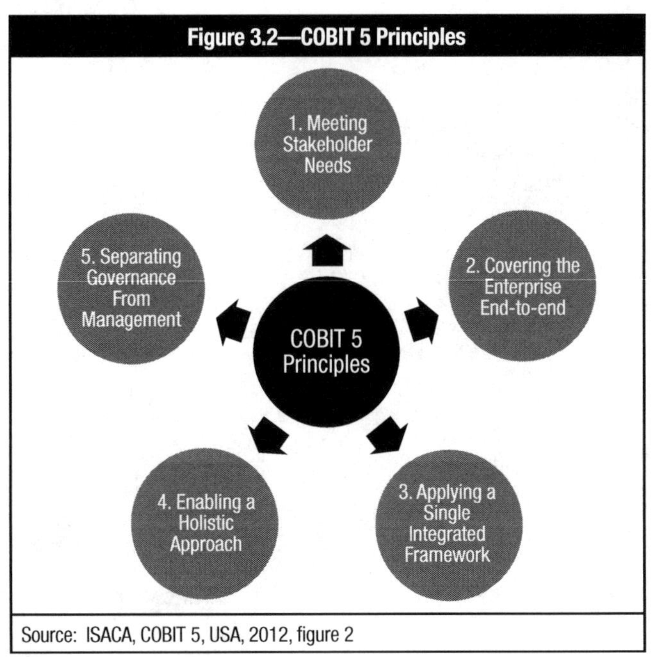

Source: ISACA, COBIT 5, USA, 2012, figure 2

3.5.2 ISO/IEC 27001:2013

The security standards *ISO/IEC 27001:2013: Information technology—Security techniques—Information security management systems—Requirements* and *ISO/IEC 27002:2013: Information technology—Security techniques—Code of practice for information security controls* provide a widely accepted framework and approach to information security management. The 114 control in the 14 domains of ISO/IEC 27001:2013 can be generally mapped to COBIT, but they are less business-oriented and comprehensive and do not provide complete tool sets. These standards do provide high-level comprehensive requirements for information security programs. Based on the British Standard (BS7799), this standard has been slightly expanded to include 14 broad control areas:

- A.5: Information security policies
- A.6: Organization of information security
- A.7: Human resource security (controls that are applied before, during or after employment)
- A.8: Asset management
- A.9: Access control
- A.10: Cryptography
- A.11: Physical and environmental security
- A.12: Operations security
- A.13: Communications security
- A.14: System acquisition, development and maintenance
- A.15: Supplier relationships
- A.16: Information security incident management
- A.17: Information security aspects of business continuity management
- A.18: Compliance (with internal requirements, such as policies, and with external requirements, such as laws)

The ISO/IEC 27000 series of standards covers virtually all aspects of security at a high level. The series includes:

- *ISO/IEC 27000 Information technology—Security techniques—Information security management systems—Overview and vocabulary*

- *ISO/IEC 27001 Information technology—Security techniques—Information security management systems—Requirements*
- *ISO/IEC 27002 Information technology—Security techniques—Code of practice for information security controls*
- *ISO/IEC 27003 Information technology—Security techniques—Information security management system implementation guidance*
- *ISO/IEC 27004 Information technology—Security techniques—Information security management—Measurement*
- *ISO/IEC 27005 Information technology—Security techniques—Information security risk management*
- *ISO/IEC 27006 Information technology—Security techniques—Requirements for bodies providing audit and certification of information security management systems*
- *ISO/IEC 27007 Information technology—Security techniques—Guidelines for information security management systems auditing*
- *ISO/IEC TR 27008 Information technology—Security techniques—Guidelines for auditors on information security controls*
- *ISO/IEC 27010 Information technology—Security techniques—Information security management for inter-sector and inter-organizational communications*
- *ISO/IEC 27011 Information technology—Security techniques—Information security management guidelines for telecommunications organizations based on ISO/IEC 27002*
- *ISO/IEC 27013 Information technology—Security techniques—Guidance on the integrated implementation of ISO/IEC 27001 and ISO/IEC 20000-1*
- *ISO/IEC 27014 Information technology—Security techniques—Governance of information security*
- *ISO/IEC TR 27015 Information technology—Security techniques—Information security management guidelines for financial services*
- *ISO/IEC TR 27016 Information technology—Security techniques—Information security management—Organizational economics*
- *ISO/IEC TR 27019 Information technology—Security techniques—Information security management guidelines based on ISO/IEC 27002 for process control systems specific to the energy utility industry*
- *ISO/IEC 27031 Information technology—Security techniques—Guidelines for information and communication technology readiness for business continuity*
- *ISO/IEC 27032 Information technology—Security techniques—Guidelines for cybersecurity*
- *ISO/IEC 27033 Parts 1 to 5 Information technology—Security techniques—Network security*
- *ISO/IEC 27034-1 Information technology—Security techniques—Application security—Part 1: Overview and concepts*
- *ISO/IEC 27035 Information technology—Security techniques—Information security incident management*
- *ISO/IEC 27036-3 Information technology—Security techniques—Information security for supplier relationships—Part 3: Guidelines for information and communication technology supply chain security*
- *ISO/IEC 27037 Information technology—Security techniques—Guidelines for identification, collection, acquisition and preservation of digital evidence*
- *ISO 27799 Health informatics—Information security management in health using ISO/IEC 27002*

3.6 INFORMATION SECURITY FRAMEWORK COMPONENTS

Most standard frameworks for information security show the development of an information security program as starting with risk assessment and the identification of control objectives and key controls defined by standards such as COBIT and ISO/IEC 27001 and 27002. It should be understood that while guidance is provided by these standards, control objectives must be based on individual organizational objectives and risk appetite and tolerance tailored to achieve the desired outcomes, as discussed in chapter 1. As the program develops and matures and conditions change, control objectives may need to be changed and additional or modified controls may be required to address new conditions, emerging risk, evolving regulations or contractual obligations.

The various components that make up the management framework can be broken down into their functional elements (i.e., technical, operational, managerial, administrative and educational components). This can be helpful in ensuring that each aspect is adequately represented in the framework and given appropriate security consideration.

3.6.1 TECHNICAL COMPONENTS

Information security is typically involved with most or all of the technical IT components in an organization. This includes providing and maintaining suitable security standards, review of procedures for policy compliance, designing and implementing appropriate security metrics, and general oversight. Many of the key controls identified in the framework will address risk associated with the technical components, including their configuration, monitoring, maintenance and operation. While IT may "own" the systems, it is typically not the data owner and serves as custodian with a requirement to provide adequate protection and compliance with the organization's information security policies by operating within the applicable standards. In other cases, the business owner may actually "own" the systems supporting the business operation but, in either case, IT will still serve as custodian. It is essential that all technology components have an identified owner and that there are no "orphan systems." This is necessary to ensure responsibility and accountability for maintaining all systems in compliance with security policies and for ownership and proper treatment of associated risk to acceptable levels.

From an information security perspective, the vast majority of the organization's information will reside with IT and will be a major focus of the information security framework. The information security function must adequately regulate the IT function and provide oversight to ensure policy compliance sufficient to achieve acceptable risk levels consistent with the information security strategy objectives.

3.6.2 OPERATIONAL COMPONENTS

Operational components of a security program are the ongoing management and administrative activities that must be performed to provide the required level of security assurance. These operational components include items such as SOPs, business operations security practices, and maintenance and administration of security technologies. They are generally conducted on a daily to weekly time line.

The information security manager must provide ongoing management of the operational information security components. Because it is common for many of these components to fall outside the direct control of the information security department (e.g., operating system patching procedures), the information security manager must work with and provide oversight of organizational groups such as IT, business units and other organizational groups to ensure that operational security needs are covered. Examples of common operational components include:
- Identity management and access control administration
- Security event monitoring and analysis
- System patching procedures and configuration management
- Change control and/or release management processes
- Security metrics collection and reporting
- Maintenance of supplemental control technologies and program support technologies
- Incident response, investigation and resolution
- Retirement and sanitization of data processing equipment and media storage

For each operational component related to information security, the information security manager needs to identify the owner and collaborate with the owner to document key information needed for management of the necessary functions. This information includes the component ownership and execution roles, activity schedule or triggers, needed information or data inputs, actual procedural steps, success criteria, failure escalation procedures, and approval/review processes. It also includes processes for providing suitable management metrics for the feedback necessary for effective ongoing management.

In addition, the information security manager should ensure that procedures for log maintenance, issue escalation, management oversight, and periodic risk assessment and quality assurance reviews are developed and implemented. It is important for the information security manager to update the roles and responsibilities documentation as new tasks arise in operational component development. For example, a new operational procedure that requires a monthly chief operating officer (COO) review of security issues related to business activities needs to be added to the appropriate task lists and schedules.

3.6.3 MANAGEMENT COMPONENTS

In addition to a variety of ongoing technical and operational security tasks, the information security manager needs to consider a number of management components. These typically include strategic implementation activities such as standards development or modification, policy reviews, and oversight of initiatives or program execution. These are activities that generally take place less frequently than operational components, perhaps on a time line measured in months, quarters or years.

Management objectives, requirements and policies are key in shaping the rest of the information security program, which, in turn, defines what must be managed. The information security manager must ensure that this process is executed with appropriate consideration to legal, regulatory, risk and resource issues as well as a suite of metrics needed for decision support.

Ongoing or periodic analysis of assets, threats, risk and organizational impacts must continue to be the basis for modifying security policies and developing or modifying standards. It should be considered that early versions are often too permissive, too restrictive or misaligned with operational realities. As a result, the information security manager is well advised to exercise flexibility in making adjustments to standards and policy interpretation during the initial stages of a security program.

Ongoing communication with business and operational units is critical to providing the feedback that can provide guidance to information security management, ensure its effectiveness, and maintain alignment with, and support of, the objectives of the organization.

During development of operational and technical management components, it is important that there is management oversight ensuring fulfillment of requirements and consistency with strategic direction. This oversight often occurs in the form of management reviews of program components (for example, the CIO, CEO, steering committee or executive committee performing a quarterly review of security operations). Topics for review might include modifications to operational or technical components, general effectiveness of program components, review of metrics and key performance indicators (KPIs), root cause analysis of detrimental events such as outages or compromises, issues hampering component effectiveness that require management attention, and/or review of action items and commitments from previous review sessions.

3.6.4 ADMINISTRATIVE COMPONENTS

As the scope and responsibilities of the information security management function grow, so do the resources, personnel and financial aspects involved. This means that information security management must address the same business administration activities as other business units. The information security manager in charge of such an organization must ensure that financial, HR and other management functions are effective.

Financial administration functions generally consist of budgeting, time line planning, total cost of ownership (TCO) analysis/management, return on investment (ROI) analysis/management, acquisition/purchasing and inventory management. These functions, particularly budgeting and time line planning, often require updates throughout the fiscal year as financial realities and organizational goals change. The information security manager should establish a working rapport with the organization's finance department to ensure a strong working relationship, support and compliance with financial policies and procedures.

HR management functions generally include job description management, organizational planning, recruitment and hiring, performance management, payroll and time tracking administration, employee education and development, and termination management. The information security program should account for the time and resources needed for these activities, particularly as program staffing grows over time. Those charged with management of larger information security programs must also address the need to develop an efficient organizational structure with appropriate layers of management and supervisory personnel. In all issues related to HR management, it is important for the information security manager to work closely with HR leadership and adhere to established procedures to prevent legal liabilities and other types of risk.

Effective management functions require the information security manager to balance project efforts and ongoing operational overhead with staff head count, utilization levels and external resources. Rarely does any information security program have an optimal number of resources, and it is always necessary to prioritize efforts. The information security manager should work with the steering committee and executive management to determine priorities and establish consensus on what project items may be delayed because of resource constraints. Spikes in activity or unexpected project efforts can often be addressed with third-party resources such as contractors. The information security manager should maintain relationships with the vendors most likely to be called on in such cases.

It is not unusual for the information security manager to be under pressure to shortcut security management, quality assurance (QA) and development processes or divert resources from daily operations to accelerate project efforts. It is the role of the information security manager to document and ensure that executive management understands the risk implications of moving an initiative ahead without full security diligence; it is up to executive management to decide if the initiative is important enough to warrant the risk. If this situation occurs, the information security manager should make every effort to utilize the first available opportunity to revisit systems or initiatives that are not certified or accredited.

To ensure that the existing security environment operates as needed, security operational resources should be diverted to project efforts only if they are not fully utilized. Even in this situation, it needs to be clearly communicated that operational security resources are provided *ad hoc*, and a spike in operational activities (e.g., an intrusion) requires the immediate attention of the operational staff.

3.6.5 EDUCATIONAL AND INFORMATIONAL COMPONENTS

Employee education and awareness regarding security risk is often integrated with employee orientation and initial training. General organizational policies and procedures, such as acceptable use policies and employee monitoring policies, should be communicated and administered at the organization's HR level. Issues and responsibilities that are specific to an employee's role or organization (e.g., call center authentication of customers) should be communicated and administered at the business unit level. Interactive education techniques such as online testing and role-playing are often more effective than a purely informational approach. Examples of these types of training include incident response and contingency plan training and exercises. In all cases, the information security manager should collaborate with HR and business units to identify information security education needs. Pertinent metrics (e.g., average employee quiz scores, average time elapsed since last employee training) should be tracked and communicated to the steering committee and executive management.

3.7 DEFINING AN INFORMATION SECURITY PROGRAM ROAD MAP

The key goals of strategic alignment, risk management, value delivery, resource management, assurance process integration and performance measurement are universal and are defined to some extent in the development of a security strategy. The process of program development requires that each of the six key goals is considered in detail and clarified in light of the evolving road map. As specific project plans for various parts of the road map develop, approaches to best achieve each key goal should become apparent.

These may be concepts that are not well understood by management and other stakeholders, which could lead to unrealistic expectations and poor outcomes. To maximize the chances for success, it may be most effective to develop a road map for the information security program in stages, starting with relatively simple objectives designed to demonstrate the value of the program and provide feedback on achievement of the key goals.

For example, the first stage might be to create that subcomponent of the program necessary to demonstrate the extent of strategic alignment and approaches to improvements. To get started, an information security manager can interview stakeholders such as department heads in HR, legal, finance and major business units to determine important organizational issues and concerns. Information taken from such interviews will point to candidates for information security steering committee members. In stage 2, that forum can be used to draft basic security policies for the implementation of an information security program for approval by upper management. Because members of the information security steering committee, by definition, represent business interests, the forum can be used to list specific business goals for security with reference to business processes (and, thus, also to systems, as depicted in **figure 3.3**). In stage 3, members of the information security steering committee hold functional roles that can promote awareness of the policy and conduct internal security reviews to see if they are in compliance. In stage 4, the compliance gaps identified in the security reviews can be used to effect change, and an approach to monitor the organizational policy compliance strategy can be simultaneously developed. From that foundation, the information security manager can begin the work of building consensus around roles and responsibilities, processes, and procedures in support of the policy.

3.7.1 ELEMENTS OF A ROAD MAP

A road map to implement the information security strategy must consider a number of factors. If a well-developed strategy exists, then a high-level road map to achieve the strategy objectives should also exist. Objectives, resources and constraints will have been defined. The work that remains is to transform the conceptual or logical architecture or design into a physical one. Construction of specific projects and initiatives must be planned, along with budgets, timetables, personnel and other tactical project management aspects that will result collectively in achieving the strategy objectives. It is similar to the differences between the architecture of a house and the tasks required to build the house.

Figure 3.3—Information Security Management System Controls Process

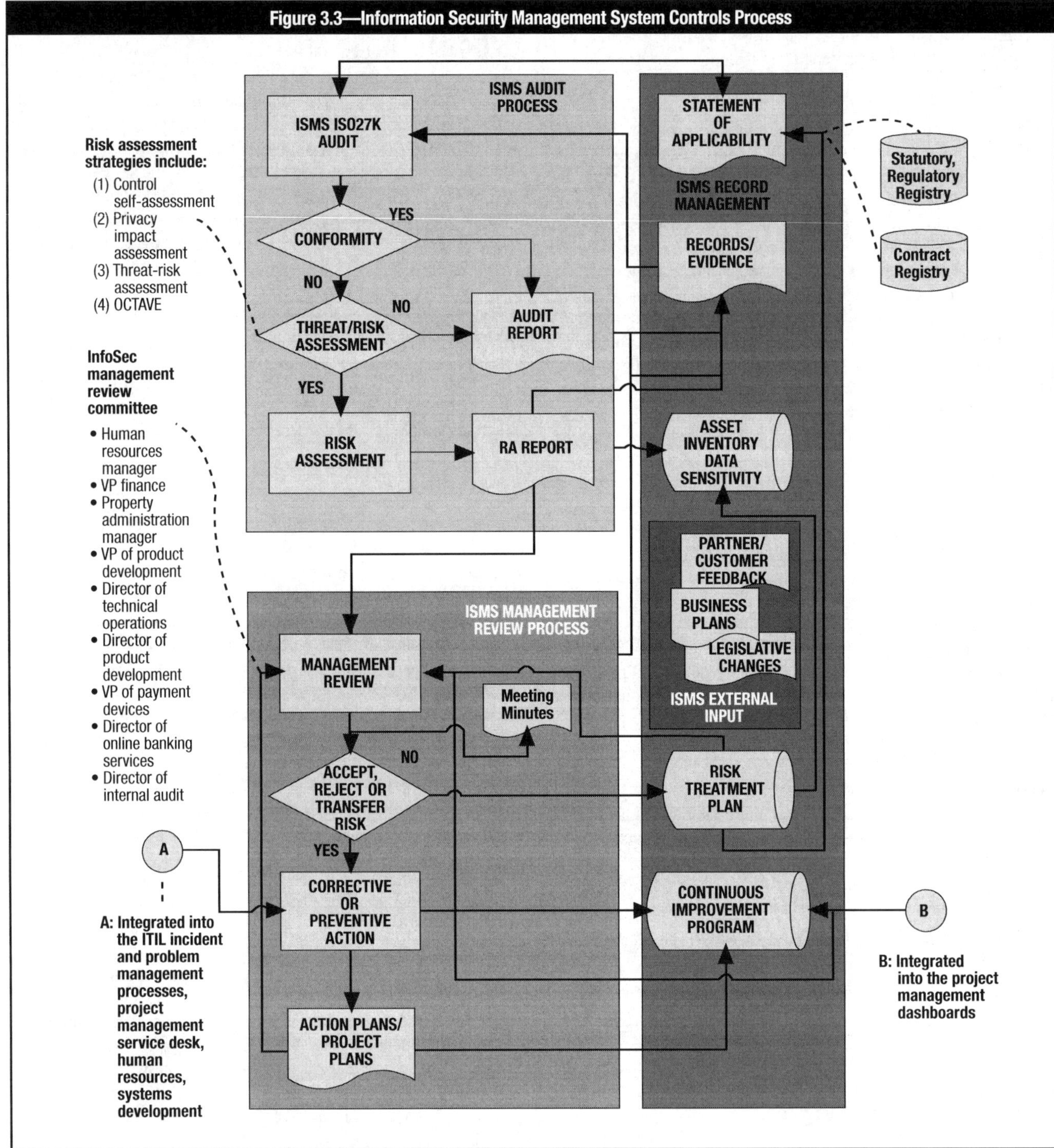

However, if a strategy is not developed and risk management objectives are not defined, there is a risk that the diverse elements that must be developed for the information security program will not be integrated or prioritized. In addition, there will be a lack of useful metrics and, over time, the results are likely to be less than optimal.

Much of an information security program development effort will involve designing controls that meet control objectives, then developing projects to implement, deploy and test the controls. One factor to consider is the ability of the organization to absorb new security activities. These activities would be initiated to address control weaknesses and meet new objectives. Consideration must be given to the extent that these activities are disruptive to other organizational activities. See section 3.12 Controls and Countermeasures.

3.7.2 DEVELOPING AN INFORMATION SECURITY PROGRAM ROAD MAP

An important skill to have in developing an information security program road map is the ability to thoroughly review the security level of existing data, applications, systems, facilities and processes. This will provide insight into the specific projects required to meet strategic objectives. Approaches to performing security reviews and evaluating the security program are discussed in section 3.11.4 Security Reviews and Audits.

An implementation road map can essentially be a high-level project plan (or set of project plans) or an architectural design that can serve the same purpose. Either can serve to define the steps necessary to achieve a particular objective of the program. The purpose is to have an overall view of the steps required as well as the sequence. In more complex projects, it can be a benefit to have both. A road map should include various milestones that will provide KGIs, indicate KPIs and define critical success factors (CSFs).

3.7.3 GAP ANALYSIS—BASIS FOR AN ACTION PLAN

Once the organizational roles and responsibilities seem appropriately established and inventory is taken of the required vs. existing technology and processes, an information security manager can identify where control objectives are not adequately supported by controls. The information security manager should work with the appropriate personnel to identify control points and assist in developing processes to monitor them. Those executing new processes and procedures should concentrate on KGIs and KPIs, frequently validating that control objectives are being met and progress toward control objectives achieves information security program goals. It is more important that the procedure for monitoring achievement of control objectives is established than it is that all processes are right on the first pass. It is this monitoring that, if effective, will provide a basis for the security program to evolve and mature.

3.8 INFORMATION SECURITY INFRASTRUCTURE AND ARCHITECTURE

Infrastructure refers to the underlying base or foundation on which information systems are deployed. Generally, infrastructure comprises the computing platforms, networks and middleware layers, and it supports a wide range of applications. In previous sections, the information security program has been presented as the foundation that enables security resources to be deployed. Infrastructure and security infrastructure refer to the same thing. When infrastructure is designed and implemented and is consistent with appropriate policies and standards, the infrastructure should essentially be secure.

3.8.1 ENTERPRISE INFORMATION SECURITY ARCHITECTURE

Considerable development of architectural approaches for security, as a part or subset of enterprise information architecture, has occurred during the past decade. There are few things as complex as the information systems in a large organization. These systems are often constructed without a comprehensive architecture or extensive design efforts. Information systems have traditionally evolved organically with bits and pieces added as needed. The result has been a lack of integration, haphazard security standardization, and a host of other weaknesses and vulnerabilities evident in most systems.

Enterprise information security architecture (EISA) was designed as an essential part of overall enterprise IT system design. As it evolved over time, EISA was developed as a stand-alone approach, although it must, by necessity, be a part of and consistent with the enterprise information architecture. Conceptually, the EISA objective is not just to manage security technology but to address the related elements of business structure, performance management and security processes as well.

The objectives of information architecture approaches include the following:

- Provide overarching structure, coherence and cohesiveness.
- Serve as a program development road map.
- Ensure strategic alignment between business and security.
- Support and enable achievement of business strategy.
- Implement security policies and strategy.
- Ensure traceability back to the business strategy, specific business requirements and key principles.
- Provide a level of abstraction independent of specific technologies and preferences.
- Establish a common language for information security within the organization.
- Allow many individual contributors to work together to achieve objectives.

Several architectural approaches have been developed for the enterprise, all of which include essential aspects of security and some of which deal exclusively with security, as discussed in section 1.11.2 Enterprise Information Security Architecture(s).

While a detailed discussion of each of these approaches is beyond the scope of this manual, it should be noted that these approaches fall into three basic categories: process approaches, frameworks and reference models. Frameworks such as COBIT, the Zachman framework, Sherwood Applied Business Security Architecture (SABSA) and The Open Group Architecture Framework (TOGAF) allow a great deal of flexibility in how each element of the architecture is developed. The essence of the frameworks is to describe the elements of architecture and how they must relate to each other. Process models are more directive in the processes used for the various elements. Reference models are a small-scale representation of the actual implementation. While the objectives of all the models are essentially the same, the approach varies widely.

In some cases, an organization has already adopted a standardized architectural approach that should be utilized to the extent possible. If no standard approach has been devised, the various methods mentioned in this manual should be evaluated for the most appropriate form, fit and function.

The TOGAF Architecture Development Method (ADM) phases, as shown in **figure 3.4**, include:

- Preliminary phase—Deals mainly with the definition of the architecture framework, as discussed earlier, as well as the architecture principles. In addition, the overall scope, constraints, objectives and assumptions are identified.
- Architecture vision—Deals with defining the vision and scope of the architecture and specific segments of work to be performed

- Business architecture—Addresses the description of the as-is business architecture domain, the development of the to-be business architecture and the gap analysis between the two
- Information systems architecture—Provides the description of the as-is and to-be data, applications domains and conducting the gap analyses
- Technology architecture—Deals with the description of the as-is and to-be technology domains and conducting a gap analysis
- Opportunities and solutions—Deals with the formulation of a high-level implementation and migration strategy to transform the as-is architectures into the to-be architecture
- Migration planning—Deals with formulation of a detailed implementation and migration road map, including the analysis of costs, benefits and risk
- Implementation governance—Ensures that the implementation projects conform to the defined architecture
- Architecture change management—Deals with keeping the architecture up to date and ensures that the architecture responds to the needs of the enterprise, as changes arise
- Requirements management—Ensures that the architecture projects are based on business requirements and the business requirements are validated against the architecture

Figure 3.4—TOGAF Architecture Development Cycle

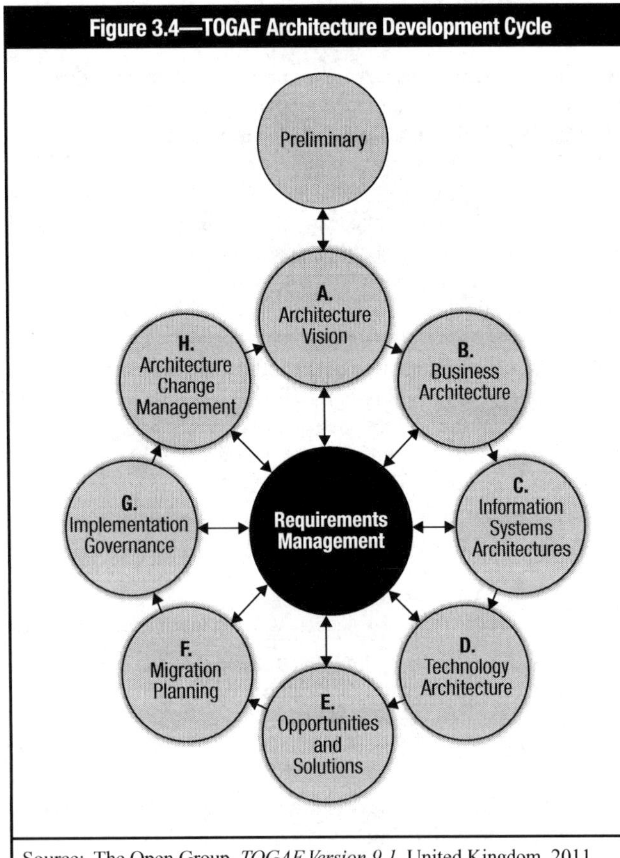

Source: The Open Group, *TOGAF Version 9.1*, United Kingdom, 2011

Enterprise Architecture Domains

There are four commonly accepted subsets of an overall enterprise architecture, as shown in **figure 3.5**, which TOGAF and COBIT are designed to support:

- A business (or business process) architecture defines the business strategy, governance, organization and key business processes.
- A data architecture describes the structure of an organization's logical and physical data assets and data management resources.
- An applications architecture provides a blueprint for the individual application systems to be deployed, their interactions and their relationships to the core business processes of the organization.
- A technology architecture describes the architectural principles, component relationships, and hardware and software infrastructure intended to support the deployment of core, mission-critical applications.

Whatever architectural approach is chosen, an effective approach must start with the enterprise business architecture. The subsequent designs, such as application, security and data architectures, function as subsets to ensure alignment with and support of business strategy and objectives.

Complexity of developing and implementing some approaches has proven to be a hindrance to adoption. For example, some notions of information security architecture frameworks include six layers, from contextual to physical. Like a building architecture, the highest or contextual level defined in SABSA and the Zachman frameworks is the "business" or utility layer, i.e., what the structure is to be used for. A theater is designed very differently from an office building because the buildings are used for different purposes. The design, or architecture, is tightly aligned with the purpose (i.e., linked to the business objective). The contextual architecture serves to define the relationships among various required business attributes. These include who, what, when, where and how, which drive the next layer, the conceptual layer, which integrates the architectural design concepts with the business requirements.

The next layer, the logical architecture, describes the same elements in terms of the relationships of logical elements. This is followed by a physical layer, which identifies the relationships among various security mechanisms that will execute the logical relationships and the component architecture consisting of the actual devices and their interconnections. Finally, there is the operational architecture, describing how security service delivery is organized.

While arguably comprehensive, organizations that have attempted to implement SABSA or Zachman frameworks have often found this framework complex and costly to develop and difficult to maintain as organizations experience rapid change and new business objectives are defined.

The important consideration, regardless of the approach, is that at the point of the development and implementation of an information security program, some form of high-level architecture or design should have been prepared. This is particularly the case for major implementations comprised of many parts and projects that must integrate well to achieve the program objectives. Although many organizations have not developed enterprise or security architectures, the adoption of an appropriate one can be essential to developing and implementing an effective information security program.

<table>
<tr><th colspan="4">Figure 3.5—Architecture Concerns of the Stakeholders</th></tr>
<tr><th>Architecture Views
May address user, planner and business management business concerns</th><th>Data Architecture Views
May address database designer administrator and system engineer concerns</th><th>Applications Architecture Views
May address system and software engineer concerns</th><th>Technology Architecture Views
May address acquires, operator, administrator and manager concerns</th></tr>
<tr><td>Business function view</td><td rowspan="4">Data entity view</td><td rowspan="4">Software engineering view</td><td rowspan="3">Networked computing/hardware view</td></tr>
<tr><td>Business services view</td></tr>
<tr><td>Business process view</td></tr>
<tr><td>Business information view</td><td rowspan="2">Communications engineering view</td></tr>
<tr><td>Business locations view</td><td rowspan="5">Data flow view (organisation data use)</td><td rowspan="5">Applications interoperability view</td></tr>
<tr><td>People view (organisation chart)</td><td rowspan="3">Processing view</td></tr>
<tr><td>Workflow view</td></tr>
<tr><td>Useability view</td></tr>
<tr><td>Business strategy and goals view</td><td rowspan="3">Cost view</td></tr>
<tr><td>Business objectives view</td><td rowspan="4">Logical data view</td><td rowspan="4">Software distribution view</td></tr>
<tr><td>Business rules view</td></tr>
<tr><td>Business events view</td><td rowspan="2">Standards view</td></tr>
<tr><td>Business performance view</td></tr>
<tr><td></td><td colspan="3">System engineering view</td></tr>
<tr><td colspan="4">Enterprise security view</td></tr>
<tr><td colspan="4">Enterprise manageability view</td></tr>
<tr><td colspan="4">Enterprise quality of service view</td></tr>
<tr><td colspan="4">Enterprise mobility view</td></tr>
<tr><td colspan="4">Source: ISACA, COBIT Mapping: Mapping of TOGAF 8.1 With COBIT® 4.0, USA, 2007, figure 7, page 24</td></tr>
</table>

3.8.2 OBJECTIVES OF INFORMATION SECURITY ARCHITECTURES

One of the key functions of architecture as a tool is to provide a framework within which complexity can be managed successfully. As the size and complexity of a project grows, many designers and design influences must all work as a team to create something that has the appearance of being created by a single design authority.

As the complexity of the business environment grows, many business processes and support functions must integrate seamlessly to provide effective services and management to the business, its customers and its partners. Architecture provides a means to manage that complexity.

Providing a Framework and Road Map

Architecture also acts as a road map for a collection of smaller projects and services that must be integrated into a single homogenous whole. It provides a framework within which many members of large design, delivery and support teams can work harmoniously and toward which tactical projects can be migrated.

Simplicity and Clarity Through Layering and Modularization

In the same way that conventional architecture defines the rules and standards for the design and construction of buildings, information systems architecture addresses these same issues for the design and construction of computers, communications networks and the distributed business systems that are required for the delivery of business services. Information systems architecture must, therefore, take account of:

- The goals that are to be achieved through the systems
- The environment in which the systems will be built and used
- The technical capabilities of the people to construct and operate the systems and their component subsystems

Business Focus Beyond the Technical Domain

Information systems architecture is concerned with much more than technical factors. It is concerned with what the enterprise wants to achieve and with the environmental factors that will influence those achievements. The word "enterprise" implies not just a large organization, but one in which all the parts of that organization exhibit a "joined up" quality and in which the organization is seen at the highest level as a single entity with an integrated mission and purpose.

In some organizations, this broad view of information systems architecture is not well understood. Technical factors are often the main influences on the architecture and, under these conditions, the architecture can fail to deliver what the business expects and needs.

Architecture and Control Objectives

Where security control objectives are considered, a systems architect can use combinations of technologies to provide control

points in a system's infrastructure. Combined with control activities and associated procedures, these control points may be used to ensure that policy compliance is preserved as new systems are deployed that use the infrastructure. For example, if a network is structured such that there is only one connection to the Internet, then all network traffic that is destined for the Internet must travel through that connection. This would allow technology to be deployed in one place that could inspect all documents destined for the Internet to ensure that the information contained in the document is authorized to be sent to an external entity. Often, no technology will be specified by the architecture; this leaves a wide range of design choices for control points that would inspect documents being sent to the Internet.

3.9 ARCHITECTURE IMPLEMENTATION

Some organizations have enough experience with combinations of technologies used for a specific purpose that they elevate architectural decisions to the policy level. That is, choices of combinations of technology used for a given purpose are mandated by policy, because certain combinations allow easy implementation of security features that accomplish specific control objectives. Often, architectural policies are warranted where potential damage from data exposure warrants redundant controls. Some examples of architecture policy domains are:

- Database management systems (to restrict application access)
- Telecommunications (to mitigate threats of phone fraud)
- Web application access

Example: Web application access is often protected from unauthorized access via user IDs and passwords. Yet inherent vulnerabilities in communicating on publicly available networks prompted security architects to require the use of transport layer security (TLS) on web servers that hosted applications that exchanged sensitive information with clients.

This configuration is often mandated by policy to ensure that traffic between client and server is encrypted to prevent the user ID and password, and the subsequent data flow, from being observed on the public network. However, such a policy does not provide any assurance that the client has not been compromised or the user password has not been stolen. Mitigation of this impersonation threat requires a technology control in addition to passwords and basic TLS encryption.

There are multiple alternatives that an application may employ to achieve the level of client identification necessary to mitigate the risk of impersonation. In financial applications such as online banking, regulators now require that banks utilize some form of multifactor authentication to provide additional assurance of the identity of the individual initiating the client request. Applying this requirement uniformly on all applications (making it part of the organization's security architecture) will allow the organization to mitigate the impersonation threat in a uniform, efficient and supportable manner.

A number of approaches to architecture for components of information systems exist (e.g., architectures for data and databases, servers, technical infrastructures, identity management). However, few organizations have developed an overall comprehensive enterprise security architecture, its management and its relationship to business objectives. This is somewhat analogous to a situation in which separate, unintegrated designs for aircraft wings, engines, navigational equipment, passenger seats, etc., exist, but there are no designs for the complete aircraft and how the various components fit together. The result would be unlikely to function well, if at all, and the outcome would be untrustworthy. Admittedly, information systems are designed to operate in a far more loosely coupled fashion, but the point is, nevertheless, relevant.

A number of architectural frameworks have been developed during the past decade to address this issue and they offer useful insights and approaches to dealing with many current design, management, implementation and monitoring issues. Some specific approaches are listed in section 3.8.1 Enterprise Information Security Architecture.

The framework approaches offer a great deal of flexibility in using a variety of standards and methods such as COBIT, ITIL and ISO/IEC 27001:2013. It is true that many of the approaches may be more sophisticated and complex than many organizations are prepared to deal with. However, with the growing reliance on increasingly complex systems, coupled with the escalating problems of manageability and security, organizations will eventually have little choice but to get organized.

For the information security manager, the approach may be helpful in developing long-term objectives or suggesting approaches to address current issues. Some organizations have adopted elements of the architectural approach, piecemeal, with the long-term objective of full implementation over time.

3.10 SECURITY PROGRAM MANAGEMENT AND ADMINISTRATIVE ACTIVITIES

Information security program management includes directing, overseeing and monitoring activities related to information security in support of organizational objectives. Management is the process of achieving the objectives of the business organization by bringing together human, physical and financial resources in an optimum combination with process and technology and making the best decision for the organization while taking into consideration its operating environment.

In the typical organization, information security program management includes short- and long-range planning and day-to-day administration in addition to directing various projects and initiatives. It also includes a variety of risk management activities and incident management and response functions and a variety of oversight and monitoring functions. Program management typically includes aspects of governance whether in terms of policy and standards development or procedural controls and general rules of use for end users. As with other aspects of information security, governance cannot remain static in a changing risk environment and must evolve to be effective.

The information security manager has the responsibility for managing information security program activities to achieve the outcomes listed in section 3.1.2 Outcomes of Information Security Program Management. It is the responsibility of senior management to support those objectives and provide adequate resources and authority to ensure that objectives are achieved. In addition, the roles and responsibilities of other assurance providers must be clearly defined to prevent gaps in protection among them.

As the scope of the information security manager's responsibilities continues to expand, there is the risk that important security elements may be overlooked or neglected. It may be useful for the information security manager to consider the following checklist for a comprehensive, well-managed security program:
- A security strategy intrinsically linked with business objectives that has senior management acceptance and support
- Security policy and supporting standards that are complete and consistent with strategy
- Complete and accurate security procedures for all important operations
- Clear assignment of roles and responsibilities
- Established method to ensure continued alignment with business goals and objectives such as a security steering committee
- Information assets that have been identified and classified by criticality and sensitivity
- Security architecture that is complete and consistent with strategy, and in line with business objectives
- Effective controls that have been well-designed, implemented and maintained
- Effective monitoring processes in place
- Tested and functional incident and emergency response capabilities
- Tested business continuity/disaster recovery plans
- Appropriate information security involvement in change management, SDLC and project management processes
- Established processes to ensure that risk is properly identified, evaluated, communicated and managed
- Established security awareness training for all users
- Established activities that create and sustain a corporate culture that values information security
- Established processes to maintain awareness of current and emerging regulatory and legal issues
- Effective integration with procurement and third-party management processes
- Resolution of noncompliance issues and other variances in a timely manner
- Processes to ensure ongoing interaction with business process owners
- Business-supported processes for risk and business impact assessments, development of risk mitigation strategies, and enforcement of policy and regulatory compliance
- Established operational, tactical and strategic metrics that monitor utilization and effectiveness of security resources
- Established methods for knowledge capture and dissemination
- Effective communication and integration with other organizational assurance providers

Note: This list is not comprehensive, nor will it fit all organizations. It is simply meant to demonstrate the range of activities and established practices that would constitute a mature security program. This list must be tailored for the size and nature of each organization.

Program Administration

Administration of a security program typically involves a series of repetitive functions similar to those required in other organizational units. There will be differences in administration of program development projects and the ongoing administration of the operations aspects of the program, which need to be considered.

Ongoing administration might include such tasks as:
- Personnel performance, time tracking and other record keeping
- Resource utilization
- Purchasing and/or acquisition
- Inventory management
- Project monitoring and tracking
- Awareness program development
- Budgeting, financial management and asset control
- Business case development and financial analysis
- HR administration and personnel management
- Project and program management
- Operations and service delivery management
- Implementation and administration of metrics and reporting
- Information technology development life cycle management

There may be a number of technical administrative and operational requirements as well. These include:
- Cryptographic key management
- Log reviews and monitoring
- Change request review and oversight
- Configuration, patch, and other life cycle management reviews and oversight
- Vulnerability scanning
- Threat monitoring
- Compliance monitoring
- Penetration testing

An effective information security manager should be familiar with existing frameworks and major international standards for IT and security management (e.g., COBIT, ISO/IEC 27001 and 27002) and be able to extract relevant elements to utilize for the management approach best suited to the organization. Some may prove to be a better fit than others, depending on organizational structure, culture, available resources, business sector, etc.

The information security manager has many responsibilities, and one may be as a facilitator to help resolve competing objectives between security and performance. As an active facilitator, the information security manager gains senior management support and organizational acceptance and is likely to achieve greater compliance with information security program policies, standards and procedures. Through a facilitative approach, the information security manager can work with the different departments to discuss information security risk and suggest solutions that address both security requirements and minimizing the impact on business activities. Through this consultative role, the information security manager also needs to ensure that the organization's life cycle processes incorporate information security. By working in a consultative role, the information security manager can facilitate the enterprise's information security program while staying informed about the organization's activities that may impact information security.

3.10.1 PERSONNEL, ROLES, SKILLS AND CULTURE

The information security manager allocates personnel resources based on the technical and administrative skills required to effectively operate the program. Staff members include security engineers, QA and testing specialists, access administrators, project managers, compliance liaisons, security architects, awareness coordinators, auditors, and policy specialists. The information security manager should develop positions in accordance with specific program needs, even merging responsibilities of multiple roles into one personnel position for small organizations.

The information security manager must ensure that personnel within the security organization as well as other responsible organizations maintain the appropriate skills needed to carry out program functions. Each organization's skill requirements will vary, generally revolving around the existing information systems and security technologies implemented.

Personnel requirements for information security program development differ after the program is implemented. For example, the personnel involved in the construction of a security program likely differ from the personnel that will administer systems once they are functioning normally. Skills that are only rarely needed are best acquired through engagement of service providers such as integrators or consulting firms. When faced with the need for a specialized skill, the information security manager should analyze the cost, timing and intellectual capital implications of hiring or training staff vs. using an external service provider. Additionally, in some organizations there may be a need for background checks, especially if classified, confidential or highly sensitive information is involved with the job. The HR department should be involved in and conduct background checks because, in some jurisdictions, there are privacy regulations that limit the extent of investigation.

Project management is a normal activity in managing the development of a security program. Larger organizations usually have a project management office (PMO), which may be available to the information security manager to assist in development and implementation projects. Existing functions within the organization should be used whenever possible both to maximize organizational involvement in the program and leverage existing capabilities.

Roles

A role is a designation assigned to an individual by virtue of a job function or other label. A responsibility is a description of some procedure or function related to the role that someone is accountable to perform. Roles are important to information security because they allow responsibilities and/or access rights to be assigned based on the fact that an individual performs a function rather than assigning them to individual people. Because there are typically fewer roles than staff members and the roles change less often, administrative costs are reduced. In addition, roles can be assigned during the onboarding process by HR, which further streamlines the process.

RACI (responsible, accountable, consulted, informed) charts can be used effectively in defining the various roles associated with aspects of developing an information security program. Clear designation of roles, responsibilities and accountability is necessary to ensure effective implementation and management.

See section 1.3 Roles and Responsibilities for more information.

Skills

Skills are the training, expertise and experience held by the personnel in a given job function. It is important to understand the proficiencies of available personnel to ensure that they map to competencies required for program implementation. Specific skills needed for program implementation can be acquired through training or utilizing external resources. External resources such as consultants are often a more cost-effective choice for skills required for only a short time and/or for specific projects.

Once it has been agreed that certain personnel will have specific information security responsibilities, formal employment agreements should be established that reference those responsibilities, and these must be considered when screening applicants for employment.

Culture

Culture represents organizational behavior, methods for navigating and influencing the organization's formal and informal structures to get work done, attitudes, norms, level of teamwork, existence or lack of turf issues, and geographic dispersion. Culture is impacted by the individual backgrounds, work ethics, values, past experiences, individual filters/blind spots and perceptions of life that individuals bring to the workplace. Every organization has a culture, whether it has been purposely designed or simply emerged over time as a reflection of the leadership.

While information security primarily involves logical and analytical activities, building relationships, fostering teamwork and influencing organizational attitudes toward a positive security culture rely more on good interpersonal skills. The astute information security program manager recognizes that developing both types of skills is essential to being an effective manager.

Building a security-aware culture depends on individuals in their respective roles performing their jobs in a way that protects information assets. Each person, no matter what level or role within the organization, should be able to articulate how information security relates to his/her role. For this to happen, the security manager must plan communications, participate in committees and projects, and provide individual attention to the end users' or managers' needs. The security department should be able to answer "What is in it for me?" or "Why should I care?" for every person in the organization. Once these questions have been answered, effective communications can be tailored to these messages.

Some outcomes that may indicate a successful security culture are: The information security department is brought into projects at the appropriate times, end users know how to identify and report incidents, the organization can identify the security manager, and people know their role in protecting the information assets of the organization and integrating information security into their daily practices.

3.10.2 SECURITY AWARENESS TRAINING AND EDUCATION

Risk that is inherent in using computing systems cannot be addressed purely through technical security mechanisms. An active security awareness program can greatly reduce risk by addressing the behavioral element of security through education and consistent application of awareness techniques. Security awareness programs should focus on common user security concerns such as password selection, appropriate use of computing resources, email and web browsing safety, and social engineering, and the programs should be tailored to specific groups. In addition, users are the front line for the detection of threats that may not be detectable by automated means (e.g., fraudulent activity and social engineering). Employees should be educated on recognizing and escalating such events to enhance loss prevention.

An important aspect of ensuring compliance with the information security program is the education and awareness of the organization regarding the importance of the program. In addition to the need for information security, all personnel must be trained in their specific responsibilities that are related to information security.

Particular attention must be paid to those job functions that require virtually unlimited data access. People whose job is to transfer data may have access to data in most systems, and those doing performance tuning can change most operating system configurations. People whose job is to schedule batch jobs have the authority to run most system jobs applications. Programmers have access to change application code. These functions are not typically managed by an information security manager. Although it is possible to set up elaborate monitoring controls, it is not technically feasible or financially prudent for an information security manager to provide oversight adequate to ensure that all data transfer jobs that transmit reports send them only to appropriately authorized recipients. Although the information security manager can ensure that there is clear policy, develop applicable standards and assist in process coordination, management in all areas must understand the security requirements and assist in providing oversight.

Employee awareness should start from the point of joining the organization (e.g., through induction training) and continue regularly. Techniques for delivery need to vary to prevent them from becoming stale or boring and may also need to be incorporated into other organizational training programs.

Security awareness programs should consist of training (often administered online); simple quizzes to gauge retention of training concepts; security awareness reminders such as posters, newsletters or screen savers; and a regular schedule of refresher training. In larger organizations, there may be a large enough population of middle and senior management to warrant special management-level training on information security awareness and operations issues.

All employees of an organization and, where relevant, third-party users must receive appropriate training and regular updates on the importance of security policies, standards and procedures in the organization. This includes security requirements, legal responsibilities and business controls, as well as training in the correct use of information processing facilities (e.g., login procedures, use of software packages). For new employees, this should be a part of new employee orientation and, therefore, occur before access to information or services is granted.

The information security manager should take a methodical approach to developing and implementing the education and awareness program, and consider such aspects as:

- Who is the intended audience (senior management, business managers, IT staff, end users)?
- What is the intended message (policies, procedures, recent events)?
- What is the intended result (improved policy compliance, behavioral change, better practices)?
- What communication method will be used (computer-based training [CBT], all-hands meeting, intranet, newsletters, etc.)?
- What is the organizational structure and culture?

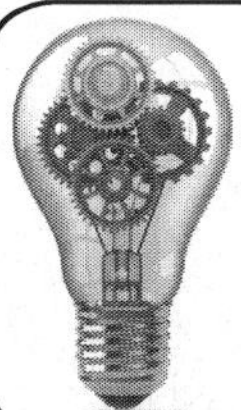

In Practice: Examine your organization's existing security awareness program. Using the questions above as a guide, identify any areas of possible improvement or see if you can successfully answer the questions based on the information in your program.

A number of different mechanisms available for raising information security awareness include:

- Computer-based security awareness and training programs
- Email reminders and security tips
- Written security policies and procedures (and updates)
- Nondisclosure statements signed by the employee
- Use of different media in promulgating security (e.g., company newsletter, web page, videos, posters, login reminders)
- Visible enforcement of security rules
- Simulated security incidents for improving security
- Rewarding employees who report suspicious events
- Job descriptions
- Performance reviews

3.10.3 GENERAL RULES OF USE/ACCEPTABLE USE POLICY

While specific procedures provide the detailed steps required for many functions at the operational level, there is still a large group of users that may benefit from a user-friendly summary of what they should and should not do to comply with policy. An effective way of assisting these general users in understanding security-related responsibilities is the development of an acceptable use policy. This policy can detail, in everyday terms and a straightforward and concise manner, the obligations and responsibilities of all users. It is then necessary to effectively communicate the use policy to all users and ensure it is read and understood. The use policy should be provided to all new personnel who will have access to information assets, regardless of employment status.

Typically, these rules of use for all personnel include the policy and standards for access control, classification, marking and handling of documents and information, reporting requirements, and disclosure constraints. They may also include rules on email and Internet usage. The rules of use provide a general security baseline for the entire organization. It is often necessary to provide supplemental or additional information to specific groups in the organization, consistent with their responsibilities.

3.10.4 ETHICS

Many organizations have implemented ethics training to provide guidance on what the organization considers legal and appropriate behavior. This approach is most common when individuals are required to engage in activities of a particularly sensitive nature such as monitoring user activities, penetration testing, and access to sensitive personal data. Information security personnel must be sensitive to potential conflicts of interest or activities that may be perceived in a manner detrimental to the organization.

Codes of ethics and conduct should be reviewed and acknowledged by each employee involved in information security management and other applicable duties. The signed acceptance of the code should be kept as a part of employee records.

3.10.5 DOCUMENTATION

Oversight of the creation and maintenance of security-related documentation is an important administrative component of an effective information security program. Documents that commonly pertain to the program include:

- Policies, standards, procedures and guidelines
- Technical diagrams of infrastructure and architectures, applications, and data flows
- Training and awareness documentation
- Risk analyses, recommendations and related documentation
- Security system designs, configuration policies and maintenance documentation
- Operational records such as shift reports and incident tracking reports
- Operational procedures and process flows
- Organizational documentation such as organization charts, staff performance objectives and RACI models

Each document should be assigned an owner who is responsible for updating documentation (or templates in the case of operational records). Changes should be made under the recommendations of management or the security steering committee; the person or group should also approve changes prior to their distribution. The owner is also responsible for ensuring that access to documentation is appropriate, controlled and auditable.

The information security manager should ensure that enablers (**figure 3.6**) are available that address creation, approval, change, maintenance, controlled distribution and retirement of documentation. In many cases, these services are provided as part of a larger enterprise document management system, an ideal situation considering that the information security manager does not necessarily maintain custodianship of all security-related documentation. The information security manager should also ensure that sensitive technical and operational documents are protected by controls and practices equal to or stronger than those implemented for other sensitive corporate information assets. Security documentation must also follow the organizational standards for appropriate classification and labeling.

Figure 3.6—COBIT 5 Enterprise Enablers

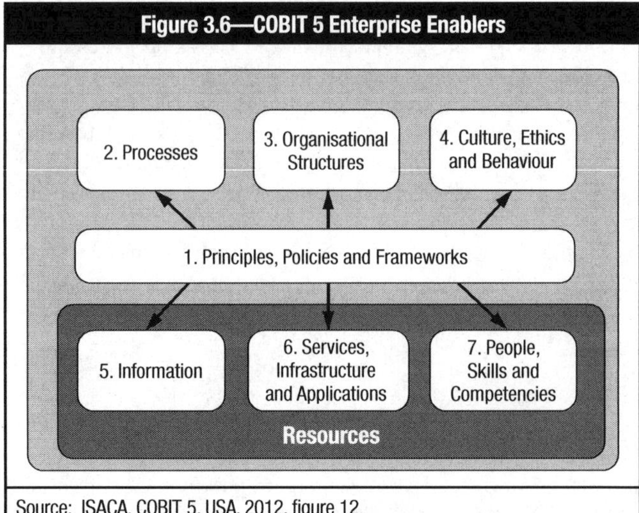

Source: ISACA, COBIT 5, USA, 2012, figure 12

A number of documents will be required to design, initiate and implement an information security program. Documentation requirements will change at various stages and it is important to make sure they are current. Standards and procedures must change as required to address changes in risk, technology, business activities or baselines, but must remain consistent with published policies, which typically change with far less frequency.

Version control is essential to ensure that all parties are operating from the same document revisions, and a reliable notification and publication process is required.

Some of the documentation required includes:

- Program objectives
- Road maps
- Business cases
- Required resources
- Controls
- Budgets
- Systems designs/architectures
- Policies, standards, procedures, guidelines
- Project plan milestones, time lines
- KGIs, KPIs, CSFs, other metrics
- Training and awareness requirements
- Business impact and risk analysis
- SLAs
- Severity criteria
- Declaration criteria

Document Maintenance

Documentation typically needs to be updated as the information security program is implemented and as the organization evolves. The information security manager must implement procedures for adding, modifying and, in some cases, retiring information security policies, standards, procedures and other documentation.

An important consideration for all documentation activities is to ensure that an appropriate version control process is in place so that all recipients are using current documentation. Processes should also be developed to retire documentation that has been superseded by later releases. Automated systems using a single source for all documentation help to ensure that only current policies, standards and procedures are being used. Also, all documentation must have standardized markings, including effective release date, version, owner and classification.

A process for change proposals can be based on policy reviews or initiated as stakeholders recognize the need for a policy change. The information security manager should track proposed changes to policies for review in the appropriate forums.

Modification of standards will probably occur more frequently than modification of policies. These changes are often driven by changes in technology such as availability of new security capabilities, changes in risk, evolution of the technical infrastructure and requirements of new business initiatives. Proposed standards should be reviewed by the steering committee and/or by departments impacted by the proposed standards. The opportunity to provide input and suggested modification should be afforded, to optimize cooperation and compliance. Standards should be reviewed periodically or as warranted by environmental changes, changing risk, request of stakeholders or audit determination of insufficiency. Changes to standards should be managed in a similar manner to policy changes and should also include technical or operational risk analysis relative to the proposed change in standards.

Modifications to policies or standards should trigger procedures to modify compliance monitoring tools and processes. Changes should be periodically shared with auditors and compliance personnel who are not already involved in the change management process to ensure awareness and acceptance of changes prior to auditing activities.

Changes in standards will necessitate a review of procedures that may require modification to be in compliance. Other documentation such as acceptable use policies, guidelines, and QA and procurement processes, may also require modification.

3.10.6 PROGRAM DEVELOPMENT AND PROJECT MANAGEMENT

Information security programs are rarely static and must undergo ongoing development to meet changing conditions and risk. The basis for development is the strategy to achieve defined objectives. These objectives include achieving the desired maturity levels and managing risk to acceptable levels while effectively supporting business operations.

The information security gap analysis, described in section 3.7.3 Gap Analysis—Basis for an Action Plan, typically identifies the need for projects that result in improvements to the information security program. The processes described in section 3.11.11 Integration With IT Processes should also be reviewed to identify projects that can fill gaps between the existing organization and the program as envisioned. Many of these projects will be technology implementations or reconfigurations that will make existing technology more secure. Each project should have time, budget and a measurable result. Each must make the environment more secure without causing control weaknesses in other areas.

An information security manager should prioritize the portfolio of projects in such a way that those that overlap are not delayed by each other, resources are appropriately allocated, and the results are smoothly integrated into or transitioned from existing operations. The information security manager should employ generally accepted project management techniques such as setting goals, measuring progress, tracking deadlines, and assigning responsibilities in a controlled and repeatable manner. This helps ensure that the security program's design and implementation will be successful.

3.10.7 RISK MANAGEMENT

Virtually all aspects of program management serve to manage risk to acceptable levels. Because the risk landscape changes continuously, it is essential for information security to change and adapt as required to ensure the business is capable of dealing effectively with current conditions.

Regardless of the effectiveness of information security, few organizations will escape the effect of security incidents. An essential aspect of program management is to ensure that the organization can respond effectively to security incidents that disrupt business operations.

Risk Management Responsibilities

Managing information risk is an integral part of the information security manager's ongoing responsibilities and a primary requirement in both program development and management. The information security manager must have a good understanding and develop the requisite skills regarding evaluation and management of risk, as detailed in chapter 2. This includes:

- Knowledge of program development life cycle risk
- Knowledge of program management risk
- Knowledge of methods for assessing the vulnerabilities in technical and operational environments
- Ability to analyze exposures, the general threat environment and threats specific to the information security manager's organization
- Knowledge of risk analysis approaches including quantitative and qualitative methods
- Knowledge of risk management processes including mitigation, elimination, transfer and informed acceptance
- Ability to understand and assess potential impacts if risk is exploited
- Knowledge of methods for tracking, documenting and communicating risk and impact issues

3.10.8 BUSINESS CASE DEVELOPMENT

In developing the information security program, the best options for major projects or initiatives will often benefit from developing a persuasive business case, as described in section 1.4.2 Information Security Roles and Responsibilities. The business case should provide a clear statement of the value proposition (or cost benefit) based on the identified needs of the organization. The need from an information security perspective must identify the risk that is to be addressed and the potential impacts if the

project is not undertaken as well as the reduction in risk and potential impacts as a result of the effort. To be effective, the business case must make it evident that there is a significant return on the proposed investment, the project is feasible and practical, and impact on productivity is acceptable.

3.10.9 PROGRAM BUDGETING

Budgeting is an essential part of information security program management and can have a significant impact on the program's success. Effective preparation and defense of a budget can mean the difference between having or not having sufficient staff and other resources to accomplish the objectives of the information security program.

As with many business activities, self-education and advance preparation are key factors in successfully navigating this frequently challenging process. Well before the budget cycle begins, the information security manager should ensure familiarity with the budgeting process and methods used by the organization. It will also be important to consider the timing of the various stages of the organization's budget cycle.

Another key consideration before beginning the budgeting process is the information security strategy. All budget expenditures for information security should be derived from and supported by the information security strategy. Ideally, detailed elements of the security strategy are laid out in a security road map, as discussed in this section. Having consensus on the strategy objectives and the strategy communicated and approved by management before entering into the budgeting process are key elements in a successful budget proposal.

Elements of an Information Security Program Budget

Many costs associated with an information security program are fairly straightforward. Expenses such as personnel (salaries, training, etc.), basic hardware and software, and subscription services fall into this category. In addition to the typical program start-up and yearly operational costs, short- and long-term projects represent various objectives on the security road map. These projects consume resources over the course of months (or years) and must be accounted for in the budget as accurately as possible. The information security manager should collaborate with the organization's PMO and the appropriate subject matter experts (SMEs) to help estimate costs for projects that will start within the fiscal year. Elements of each project that should be considered include:

- Employee time
- Contractor and consultant fees
- Equipment (hardware, software) costs
- Space requirements (data center rack space, etc.)
- Testing resources (personnel, system time, etc.)
- Training costs (staff, users, etc.)
- Travel
- Creation of supporting documentation
- Ongoing maintenance
- Contingencies for unexpected costs

While it will be impossible to be completely accurate, engaging the appropriate SMEs and skilled project managers can be invaluable in arriving at a fairly accurate estimate.

Finally, it is important to keep in mind that some aspects of an information security program are not entirely predictable and can incur unanticipated costs—particularly in the area of incident response. These costs are often the result of the need for external resources to assist with a security event that exceeds the skills or bandwidth of the organization's staff. One approach to budget for these situations is to use historical data of incidents and remediation costs of previous security events that required unbudgeted external resources.

Even if the organization has recently initiated its security program, there may have been prior events that necessitated external resources. If the planned program does not anticipate having these skills on staff, it is likely external resources will be needed again in the future. For example, if during the past three years the organization has engaged external forensic consultants to assist with security events because the required skill sets were not on staff, the security manager should consider adding a line item in the budget to account for this situation. Costs can be estimated based on the average costs over the previous years as a starting point. If this information is not available, industry statistics from peer organizations may be useful as an indicator of risk and remediation costs.

3.10.10 INFORMATION SECURITY PROBLEM MANAGEMENT PRACTICES

In addition to crisis or event management practices, the information security manager needs to understand the various aspects of an effective problem management approach. Problem management is focused on ascertaining the root causes of issues. This typically requires a systematic approach to understanding the various aspects of the issue, defining the problem, designing an action program, assigning responsibility and assigning due dates for resolution. A reporting process should also be implemented for tracking the results and ensuring that the problem is resolved.

As the information systems environment continually undergoes changes via updates and additions, it is not unusual for the security controls in place to occasionally develop a problem and not work as intended. It is at this point that the information security manager must identify the problem and assign a priority to it.

The information security manager should also be familiar with mitigating controls that may have to be employed if the primary security control fails. Rather than allowing the security vulnerability to put the organization at risk, it may be necessary for the information security manager to take alternative actions to protect the information resources until the problem is resolved. For example, if a firewall fails, the information security manager may decide to disconnect the system from the outside until the firewall problem is corrected. While this would protect the information resources from outside risk, it would likely affect the organization's ability to perform business. Therefore, it is important that specific authority and limits are established by management.

3.10.11 VENDOR MANAGEMENT

An ongoing management and administrative responsibility for the typical information security manager is the oversight and monitoring of external providers of hardware and software, general supplies, and various services. This is to provide assurance that risk introduced in acquisition processes, implementation and service delivery is managed appropriately. Other organizational units involved in aspects of managing vendors include legal, finance, procurement and some business units.

Security service providers are a common feature of many information security programs and often the most cost-effective approach to various monitoring and administrative functions. They can provide the information security manager with specialist skills as needed, longer-term staff augmentation while recruiting for open positions and even offloading of routine tasks. Outsourced security service providers can deliver a range of services such as assessment and audit, engineering, operational support, security architecture and design, advisory services, and forensics support.

Security service providers can free up internal resources to focus on projects or operations where preservation of intellectual capital is at a premium. The use of external security resources can also provide an objective, fresh perspective on the information security program. If the information security manager uses an outsourced security provider, the capabilities and approaches the vendor takes should align with the organization's information security program.

The use of external parties to provide security-related functions usually creates risk that must be managed by the information security manager. Issues such as financial viability, quality of service, adequate staffing, adherence to the organization's security policies and right to audit must be addressed.

3.10.12 PROGRAM MANAGEMENT EVALUATION

Certain situations call for the information security manager to assess the current state of an existing information security program (e.g., if promoted or hired into an existing CSO role). It is also important for the information security manager to periodically reevaluate the effectiveness of the program relative to the changes in organizational demands, environment and constraints. The results of such analysis should be shared with the information security steering committee or other stakeholders for review and development of needed program modifications.

While the information security manager must determine the most appropriate scope for assessing current state, the following section outlines several critical areas for evaluation.

Program Objectives

The information security manager must evaluate the documented security objectives established for the program. Key considerations include:

- Has an information security strategy and development road map been developed?
- Have criteria for acceptable risk and impact been determined?
- Do complete and current policies, standards and procedures exist?
- Are program goals aligned with governance objectives?
- Are objectives measurable, realistic and associated with specific time lines?
- Do program objectives align with organizational goals, initiatives, compliance needs and operational environment?
- Is there consensus on program objectives? Were objectives developed collaboratively?
- Have metrics been implemented to measure program objective success and shortfalls?
- Are there regular management reviews of objectives and accomplishments?

Compliance Requirements

Alignment with and fulfillment of compliance requirements are among the most visible indications of security management status. Because many standards establish program management requirements, the information security manager must evaluate the management program itself—framework and components—against compulsory and/or voluntary compliance standards. Key considerations include:

- Has management determined the level of compliance the organization will undertake as well as time lines and milestones?
- Is there facilitation of close communication between compliance and information security groups? Are information security compliance requirements clearly defined?
- Does the information security program specifically integrate compliance requirements into policies, standards, procedures, operations and success metrics?
- Do the program's technical, operational and managerial components align with the components required by regulatory standards?
- What have been the results of recent audit and compliance reviews of the information security program?
- Are program compliance deficiencies tracked, reported and addressed timely?
- Are compliance management technologies used to increase the efficiency of fulfilling security compliance demands?

Program Management

Evaluation of program management components reveals the extent of management support and the overall depth of the existing program. Very technical, tactically driven security programs tend to implement fewer management components, whereas strategic programs driven by standards, compliance and governance implement a more comprehensive set of management activities to ensure that requirements are established and fulfilled. Considerations of program management components include:

- Is there thorough documentation of the program itself? Have key policies, standards and procedures been reduced to accessible operating guidelines and distributed to responsible parties?
- Do responsible individuals understand their roles and responsibilities?
- Are roles and responsibilities defined for members of senior management, boards, etc.? Do these organizations understand and engage in their responsibilities?
- Are responsibilities for information security represented in business managers' individual objectives and part of their individual performance rating?
- Are policies and standards complete, formally approved and distributed?
- Are business unit managers involved in guiding and supporting information security program activities? Is there a formal steering committee?
- How is the program positioned within the organization? To whom is the program accountable? Does this positioning impart

an appropriate level of authority and visibility for the objectives that the program must fulfill?
- Does the program implement effective administration functions (e.g., budgeting, financial management, HR management, knowledge management)?
- Are meaningful metrics used to evaluate program performance? Are these metrics regularly collected and reported?
- Are there forums and mechanisms for regular management oversight of program activities? Does management regularly reassess program effectiveness?

Security Operations Management

The information security manager must evaluate the effectiveness with which the information security program implements security operational activities, both within the security organization and in other organizational units. Some key considerations include:
- Are security requirements and processes included in security, technology and business unit standard operating procedures (SOPs)?
- Do security-related SOPs provide for accountability, process visibility and management oversight?
- Are there documented SOPs for security-related activities such as configuration management, access management, security systems maintenance, event analysis and incident response?
- Is there a schedule of regularly performed procedures (e.g., technical configuration review)? Does the program provide for records of scheduled activities?
- Is there segregation of duties (SoD) among system implementers, security administrators and compliance personnel?
- Does the program provide for effective operational, tactical and strategic metrics reporting that provides management with needed information for oversight? Are other oversight mechanisms in place?
- Does management regularly review security operations? Is there a forum for operational issues to be escalated to management for resolution?

Technical Security Management

Management of the technical security environment is critical to ensuring that information processing systems and security mechanisms are implemented effectively. In addition to evaluating the current technical environment itself, the information security manager should consider the following issues regarding management of technical security concerns:
- Are there technical standards for the security configuration of individual network, system, application and other technology components?
- Do standards exist that address architectural security issues such as topology, communication protocols and compartmentalization of critical systems?
- Do standards support and enforce high-level policies and requirements? Are standards a collaborative effort among technology, operations and security staff?
- Are technical standards uniformly implemented? Do procedures exist to regularly evaluate and report on compliance with technical standards? Is there a formal process to manage exceptions?
- Is there continuous monitoring of key controls? Do controls provide notification on failure?
- Is separation of development, test and production environments enforced?
- Do systems enforce SoD, especially where high levels of administrative access are concerned?
- Is there reliable and comprehensive visibility (logging) into system activities, configurations, accessibility and security-related events? Is this visibility continual or intermittent?
- Are proper decommissioning processes in place to prevent data leakage?

Resource Levels

The information security manager must assess the level of financial, human and technical resources allocated to the program. Areas of deficiency must be identified and escalated to senior management and/or the steering committee. Considerations include the following areas:
- Financial resources:
  - What is the current funding level for the program?
  - Is a comprehensive capital and operating budget maintained?
  - Do financial allocations align with program budget expectations?
  - Are there linkages between resource allocation and business objectives?
  - Are functions within the program prioritized in terms of financial allocation?
  - Which functions are likely to suffer from underfunding?
- HR:
  - Does the program implement a workload management methodology?
  - What is the current staffing level for the program?
  - Are existing resources fully utilized in terms of time and skills?
  - Are existing resources adequately skilled for the roles they are in?
  - Are there low-value tasks that other resources could be leveraged to complete?
  - What other human resources (e.g., IT staff) is the program dependent on to operate effectively? Is information security a formal part of these resources' job descriptions and activity plans?
- Technical resources:
  - What technologies currently support information security program objectives?
  - Is the capacity of supporting technologies sufficient to support current demands? Will these technologies scale to meet future needs?
  - Does the program account for maintenance, administration and eventual replacement of supporting technologies?
  - Are there other technologies that could make the program more efficient or effective?

3.10.13 PLAN-DO-CHECK-ACT

The information security program is based on the effective, efficient management of controls designed and implemented to treat or mitigate threats, risk, vulnerabilities and impacts. The unique dependency on the effective, efficient management of a business process such as information security lends itself to the concepts and methodologies encompassed within the total quality management (TQM) system. TQM is based on cycle made up of four primary processes, plan-do-check-act (PDCA), as depicted in **figure 3.7**.

The TQM approach, combined with a governance methodology that focuses on strategic program alignment with organizational goals, will provide the information security manager with tools that he/she can use to implement and maintain a highly effective, efficient program. As described in chapter 1, the basic elements of a governance methodology include a strategic vision, objectives, KGIs, CSFs, KPIs, and key actions or tactical and annual action plans. These elements are defined as follows and depicted in **figure 3.8:**

Figure 3.7—PDCA Methodology

The PDCA methodology is an iterative process model.

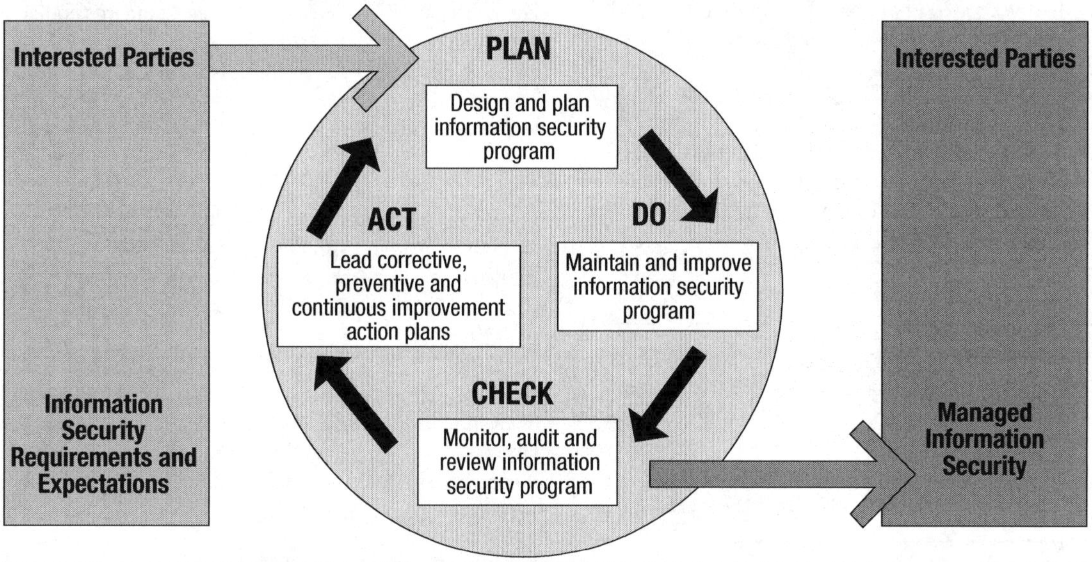

Interested Parties

Information Security Requirements and Expectations

PLAN

Design and plan information security program

DO

Maintain and improve information security program

CHECK

Monitor, audit and review information security program

ACT

Lead corrective, preventive and continuous improvement action plans

Interested Parties

Managed Information Security

PLAN	Design, plan and initiate the information security program. These activities include creating a strategy; socialization concepts; and policies, goals, objectives and practices as necessary to manage risk. The activities include creating a strategy; socialization concepts; and policies, goals, objectives and practices as necessary to manage risk."
DO	Execute and control the information security strategy including the integration into organizational practices.
CHECK	Facilitate semiannual audits to determine conformance to the statement of applicability and identify opportunities for improvement. Wherever appropriate, develop and integrate performance matrices which support information security program goals and objectives.
ACT	Upon the discovery of nonconformities and/or opportunities, create and track corrective, preventive and continuous improvement action plans. Present findings from internal/external audit and risk assessments to the management review committee for decisions regarding the acceptance, rejection or transfer of risk and the commitment of resources and capital to facilitate subsequent efforts.

Figure 3.8—Strategic Objectives, CSFs, KPIs, Key Actions

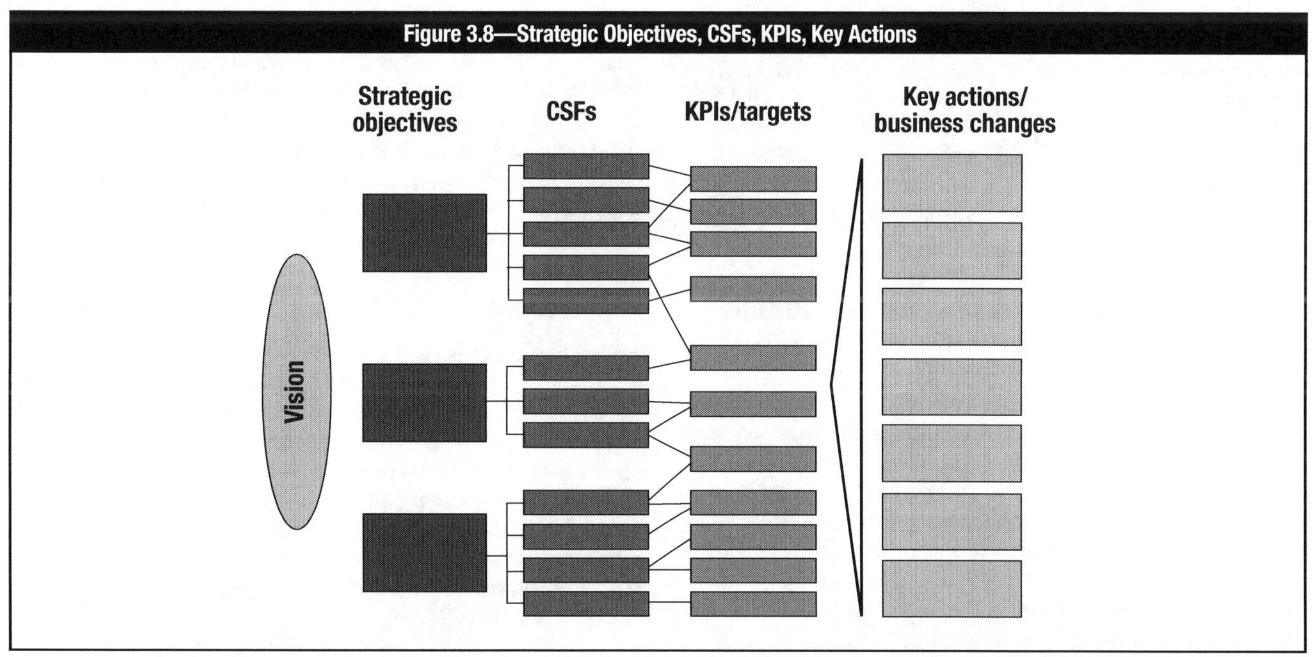

- **Vision** is a broadly defined, clear and compelling statement about the organization's purpose. This should include the desired outcomes of the information security program.
- **Strategic objectives** are a set of goals that are necessary and sufficient to move the organization toward its vision. These goals should be reflected in KGIs.
- **CSFs** are a set of circumstances or events that are necessary to achieve the strategic objectives.
- **KPIs** are concrete metrics tracked to ensure that the CSFs are being achieved.
- **Key actions, including tactical and annual action plans**, are the initiatives to be delivered to achieve the strategic objectives and KGIs.

In Practice: Using **figure 3.8** on page 167 as a guide, map the items to an objective in your organization. Identify a strategic objective and the related CSFs and KPIs to help to guide key actions to take to achieve these goals. An example is available on page 205.

3.10.14 LEGAL AND REGULATORY REQUIREMENTS

Corporate legal departments are often primarily focused on contracts and securities or company stock-related matters. The result is that they are, in many cases, not aware of regulatory requirements, and the information security manager should not rely on the legal department to identify them. Typically, the impacted department will be the most knowledgeable about legal and regulatory issues. It is, however, prudent for the information security manager to request legal review and interpretation of legal requirements that have security implications to ensure clarity on the organization's official position on the matter.

In addition, the information security manager may be required to support legal standards related to privacy of information and transactions, the collection and handling of audit records, email retention policies, incident investigation procedures, and cooperation with legal authorities. Legal issues must also be taken into careful consideration whenever an employee of the organization is investigated or monitored, or when disciplinary action is to be taken for inappropriate behavior. Because legal and regulatory requirements differ considerably in different regions and jurisdictions, it is prudent to engage the HR and legal departments prior to such actions being taken.

3.10.15 PHYSICAL AND ENVIRONMENTAL FACTORS

The confidentiality, integrity and availability of information can be compromised through unauthorized physical access and damage or destruction to physical components. The level of security surrounding any IT hardware and software, or any physical information assets such as documents or other media, should depend on the criticality of the systems, sensitivity of the information that can be accessed, the significance of applications processed, the cost of the equipment and the availability of backup equipment. Control of physical security may reside with a separate security department or sometimes with facilities management, but increasingly, this function falls under the information security group. Regardless of who is responsible for physical security, the information security manager must ensure that physical security policies, standards and activities are sufficient to not jeopardize information security efforts. A wide range of physical security controls—such as electronic locks, motion detectors, cameras, steel wire caging and radio-frequency tracking devices—are available to implement physical security.

Physical security policies and standards must be established in addition to control processes. Physical control of access to computing resources should be applied based on the sensitivity of information accessed, processed and stored there. Access should be provided on an as-needed basis.

Location within facilities and environmental factors are also concerns. Locating critical systems in areas with unstable environments or in proximity to water pipes or other potential hazards should be avoided. Computing environments should incorporate systems to monitor and control environmental factors such as temperature, humidity and electrical power quality.

Personal computers are often used in less secure user areas and may require special consideration to ensure adequate levels of security. If a workstation has a particularly sensitive function or is required to store sensitive information, isolating it may be desirable. Other physical workstation controls can include physically securing the device to prevent theft, locking the chassis to prevent tampering, removing or disabling external device interfaces (e.g., universal serial bus [USB], serial ports), and enforcing local area network (LAN)-based storage to minimize sensitive data on workstation disks.

Laptops and portable devices require special consideration, particularly given the significant risk of theft or loss. The information security manager should consider the use of whole disk encryption to protect sensitive information in the event of loss.

Security of electronic and print media should also be protected. Disclosure of sensitive information on paper, microfilm, tape, CD-ROM or other physical media is as great a risk to the organization as an online compromise and the media must be stored in secure locations. The transport and storage of backup tapes, if not encrypted, can also present significant risk, particularly when stored offsite. The information security manager should also consider a clean desk policy to prevent unauthorized access to sensitive information in less secure office areas.

Geographic concerns need to be considered, particularly where organizational facilities and disaster recovery sites are involved. Regions with a risk of earthquakes, hurricanes, flooding or other natural disasters should be avoided when selecting facility sites. Proximity to facilities that present special risk (e.g., nuclear power plants, chemical production facilities, airports) should be considered. Primary processing facilities, disaster recovery sites and offsite data storage facilities should be located far enough from one another to ensure that a disaster event does not impact more than one site.

3.10.16 CULTURE AND REGIONAL VARIANCES

The information security manager should be aware of differences in perceptions, customs and appropriate behavior across different regions and cultures and recognize that what is viewed as reasonable in one culture may not be acceptable in another. The information security manager needs to identify the audience and those who will be affected by the information security activities.

In addition, laws in different countries restrict certain sharing of personal information. The information security manager needs to be aware of these complications and work to develop an information security program that meets the individual needs of the organization.

Policies, controls and procedures should be developed and implemented with respect to these differences. Elements that might be culturally offensive to others must be avoided, particularly if alternate elements are available that meet control requirements. If in doubt, the information security manager should work with the legal and HR department to develop appropriate strategies for addressing differences across the regions and cultures represented within the organization to identify potential conflicts and work toward solutions.

3.10.17 LOGISTICS

The information security manager must address logistical issues effectively, particularly given the significant amount of interaction with other business units and individuals that is required by an effective information security program. Some of the logistic issues that the information security manager needs to be able to manage include:

- Cross-organizational strategic planning and execution
- Project and task management
- Coordination of committee meetings and activities
- Development of schedules of regularly performed procedures
- Resource prioritization and workload management
- Coordination of security resources and activities with larger projects and operations

Existing corporate resources, such as online scheduling and resource management systems, can assist with these concerns. In addition, the information security manager should develop logistics management skills through training, self-study or mentoring.

3.11 SECURITY PROGRAM SERVICES AND OPERATIONAL ACTIVITIES

The typical information security program has a number of operational responsibilities in addition to providing various security-related services. While the nature of these operations and services varies considerably among organizations, the following section describes those most typically found in larger organizations.

3.11.1 INFORMATION SECURITY LIAISON RESPONSIBILITIES

In addition to the roles discussed in the forgoing section, it is essential for an effective information security manager to maintain ongoing relationships with a number of other groups and departments in the organization. These other organizational functions have a great impact on the ability of the information security manager to be effective in implementing and managing the security program. While the names of these departments may vary, most organizations have the majority of these functions.

While the information security program will achieve greater success by integrating assurance activities with the following departments to the extent possible, effective integration with IT is essential. Integration with IT is covered in greater detail in section 3.11.11.

Physical/Corporate Security

Most large organizations have a corporate security department charged with physical security responsibilities. These departments are typically managed by individuals from law enforcement and often have limited exposure to information security. In many small organizations, physical security is handled as a part of facilities management. In either case, physical security issues invariably impact information security so a close working relationship between the two departments is important. It is essential for the information security manager to understand the physical security operation, including the relevant policies, standards, procedures and practices, to avoid a situation where inadequate physical security undermines the information security program.

IT Audit

IT or internal audit is generally charged with providing assurance of policy compliance and identifying risk. Often, especially in the absence of complete policies and standards, these auditors will have findings on information security based on what they consider good or acceptable practices. Depending on the expertise of the auditors, these findings may or may not agree with the information security manager's perspective, and this underscores the necessity for complete governance documentation. Because internal audit activities invariably impact the information security program, it is essential for the information security manager to develop and maintain a good working relationship with internal audit. A good relationship with internal audit can also provide considerable support for achieving information security objectives.

Information Technology

As the hands-on implementers and operators of information processing systems, an organization's IT department has a critical role in information security program development and management. It is important for the information security manager to develop a strong working relationship with the IT department to foster rapport, trust, understanding of common goals and open communication. This can be a challenge because IT often perceives security as an impediment to its efforts.

IT often has conflicting requirements to ensure that policies and standards are met at the same time that performance and efficiency requirements are being addressed. This conflict between safety and performance can result in sacrificing security to meet operational objectives.

Typically, security requirements encompass implementation of control mechanisms in the network, systems and application environments as well as ensuring that technology operations address security requirements. This can be a challenging task for

IT because it must fulfill security needs while also addressing issues such as functionality, accessibility, performance, capacity and scalability. The information security manager should work with IT to determine solutions that fulfill security requirements and still meet performance requirements. IT management must understand that a serious compromise will severely affect its ability to meet those performance requirements

In addition to configuring security within the actual technical environment, many organizations use IT to design, deploy and operate security systems such as firewalls, identity management systems and encryption technologies (e.g., VPNs, secure sockets layer [SSL] accelerators). Effective oversight, monitoring and good communication are critical to achieve information security objectives.

The separation of operational responsibilities between the information security department and IT is often based on the impact that a particular security system has on the production technology environment. Typically, any system that directly affects production will be operated by IT (e.g., while failure of a passive control such as an NIDS is unlikely to create a systemwide outage, the failure of an in-line system such as a firewall could cause the Internet or critical business partner connections to go down).

Business Unit Managers

The information security manager should engage business unit management when developing the information security program and continue to develop those relationships in ongoing security management activities. This provides the basis for ensuring the ongoing alignment of information security with business objectives. It will also provide assurance that the business unit managers' responsibility for front-line business operations meets security requirements and they understand their responsibility for identifying and escalating security incidents and other risk concerns.

Most organizations benefit from having an information security steering group composed of representatives of the business units and other main organizational units (i.e., finance, legal, IT, information and physical security, etc.). This forum can provide the information security manager consistent information on such things as business direction, emerging risks, control effectiveness and other matters related to information security. It is also useful in ensuring the principal stakeholders are informed about information security activities and issues.

In addition to daily operational business units, most organizations have business units responsible for the development of new products or services targeted at internal users, an external marketplace or both. It is important that the information security manager engages in the development process for any products or services related to the organization's information resources—in other words, virtually all such initiatives. The product development business unit should use an established baseline of standing security requirements (e.g., authentication controls, activity logging) for any new development project and work with the information security manager to develop additional controls to safeguard against application-specific risk. Early involvement in the product development cycle and standard baseline information security requirements helps the information security manager ensure that resources and time are allocated for effective controls implementation. If the information security group has a security architecture team, this is an appropriate area for its involvement to understand the effects on the architecture and possible requirements for modifications.

Human Resources

The HR department within most organizations has significant information security responsibilities with regard to employee policy distribution, background checks, education and enforcement. The information security manager should work with HR leadership to establish the best means to administer employee education on, and in agreement with, computer resource usage policies and procedures.

The information security manager must ensure that the HR and legal departments are intimately involved in any action involving monitoring of an employee's actions or suspected abuse of computing resources. Because of the legal implications involved with personnel actions in most localities, this cooperation is crucial. The information security manager should establish procedures and escalation as part of the incident management procedures so all involved parties understand their roles and responsibilities and are prepared for immediate action when an event does occur. A senior representative of the HR department should be assigned to the information security steering committee.

Legal Department

Information security issues are frequently related to compliance, liability, corporate responsibility and due diligence. In most organizations, these areas are the domain of the legal department. Another major area that falls to the legal department includes contracts for outsourcing and service providers. While this is primarily in the purview of the legal department, it is essential that the information security manager is in the loop to ensure inclusion of adequate security considerations. The information security manager should liaise with a representative of the legal department, who should also be on the security steering committee. By ensuring that the legal department has ongoing awareness of information security issues and acting with their consensus, the information security manager can help protect the organization from legal liability.

Employees

Employees serve as the first line of defense in the security of information. It is essential that appropriate security training on relevant policies, standards and applicable procedures is provided periodically or as needed. Testing should be used to confirm understanding and signed acknowledgment of the training should be kept with personnel records.

Once trained, it is the responsibility of all employees to follow policies, standards and procedures. Employees should be trained to report potential threats and incidents and offer suggestions for improvement to the information security program, based on their day-to-day involvement with the program.

Procurement

Most organizations use a formal procurement process that can have consequences for information security in terms of product acquisitions. The information security manager must have visibility into the process and be able to provide input into acquisition practices. Mature organizations have an approved equipment list that has been evaluated for policy and standards compliance to manage and minimize vulnerabilities introduced as a result of new equipment acquisitions. In the absence of such a process, it is important for the information security manager to be advised of proposed acquisitions and provided the opportunity to determine what risk may be introduced as a result.

Compliance

As a consequence of the increasingly complex legal and regulatory landscape that organizations must navigate, many organizations have implemented a compliance office that may be an independent department or a part of the legal department. Because there may need to be policy, standards and procedural recognition of applicable legal and regulatory requirements, it is necessary for the information security manager to establish a working relationship with the compliance office.

Privacy

Privacy regulations have become increasingly common and restrictive in many parts of the world. In response, many organizations have instituted a privacy office or a privacy officer. In some cases, this is a part of the compliance office or it may be a separate function. In any event, privacy requirements in some jurisdictions are vigorously enforced and compliance must be a major focus of information security. To ensure requirements are met, the information security manager must maintain coordination with the privacy office to avoid sanctions that have grown increasingly severe.

Training

Many larger organizations have a separate training and education department. The information security manager should have contact with this function for assistance in providing security awareness training and education in needed security skills.

Quality Assurance

Quality assurance must include acceptable levels of security-related controls. Understanding the QA process and ensuring that it includes testing of security-related aspects are important consideration for information security. The information security manager should maintain a relationship with the QA department to ensure that risk is addressed as a standard part of the process.

Insurance

Most organizations maintain various insurance policies, such as business interruption coverage, that have relevance for information security activities related to incident response, business continuity and disaster recovery. It is incumbent on the information security manager to understand the kinds and extent of insurance the organization has in order to include it in risk analysis and management and recovery planning because it serves as a compensating control.

Third-party Management

Third-party management includes outsourced functions and services. Some services, such as outsourced IDS monitoring, may be under the direct control of information security but most are likely to be managed by other departments. It is important that the information security manager understands what functions or services are provided by external parties and the associated risk. Managing risk to acceptable levels in these situations can pose a challenge and may require a variety of preventive, detective and compensatory controls including oversight and monitoring.

Project Management Office

The information security organization should maintain a strong relationship with any PMOs within the enterprise. It is important that the information security manager or representatives have an awareness of all projects, particularly IT projects, across the organization. Creating and maintaining a relationship with the PMO help to ensure that the information security team will be able to review projects to provide insight into any potential risk and/or required security measures. This relationship also provides insight into which project managers have the best qualifications for particular information security projects as the program develops.

3.11.2 CROSS-ORGANIZATIONAL RESPONSIBILITIES

The information security manager is directly responsible for many critical aspects of the information security program. It is important to be aware that there can be SoD issues if the same manager is responsible for overlapping aspects of policy, implementation and monitoring. Although in smaller organizations it may be difficult to achieve an optimal level of separation, the information security manager must be aware of the risk presented and endeavor to provide compensating controls to the extent feasible. It also generally falls to the information security manager to provide assurance that these areas of responsibility are appropriately assigned across senior managers within the organization to avoid possible conflicts of interest. A typical set of roles and responsibilities, with a sample approach for KPIs, is illustrated in **figure 3.9**. There a many other possible KPIs that can be considered. However, for the KPIs to be useful as a management tool, KGIs must also be defined. This provides the most useful metric as it shows the extent to which the goal or objective has been achieved. This will show whether there is adequate progress or additional efforts are required to achieve the goal.

As each phase of a security program is developed, executive management, managers with risk management responsibilities and department management should be made aware of the content of the information security program so activities can be coordinated and specific areas of responsibility confirmed. Information security programs typically cross numerous department boundaries; therefore, fostering awareness and getting consensus early in the process are important. To achieve this, the information security manager often becomes an ambassador for the information security program.

The information security manager must work closely with management to ensure that those in various departments and business units understand, accept and have the resources necessary to implement their part of the information security program.

Figure 3.9—Information Security Roles and Responsibilities		
Role	**Associated Information Security Process Responsibility**	**Sample Key Performance Indicator**
Executive management	Security strategy oversight and alignment	Organizational responsibility for executing all security program elements is assigned.
Business risk management	IT risk assessment	Prioritized list of IT risks to be addressed is maintained and periodically updated.
Department management	Security requirements sign-off and acceptance testing	Security features to be incorporated into application are formally approved.
	Access authorization	Individuals or groups to have access to data are formally approved.
Legal advisor to executive management	Information protection counsel	Information protection policies are consistent with applicable laws and regulations, formally approved, and those affected are aware of them.
IT operations management	Security monitoring	Security incidents are identified before they cause damage.
	Incident response	Appropriate responses to security incidents are embedded in operational procedures.
	Crisis management	Recovery procedures are periodically and successfully tested.
	Site inventory	All purchased computing devices are accounted for and correlate with a business purpose.
Quality manager	Security review participation	Security-policy-compliant systems are configured.
	Security requirements capture	Business requirements for confidentiality, integrity, and availability are documented.
	Application security design	Technical implementation plans for meeting business process security requirements are established.
	Change control	Secure archival, retrieval and compilation plans of organization-maintained source code and product customizations are established.
	Security upgrade management	Ensure testing and application of security software fixes
Purchasing	Security requirements capture	Formal requirements for security in all requests for product information and proposals are established.
	Contract requirements	Business requirements for confidentiality, integrity and availability in information service provider and technology maintenance contracts are established.

One strategy for incorporating the ideas and support of the organization's management includes the formation of an information security steering committee or an executive security council, as discussed in chapter 1. These committees can also serve to coordinate activities among groups involved in other aspects of risk management and assurance functions, thereby helping to achieve assurance process integration and strategic alignment. The members of these committees are selected for their ability to support the information security program and represent the organization's interests. They help ensure that information security requirements are identified and organizationwide support is achieved. The steering committee typically owns the information security strategy and is commonly designated as the group empowered to approve changes to policy or standards.

An important part of program development is the review, modification and/or creation of policies required to establish a framework for the development of organizational standards with respect to security. Documents that reflect the decisions set by security strategy should be clearly set forth in the form of policy mandates. Policy documents identify management intent and direction and form the basis for the organizational standards that comply with management and regulatory objectives for data confidentiality, integrity and availability. See section 1.13 Action Plan to Implement Strategy.

Next, awareness is required to educate those affected by security policy on their roles and responsibilities. Awareness activities should be conducted by all business areas that are responsible for maintaining processes in conformance with security policy. These do not need to be formal training classes but should fit in with the culture of the organization and management's preferred method of communication. Depending on the organization, executives may fulfill their awareness responsibilities with a variety of alternatives, including videos, memos, email reminders, posters, seminars and formal security training classes. If the organization has a training department, some or all of the awareness requirements may be part of its responsibilities. In any event, it is incumbent on the information security manager to initiate, coordinate and monitor the delivery of suitable information security awareness materials tailored to the various groups in the organization. Experience has shown that a variety of parallel approaches are likely to be most effective.

Considerations for implementation of security measures are rarely limited to a few central security architecture projects or major initiatives that the information security manager personally manages. Security implementation responsibilities might range from protecting a personal laptop from theft to configuring telecommunications equipment. For this reason, most managers and executives will have some aspect of security implementation within their scope of responsibilities. The manager in each functional unit the manager must understand that he/she serves as the policy compliance owner for his/her area of responsibility and provides appropriate monitoring and oversight to ensure compliance.

Those with monitoring responsibility should establish processes that create and maintain alerts, logs and metrics on system security configuration and activity. The information security manager is often the point of escalation for security issues identified by monitoring processes as well as the primary contact for incidents that may require investigation. Security monitoring must be implemented in a manner that ensures SoD. This is important because in order to be effective, regulatory functions such as security, audit and quality control cannot be under the control of those being monitored. This is also true for compliance and enforcement.

Finally, compliance includes any activity that tracks security issues and helps ensure that resources facilitate the resolution of security issues. Executives with responsibility for security compliance should establish programs that track trends in metrics and investigate anomalies and known security violations. A strong compliance program will further ensure that conclusions of these investigations are reported to others in executive management in such a way that risk is well understood. These reports should, if possible, be accompanied by recommendations for changes necessary to achieve satisfactory compliance levels. The information security manager will often be an SME within this process or may lead the program.

3.11.3 INCIDENT RESPONSE

Incident response is typically an operational requirement for the information security department. Incident response, discussed in chapter 4, provides first responders for the inevitable security incidents experienced in virtually all organizations. The objective is to quickly identify and contain incidents to prevent significant interruptions to business activities, restore affected services, and determine root causes so improvements can be implemented to prevent recurrence.

3.11.4 SECURITY REVIEWS AND AUDITS

During the development and management of an information security program, it is essential for the manager to have a consistent standardized approach to assessing and evaluating the state of various aspects of the program. Using a consistent approach will provide trend information over time and can serve as a metric for improvements to the information security program. This can be accomplished using a security review process similar to an audit. As with standard approaches to auditing, security reviews have:

- An objective
- A scope
- Constraints
- An approach
- A result

A review objective is a statement of what is to be determined in the course of a review. For example, the objective of a review may be to determine whether an Internet banking application can be exploited to gain access to internal systems.

Objective defines what the information security manager wants to get out of the review. Usually, it is to determine whether or not a given systems environment meets some security standard. In the above example, the review objective is that of a typical external penetration study (i.e., to make sure that users of web services cannot exploit system vulnerabilities to gain access to the systems that host those services).

Scope refers to the mapping of the objective to the aspect that is to be reviewed. Thus, the review objective dictates scope. For example, the review objective in the previous example dictates that the scope includes the Internet access points of the application and all the underlying technology that enables that access. If the scope is hard to describe, the review objective should be clarified to ensure that the result of the review will be well defined and actionable.

Constraint is the situation within which a reviewer operates that may impact aspects of conducting the review. It may or may not hinder his/her ability to review the entire scope and complete the review objective. In the example, a constraint may be a prohibition on accessing the application during business hours. An information security manager must evaluate his/her ability to fulfill the objective of the review in the context of constraints.

Approach is a set of activities that cover the scope in a way that meets the objective of the review, given the constraints. There are usually alternative sets of activities that can cover the scope and objective. The idea is to identify the set that is hampered by the fewest constraints. In the previous example, a constraint could be that the information security manager would not be given the credentials necessary to create a web session as an authorized user of the application under review. (In this example, the objective requires the reviewer to provide a determination on whether application Internet access can be exploited to gain access to internal systems.) To meet this objective, the review must have application Internet access as part of the scope. An information security manager, in this situation, should identify the lack of authorization for application access as a constraint and find some other way to achieve the same objective. One approach could be to set up the system in a test environment where access authorization is not an issue.

Result is an assessment of whether the review objective was met. It is an answer to the question, "Is this secure?" If it is not possible to answer the question with any level of assurance, the review should be declared incomplete. This would occur in the case above if the application access required for covering the scope was not achieved.

Figure 3.10 depicts some common types of reviews and the objective, scope, constraint and approach results of each.

In the course of performing security reviews, an information security manager can gather data about not only policy and process at various levels of the organization, but also specific control weaknesses that may put information at risk. This data can be used to help prioritize program development efforts.

Audits

Like security reviews, audits have objectives, scope, constraints, approach and results. The professional practice of information systems audit is based on an approach in which auditors identify, evaluate, test and assess the effectiveness of controls. Effectiveness is judged on the basis of whether controls meet a given set of control objectives such as compliance with policies and standards. In performing an audit, an audit team assembles documentation that: 1) maps controls to control objectives, 2) states what the team did to test those controls and 3) links those test results to the final assessment. This documentation, called "work papers," may or may not be delivered with the final report.

When an information security program has established policies and standards, an audit is extremely useful in identifying whether those policies and standards have been fully implemented. However, when an information security program is under development, those policies and standards may not yet be finalized. In this situation, an information security manager may select an externally published standard and engage an audit team to determine the extent to which the organization is in compliance. An external standard or framework, such as COBIT or ISO/IEC 27001 and 27002, provides a structure for control objectives that enables an audit team to organize its examination of existing controls. The work papers from such an audit can often be more useful than the final report. The value in the analysis of existing controls and their mapping to a set of external standards is most useful if the set of external standards closely resembles those that the information security manager intends to put in place as part of the information security program.

Different standards focus on different aspects of controls. The following examples start at the most comprehensive in scope, followed by those whose scope narrows into specific technology domains:
- COBIT lists control objectives and, for each control objective, control practices.
- The Standard of Good Practice for Information Security catalogs information security management practices and lists corresponding requirements for resources and responsibilities.
- ISO/IEC 27001 and 27002 list control practices in the domain of IT security, followed by an appendix listing security-related control objectives and, for each control objective, control practices.

Auditors

It is essential for the information security manager to develop a good working relationship with the auditors, whether internal or external. While it is not unusual for auditors to be viewed in a negative light by IT and information security staff and others in the organization, effective information security managers understand that audits are both an essential assurance process and a critical and influential ally in achieving good security governance and compliance. They can be instrumental in implementing security standards by providing feedback to senior management through audit findings that can serve to influence the tone at the top and create high-level support for security activities. Involving auditors in overall security management can be a powerful tool for improving an organization's security culture.

Figure 3.10—Security Review Alternatives

Review Type	Control Self-assessment	Security Architecture or Design Review	Security Spot Check
Objective	To establish that the controls implemented to maintain security are sufficient to do so in the systems environment required to effect security	To establish that a system is capable of securing data, and identify configuration parameters	To decide whether a given security process is working
Scope	The systems environment housing the data that an organization is charged to secure documents on system security mechanisms	Network and operating system placement diagrams, as well as detailed technical design	Process description, system security parameters of system directly supporting the process
Constraint	• Unknowns or lack of expertise in security mechanisms in third-party products • Time • The possible bias of participants who are also responsible for system maintenance	• Unknowns or lack of expertise in security mechanisms in third-party products • Time	Reliance on assumptions with respect to systems interfaces and supporting systems (e.g., data feeds, network, OS)
Approach	• Identify risks, exposures and potential perpetrators. • Evaluate ability of controls to protect, detect or recover from exploits.	Compare settable parameters of all system components to known secure configurations and/or security policy.	• Review all system security procedures and settings. • Identify expected user community. • Evaluate whether expected controls are in place.
Result	List of control weaknesses	List of issues to address, iterative process	Yes or no

The information security program must integrate with internal and external auditing activities. Some audits are compulsory, such as those required to establish compliance with a regulatory standard. Others are voluntary, as when an independent auditor makes an attestation to compliance with an industry standard.

The information security manager should coordinate with organizational audit coordinators to ensure that time and resources are allocated to address audit activities. Procedures should be established in advance for scheduling, observation of employee activities and provision of configuration data from technical systems.

In some cases, a deficiency identified by an auditor may not be applicable to the information security manager's specific organization. If concerns are identified during an audit, the information security manager should work with the auditors to agree on associated risk, mitigating factors and satisfactory control objectives. With this information in hand, the information security manager may craft one or more potential solutions that fit the organization's operational, financial and technical environment. Any combination of mitigating or compensating controls that enforce the agreed-on control objectives should satisfy the issue.

Audit findings provide strong, independent feedback for the steering committee and/or management to utilize in assessing the effectiveness of the information security program.

3.11.5 MANAGEMENT OF SECURITY TECHNOLOGY

The typical security program employs a number of technologies that require effective management and operation if optimal value delivery and resource management are to be achieved. While a newer organization may use contemporary technologies, more mature organizations are often constrained by the legacy architecture of the organization. However, these constraints can be minimized because there is usually a wide range of technology alternatives available to address a given control objective. Features that are available within a given set of legacy devices will differ depending on the technical footprint of the organization. Yet, through decades of development of alternative preventive, detective and recovery controls, solid tools and techniques to achieve information security goals are usually available.

Technology Competencies

Although information security spans technical, operational and managerial domains, a significant portion of the actual implementation of the information security program is likely to be technical. The information security manager and security department personnel are often considered the primary sources for technical security subject matter expertise within an organization. It is important for the information security manager to work with the security steering committee, senior management and other security stakeholders to establish the scope and approach of technical skills delivery in which the information security manager and security organization are expected to engage.

Organizations differ with regard to the technical scope of the information security department. At one end of the spectrum is the approach in which the information security program operates at the corporate level and primarily sets security standards at a high level. Another common approach is to use information security personnel as technical SMEs, providing consultative services to system administrators and other information technologists who, in turn, implement technical controls and security systems. At the other end of the spectrum are organizations in which the information security group takes ownership of specific pieces of infrastructure, such as access control systems, intrusion detection and monitoring systems, and compliance and vulnerability assessment automation tools.

For the individual information security manager, the technology skills that are needed vary based on his/her operational role, the organizational structure and technical scope. More technically focused information security managers (e.g., those in system administration roles) will obviously need more in-depth technical knowledge and should make training and educational arrangements accordingly. Information security managers operating at a higher level (e.g., as a CISO) may not require hands-on technical skills, but should be knowledgeable about the information technologies implemented by their organization from architectural and data flow perspectives and how and to what extent technology contributes to achieving control objectives. Regardless of operating level, all information systems managers should have a thorough understanding of security architecture, control implementation principles, and commonly implemented security processes and mechanisms. This understanding should include the strengths, limitations, opportunities and risk of common security controls in addition to the financial and operational implications of deployment.

It is important for information security managers to take into consideration all levels of technology as they plan for skills development, both personally and for the overall information security program. While traditional perimeter, network and systems security are still crucial to a strong technical security environment, information security managers are increasingly expected to address issues of application security (e.g., coding practices, functional application security mechanisms, data access control mechanisms), database security (e.g., data access control methods, application integration, content protection) and, increasingly, elements of physical, operational and environmental security issues. Highly integrated and tightly coupled systems, such as enterprise resource planning (ERP) implementations, can create an additional challenge; the entire system has to be considered from a security perspective because compromise of one element can disrupt the operations of the entire enterprise. It is important that the information security manager understand and plan for the potential domino effect of cascading risk.

3.11.6 DUE DILIGENCE

Due diligence is essentially a term related to the notion of the "standard of due care." It is the idea that there are steps that should be taken by a reasonable person of similar competency in similar circumstances. In the case of an information security manager, this means ensuring that the basic components of a reasonable security program are in place. Some of these components might include:
- Senior management support
- Comprehensive policies, standards and procedures

- Appropriate security education, training and awareness throughout the organization
- Periodic risk assessments
- Effective backup and recovery processes
- Implementation of adequate security controls
- Effective monitoring and metrics of the security program
- Effective compliance efforts
- Tested business continuity and disaster recovery plans
- Protection of data (in transit and at rest)

It is also important to take into consideration that the third parties the organization uses and relies on can present significant risk to an organization's information resources. Due diligence regarding placement of appropriate security language into contracts and agreements with third parties, as well as subsequent third-party performance against security requirements, must also take place. Because contracts are usually prepared by the legal department, it is important to collaborate to ensure inclusion of language needed to provide adequate levels of protection. An organization's information must be protected as specified by its policies, regardless of its location.

Periodic reviews of the infrastructure, preferably by an independent and knowledgeable third party, may be a reasonable requirement both internally and of an outsource service provider as well. The infrastructure is a critical component that the organization relies on to meet its business objectives. Risk must be identified and, to the extent is not at an acceptable level, must be reasonably addressed.

Managing and Controlling Access to Information Resources

The information security manager must be aware of the various standards for managing and controlling access to information resources. It should also be considered that, depending on the organization's industry sector, specific regulatory bodies may have defined standards that must be addressed.

Increasingly, information security management is defined by the need to satisfy regulatory requirements. While these regulatory requirements may establish specific protection measures that need to be in place, they are not necessarily comprehensive in their approach. More broadly defined guidance for information security program development and administration is provided by various standards bodies and by not-for-profit organizations whose members are involved with governance, assurance or information protection. The following list, while not meant to be complete, identifies some of the more widely recognized organizations that provide reference materials of interest to information security managers:

- American Institute of Certified Public Accountants (AICPA)
- Canadian Institute of Chartered Accountants (CICA)
- The Committee of Sponsoring Organizations of the Treadway Commission (COSO)
- German Federal Office for Information Security (BSI)
- International Organization for Standardization (ISO)
- ISACA
- International Information Systems Security Certification Consortium, Inc. $(ISC)^2$
- IT Governance Institute
- National Fire Protection Association (NFPA)
- Organisation for Economic Co-operation and Development (OECD)
- US Federal Energy Regulatory Commission (FERC)
- US Federal Financial Institution Examination Council (FFIEC)
- US National Institute of Standards and Technology (NIST)
- US Office of the Comptroller of the Currency (OCC)

Most countries have governmental regulatory organizations that deal with medical and financial information and other privacy issues. It is incumbent on the information security manager to understand and consider these regulations in all the jurisdictions where the organization operates.

Vulnerability Reporting Sources

Threats to information systems are global. Requirements for rapid time to market and other issues have resulted in a variety of vulnerabilities in both hardware and software. These vulnerabilities are constantly being discovered and reported by a variety of organizations. It is an important part of any effective security program to maintain daily monitoring of relevant entities that publish this information, which include the US Computer Emergency Readiness Team, MITRE's Common Vulnerabilities and Exposures database, Security Focus' BUGTRAQ mailing list, SANS Institute, OEMS and numerous software vendors. Having the most current possible vulnerability information makes it possible for the information security manager to respond promptly with appropriate mitigation, compensation or elimination action to address newly discovered software and system flaws.

3.11.7 COMPLIANCE MONITORING AND ENFORCEMENT

Compliance enforcement processes must be considered during program development to ensure subsequent effectiveness and manageability once the program is implemented. Compliance enforcement refers to any activity within the information security program that is designed to ensure compliance with the organization's security policies, standards and procedures.

Compliance, especially with procedural controls, may pose one of the major challenges for managing a security program and must be given careful attention when designing controls during program development. Ease of monitoring and enforcement is often one of the essential factors in control selection. Complex control processes that are not readily enforceable or are difficult to monitor for compliance are generally of little value and may pose a considerable risk themselves.

Enforcement procedures should be designed to assume that control activities are in place in support of control objectives. These procedures are an added layer of control that checks that the procedures established by management are actually followed. For example, in a password reset procedure, an enforcement procedure may consist of a supervisor listening to randomly selected help desk calls and listing any help desk staff who neglect to ask a user for a security identification code prior to resetting the passwords. The enforcement procedure would be to use the list first to warn and then to discipline help desk staff who did not follow the password reset procedure.

Policy Compliance

Policies form the basis for all accountability with respect to security responsibilities throughout the organization. Policies must be comprehensive enough to cover all situations in which information is handled, yet flexible enough to allow different processes and procedures to evolve for different technologies and still be in compliance. Except in very small organizations, an information security manager will not have direct control over SDLC activities of all information systems in the organization. Consequently, it is necessary to designate formal security roles that establish which department head is responsible for putting processes in place that maintain security policy compliance and meet the appropriate standards for a given set of information systems.

It is the responsibility of the information security manager to ensure that, in the assignment process, there are no orphan systems or systems without policy compliance owners. It is also the responsibility of the information security manager to provide oversight and ensure that policy compliance processes are properly designed. An information security manager can accomplish this oversight via a combination of security review, metrics gathering and reporting processes.

Information security management literature often refers to a policy exception process. This is a method by which business units or departments can review policy and decide not to follow it based on a various factors. Several justifications can exist for policy exceptions. It may be based on a risk/reward decision where the benefit of not following the policy justifies the risk. It may be financially or technically infeasible to comply with specific policies or standards. Such trade-offs should be considered in the policy development process, when possible, to minimize the need for subsequent exceptions. As part of the program development, a formal waiver process should be implemented to manage the life cycle of these exceptions to ensure that they are periodically reviewed and, when possible, closed. Any such policy exceptions must first be assessed for risk and impact prior to implementation and the identified risk accepted by appropriate levels of management.

Standards Compliance

Standards provide the boundaries of options for systems, processes and actions that still comply with policy. The standards must be designed to ensure that all systems of the same type within the same security domain are configured and operated in the same way based on criticality and sensitivity of the resources. These will allow platform administration procedures to be developed using standards documents as reference to ensure that policy compliance is maintained. Standards also provide economy of scale; the configuration mapping to policy needs to be done only once for each security domain, and technology and process engineering efforts are reused for systems of the same type.

As much as possible, compliance with standards should be automated to ensure that system configurations do not, either intentionally or unintentionally, deviate from policy compliance. However, as policy should state only management intent, direction and expectations to allow flexibility for different standards to develop, and provide many options to comply with policy, exceptions to standards should always be reviewed to see if they deviate from the intent of policy. It may also be that a business situation justifies a deviation from existing standards but, nevertheless, may be determined to fall within the intent of policy. As with policy exceptions, standards exceptions must include risk assessment and acceptance by appropriate management. If exceptions are required to go through the change management process (assuming one is in place), assessing the risk of the change will be a standard part of the process.

Resolution of Noncompliance Issues

Noncompliance issues may result in risk to the organization, so it is important to develop specific processes to deal with these issues in an effective and timely manner. Depending on how significant the risk is, various approaches can be taken to address it. If a particular noncompliance event is a serious risk, then resolution needs to occur quickly. The security manager benefits from a method of determining criticality and then having a risk-based response process.

Typically, a timetable is developed to document each noncompliance item and responsibility for addressing it is assigned and recorded. Regular follow-up is important to ensure that the noncompliance issue and other variances are satisfactorily addressed in a timely manner. Noncompliance issues and other variances can be identified through a number of different mechanisms, including:

- Normal monitoring
- Audit reports
- Security reviews
- Vulnerability scans
- Due diligence work

Compliance Enforcement

Compliance enforcement is an ongoing set of activities that endeavor to ensure policy and, by default, standards requirements that are not being met are brought into compliance. Audits are a snapshot of compliance in time; compliance enforcement is an ongoing process that helps reduce risk and ensure positive audit opinions.

Compliance enforcement responsibilities are usually shared across organizational units and the results are commonly shared with executive management and the board's audit or compliance committees. The legal and internal audit departments often have responsibility for assessment of business strategies and operations, respectively. The information security unit is often responsible for implementing independent evaluation of technical standards, preferably using automated tools. In larger organizations and heavily regulated industries, an independent compliance organization may be established to handle and coordinate these activities.

The information security program itself is also a target of compliance evaluation and performance. The information security manager should be prepared to work closely with compliance and/or internal audit personnel to demonstrate compliance of the information security program with pertinent standards and regulations. As with a formal audit, issues identified should be defined in terms of risk, mitigating factors and acceptable control objectives. Depending on the magnitude of the issue, the information security manager may address the concern

independently or may collaborate with executive management and/or the security steering committee to create acceptable solutions.

3.11.8 ASSESSMENT OF RISK AND IMPACT

A primary operational responsibility for the information security manager and the fundamental purpose of the program is to manage risk to acceptable levels. The objective is to minimize disruptions to organizational activities balanced against an acceptable cost. A number of ongoing tasks are required to achieve this objective. See chapter 2 for a more detailed discussion.

Vulnerability Assessment

The organization's information systems environment should be constantly monitored for development of vulnerabilities that could threaten confidentiality, integrity or availability. In addition to searching for known vulnerabilities, this process should also detect unexpected changes to technical systems. This process is best implemented using automated, network- or host-based tools that deliver concise reports to information security management, including immediate alerts if severe vulnerabilities are noted.

In addition to regularly scheduled scanning, the information security manager should ensure that scheduled changes to existing technical environments (e.g., installation of a new service, hosts being relocated, firewall upgrades) do not inadvertently create architectural vulnerabilities and all such changes are handled through a change management process. Human error and unexpected system behaviors can cause the enforcement policies of technical security control mechanisms to change, creating opportunities for exploit and organizational information systems impact.

Threat Assessment

Technical and behavioral threats to an organization evolve as a result of internal and external factors. Implementation of new technologies, granting broader network and application access to partners and customers, and the ever-growing capabilities of attackers warrant periodic reassessment of the threat landscape an organization faces. This activity is particularly important for organizations too small or resource-constrained to adopt a continuous assessment approach to threat management.

The information security manager should perform this analysis at least annually by evaluating changes in the technical and operating environments of the organization, particularly where external entities are granted access to organizational resources. Internal factors such as new business units, new or upgraded technologies, changes to products and services, and changes in roles and responsibilities represent areas where the level of threat may increase or new threats may emerge.

As new threats or changes in threat levels are identified and prioritized in terms of impact, the information security manager must evaluate the ability of existing controls to mitigate risk associated with these threats. In some cases, the technical security architecture may need to be modified, a threat-specific countermeasure may be deployed, or a compensating mechanism or process may be implemented until mitigating controls are developed. Threat sources can include technical/cyber, human, facility-based, natural and environmental, and pandemic events.

Numerous threats exist that may impact security program development efforts and objectives. The range of possible threats must be evaluated to determine if they are viable; the likelihood they will materialize; their potential magnitude; and the potential impact to systems, operations, personnel or facilities.

Risk Assessment and Business Impact Analysis

Risk assessment is a part of the risk assessment process and is used to identify, analyze and evaluate risk; the probability of compromise; and its potential impact on an organization in quantitative or qualitative terms. A BIA is an exercise that determines the impact of losing the availability of any resource to an organization; it establishes the escalation of that loss over time, identifies the minimum resources needed to recover, and prioritizes the recovery of processes and supporting systems. While a BIA is often thought of in the context of business continuity and disaster recovery, whether potential impact is determined by this process or another process is not important. Impact is the bottom line of risk and the range of severity in terms of the organization must be determined to provide the information needed and guide risk management activities. Potential impact is also the basis for asset classification. Obviously, even high risk with little or no impact is not of concern.

In developing KGIs necessary for achieving control objectives, an information security manager must make choices about the relative benefits compared to costs that achieving control objectives will have on the confidentiality, integrity and availability of information resources. However, even if a risk management process has been developed, as described in chapter 2, mapping control benefits and impact onto key business goals for security is not necessarily a straightforward process. The relationship of developing and implementing security controls to achieving organizational objectives usually requires a well-developed business case (see section 1.4.2 Information Security Roles and Responsibilities) to achieve the level of buy-in needed to achieve success.

The business case must address the fact that residual risk will always remain regardless of the level of control, and it must address the fact that risk may aggregate into levels that are unacceptable. In other words, even with effective controls apparently managing risk to acceptable levels, the aggregate effect of a number of types of acceptable risk may not be acceptable and may pose a serious threat to the organization.

Numerous studies show that disasters are typically not a single calamitous occurrence but rather the result of a number of small incidents and mistakes that contribute to a major event. The lesson is that, while, individually, types of residual risk might be low, collectively, they can be disastrous.

As threats and vulnerabilities emerge, the information security manager must take steps to analyze and communicate the effect on the organization's risk posture. This process is critical to ensuring that security stakeholders are aware of potential business impact and can take actions to mitigate risk accordingly. This entire process should be completed annually, or the information security manager may choose to take an incremental approach, analyzing portions of the enterprise monthly or quarterly.

The information security manager should also recognize that asset values and risk characteristics can change, requiring reanalysis of risk posture. For example, a company can grow increasingly revenue-dependent on an application that was initially not considered to be critical to the enterprise. Asset value can increase or decrease over time in terms of real monetary value or strategic value to the organization. In addition, the risk associated with an asset can grow. A small database may initially contain only a few dozen personal information records; the same database five years later may contain 10,000, representing a much higher impact if compromised.

Periodic risk assessment results should be provided to the steering committee and/or senior management for use in guiding information security priorities and activities. The information security manager should manage this process and guide the committee through making appropriate decisions based on risk analysis results. More details on risk assessment are in section 2.7 Risk Assessment.

Resource Dependency Assessment

If resources or other constraints do not allow for comprehensive BIAs, a business resource dependency assessment is a less expensive alternative to provide the basis for allocating available resources based on the criticality of the function. A business resource dependency assessment reviews the resources that are used to conduct business (i.e., servers, databases, etc.). Depending on the criticality of the business function, the assets and resources needed for that function are identified, providing a basis for prioritizing protection efforts. In other words, a business resource dependency assessment is based on determining the various applications and infrastructure used by a business for day-to-day operations. While it should also identify interdependencies and other resources needed to perform the required functions, it does not capture the financial and operational impact of potential disruptions and does not replace a BIA.

3.11.9 OUTSOURCING AND SERVICE PROVIDERS

The two types of outsourcing that an information security manager may be required to deal with are third-party providers of security services and outsourced IT or business processes that must be integrated into the overall information security program. Most of the security requirements are similar, depending on criticality and sensitivity of assets and extent of services involved, but the ownership is different (i.e., the information security manager is normally the process owner for outsourced security services whereas other outsourced services are typically the responsibility of the process owner). The risk posed by third parties connected to the organization's internal network can be substantial and must be carefully considered.

Economics are the primary driver of outsourcing. As a result, early engagement by the information security manager is essential to ensure that those making the decisions do not unduly compromise security for the sake of cost. It is also important to understand the TCO of outsourcing over the duration of the contract. Typically, over an extended contract, there is very little, if any, overall economic benefit. This can be the result of the fact that service levels are usually contracted for at a specific level while circumstances may result in a contraction of the business and require less services with no commensurate reduction of costs. It is also possible that a growth in the business will require additional services with the potential result that the outsourcer demands much higher fees. Either event can increase cost per unit of service provided, reducing or eliminating any initial cost savings. Some organizations have also found that it takes a considerable amount of time to negotiate increased service levels, which adversely affects the organization's agility to meet changing business needs. Finally, the potentially considerable costs of repatriation of outsourced services at the end of the contract period, in the event the organization chooses to insource, must be considered as well. This may occur as a result of the organization finding the constrictions imposed by outsourcing not acceptable or the costs associated with a new agreement too high and choose not to renew the outsourcing contract.

Other important considerations when evaluating outsourcing option that can adversely affect the organization include:

- Loss of essential skills
- Lack of visibility into security processes
- New access and other control risk
- Viability of the third-party vendor
- Complexity of incident management
- Cultural and ethical differences
- Unanticipated costs and service inadequacies

Whether considering outsourcing entire operations or evaluating service providers, the adequacy of the vendor's controls needs to be audited and monitored through the life of the contract to ensure that security measures are not marginalized over time as a result of cost pressures. This can be accomplished by independent audit or onsite visits at the third-party facility to ensure that the proper controls are in place.

The existence and enforcement of privacy laws to protect a company's data also need to be considered. This is particularly true as individuals in different cultures may treat information differently (i.e., what a firm in one country considers sensitive information may not be considered sensitive in another country).

While technical controls for the IT component of outsourced processes may seem obvious, arrangements that deal with business processes require that training, awareness, manual controls and monitoring be put in place to ensure that employees in third-party facilities are treating the processes and physical and electronic data to acceptable standards.

Third-party security providers are a viable strategy that the information security manager can use to assist in the design and operation of the organization's information security program or provide other IT services. One of the main concerns for security is the extent that the third-party provider can and will meet the organization's security policies and standards on an ongoing, verifiable basis. This can result in security breaches.

The maturity of a vendor's security program and the vendor's assurance of compliance with the contracting organization's security policies should be high on the list of decision factors when selecting a vendor. This proposal and evaluation process is also applicable when evaluating whether to use an organization's autonomous division or subsidiary for certain security services.

The issues associated with outsourcing that require attention are broad and must be enumerated based on the scope, type and risk associated with the initiative. Some common issues include:
- Isolation of external party access to resources
- Integrity and authenticity of data and transactions
- Protection against malicious code or content
- Privacy/confidentiality agreements and procedures
- Security standards for transacting systems
- Data transmission confidentiality
- Identity and access management of the third party
- Incident contact and escalation procedures
- Compliance enforcement of critical sole source vendors

Outsourcing Contracts

The fundamental purpose of contracts is two-fold: 1) to ensure that the parties to the agreement are aware of their responsibilities and rights within the relationship and 2) to provide the means to address disagreements once the contract is in force. Within that framework, the information security manager should be familiar with certain provisions for security and information protection. While the information security manager may or may not be included in contract reviews, the following issues should be addressed with the legal department to ensure an understanding of security requirements that should be covered.

The most common security provision addresses confidentiality or nondisclosure. Each party will typically agree that any confidential or sensitive information it receives as part of the agreement, or about the other party, will be kept confidential through appropriate measures. This may also include the requirement to return or destroy any proprietary or confidential information upon termination of the contract or after a specified period of time. The information security manager will need to determine the specific level of destruction that will be required (e.g., document shredding, disk and tape degaussing).

The contract may also stipulate that either or both parties must maintain appropriate security controls to ensure that the systems and information used under the agreement are protected by appropriate means. The contract should explicitly define what is meant by "appropriate" and the requirements for demonstrating the effectiveness of those protections. Whether it is through the production of a current third-party audit report (e.g., SSAE16) or compliance with an industry-standard security framework (such as ISO/IEC 27001:2013, 27002:2013 or COBIT 5), the standards by which the program will be judged should be defined in the contract. Additionally, if the contracted product or service includes network connectivity between the buyer and the seller, the contract should address responsibility for the security of that connection. Specifications as to the level of security expected (e.g., firewalls, intrusion detection/prevention, monitoring) or specific technical requirements should also be addressed.

A contract with a service provider that has been determined (through a risk assessment) to be beyond a predetermined risk threshold should always contain provisions concerning right-to-audit and a right-to-inspect without notice. This would typically be for any third party that accesses, stores or processes any sensitive or otherwise business-critical information; provides mission-critical services; or connects to the network infrastructure of the contracting organization. The right-to-audit clause should allow the customer, upon proper notice, the right to conduct an in-depth audit of the third party's security program and processes to verify the effectiveness of all associated controls. The right-to-inspect without notice, although not generally favored by service providers, can provide greater incentive to adhere to contractual requirements on an ongoing basis. If these provisions are included in the contract, the parameters for an audit, such as compliance criteria, notification, scope limitations, frequency and responsibility for incurred costs, should be explicit. The right to inspect should ideally have few, if any, constraints.

In the event that a security breach happens at either party, the contract should specify the roles each party will play in the investigation and remediation process. Issues such as which party will lead the investigation, notification procedures and responsibilities (including law enforcement or regulatory notifications), and timing must all be addressed. During an active incident, pressure is high, tempers can flare, and fairness and equity are much harder to come by, so addressing these issues in the contract is easier and can be done much more equitably.

Finally, the contract should contain indemnity clauses that ensure compensation for impacts caused by the service provider. Service providers will typically attempt to limit compensation for their failure to specific amounts or, in some cases, to the amount of fees paid to them for a specific period, such as the past 12 months. It is obvious that such limits are not in the best interests of the contracting organization. Another consideration is what is called "choice of law" provisions, which require that any potential litigation take place in certain jurisdictions typically favorable to the service provider. While these elements are in the purview of the legal department (if there is one, which may not be the case in a small organization), it is useful for the information security manager to have an understanding of these issues.

The contract between the parties is a key element of establishing an appropriate level of control of the organization's information processing facilities (IPFs). Depending on the business processes and operational needs of the organization requiring the services of third-party vendor outsourcing, contracts may need to deal with a number of complex security questions. Points that should be covered in the contract (from the perspective of the information security manager) can include, but are not limited to:
- Detailed specification of outsourced service
- Specific security requirements
- Restrictions on copying information and securing assets
- Prohibiting access without explicit authorization and maintaining a list of individuals who have access
- Right to audit and/or inspect
- Indemnity clauses to mitigate impacts caused by the service provider
- Requirements for incident response plans (IRPs) and business continuity plans (BCPs)
- Level of service quality
- Integrity and confidentiality of business assets
- Nondisclosure agreements to be signed by the employees/agents of third parties
- Protection of intellectual property
- Ownership of information

- Requirement that applicable legal and regulatory requirements are met
- Return and/or destruction of information/assets at the end of the contract
- Duration up to which confidentiality shall be maintained
- Employees or agents of the third party required to comply with security policies of the organization
- Escalation processes
- Predetermined reduction or increases in costs associated with changing service levels
- Response and implementation times for changing service requirements
- Conditions and circumstances for contract cancelation and associated costs

Third-party Access

Third-party access to the information security manager's organization's processing facilities under any circumstances should be controlled based on risk assessment and must be clearly defined in an SLA. Access should be granted based on the principles of least privilege, need-to-know and need-to-do. It is important to bear in mind that third parties may have a different set of ethics and business culture that must be considered in terms of risk.

Providing access to third parties must be based on clearly defined methods of access, access rights and level of functionality, and access must require the approval of the asset owner.

Access usage should be fully logged and reviewed by the security manager on a regular basis. The review frequency should be decided based on factors such as:
- Criticality of information to which access rights are given
- Criticality of privileges given
- Period of contract

Anomalies noticed should be immediately reported to the asset owner and escalation conditions specified wherever required. The access rights given to third parties should be removed immediately after the contract expires.

Network and information access should not be granted to a third party until the contract has been signed. The contract should define the terms for access and control requirements and make allowances for assurance that appropriate safeguards are in place and will remain in place for the duration of the contract.

3.11.10 CLOUD COMPUTING

Cloud computing is the evolution of a concept that dates back to the 1960s when the notion of "utility computing" was first suggested. The idea was based on the notion of an electric utility or telephone service provider. The intervening decades have provided the technological basis to make the concept practical through increased bandwidth and near-universal Internet availability.

While current offerings vary considerably in scope and capabilities, there is growing consensus on the definition. NIST defines cloud computing as "a model for enabling convenient, on-demand network access to a shared pool of configurable computing resources (e.g., networks, servers, storage, applications, and services) that can be rapidly provisioned and released with minimal management effort or service provider interaction."

The defining characteristic of cloud computing is that processing and data are somewhere in "the cloud" as opposed to being in a specific known location. Cloud computing can be provided as either public hosting for a number of unrelated entities or private hosting, in the case of large organizations wanting greater control over the environment.

The five essential characteristics of the cloud include:
- **On-demand self-service**—Computing capabilities can be provisioned without human interaction from the service provider.
- **Broad network access**—Computing capabilities are available over the network and can be accessed by diverse client platforms.
- **Resource pooling**—Computer resources are pooled to support a multitenant model.
- **Elasticity**—Resources can scale up or down rapidly and, in some cases, automatically, in response to business demands.
- **Measured service**—Resource utilization can be optimized by leveraging charge-per-use capabilities.

The cloud model can be thought of as being composed of three primary service models (**figure 3.11**) and four deployment models (**figure 3.12**). As cloud services have evolved, a variety of other services and permutations have become available. Overall risk and benefits differ per model, so when enterprises consider different types of service and deployment models, they should consider the risk that accompanies them. It should be noted that a large variety of services, platforms and software is currently available under each of areas.

Subsequent evolution of cloud computing has led to an emerging range of options and aspects of information technology offered as a service.

Security as a service (SecaaS) comes in two major forms:
- The cloud service provider (CSP) provides stand-alone managed security services ranging from antivirus scanning and mail security to full deployment of end-point security.
- The CSP offloads appliance utilization for the client, and CPU- and memory-intensive activities are moved to cloud services. For example, antivirus activities on unified threat management (UTM) devices are often offloaded to a SecaaS provider to reduce the number of chassis at the client site. The advantage to clients is minimized risk when applying patches or updates, because they are no longer directly linked to the device.

> **Note:** The Cloud Security Alliance (CSA) is a nonprofit organization with a mission to promote best practices for providing security assurance with cloud computing and can be considered a main source of information on this topic.

With **disaster recovery as a service (DRaaS)**, the CSP offers its cloud infrastructure to provide an enterprise with a disaster recovery (DR) solution. In most cases, the CSP not only provides

Figure 3.11—Cloud Computing Service Models		
Service Model	**Definition**	**To Be Considered**
Infrastructure as a Service (IaaS)	Capability to provision processing, storage, networks and other fundamental computing resources, offering the customer the ability to deploy and run arbitrary software, which can include operating systems and applications. IaaS puts these IT operations into the hands of a third party.	Options to minimize the impact if the cloud provider has a service interruption
Platform as a Service (PaaS)	Capability to deploy onto the cloud infrastructure customer-created or acquired applications created using programming languages and tools supported by the provider	• Availability • Confidentiality • Privacy and legal liability in the event of a security breach (as databases housing sensitive information will now be hosted offsite) • Data ownership • Concerns around e-discovery
Software as a Service (SaaS)	Capability to use the provider's applications running on cloud infrastructure. The applications are accessible from various client devices through a thin client interface such as a web browser (e.g., web-based e-mail).	• Who owns the applications? • Where do the applications reside?
ISACA, *Cloud Computing: Business Benefits With Security, Governance and Assurance Perspectives*, USA, 2009, fig. 1, p. 5, *www.isaca.org/Knowledge-Center/Research/ResearchDeliverables/Pages/Cloud-Computing-Business-Benefits-With-Security-Governance-and-Assurance-Perspective.aspx*		

Figure 3.12—Cloud Computing Deployment Models		
Deployment Model	**Description of Cloud Infrastructure**	**To Be Considered**
Private cloud	• Operated solely for an organization • May be managed by the organization or a third party • May exist on-premise or off-premise	• Cloud services with minimum risk • May not provide the scalability and agility of public cloud services
Community cloud	• Shared by several organizations • Supports a specific community that has shared mission or interest. • May be managed by the organizations or a third party • May reside on-premise or off-premise	• Same as private cloud, plus: • Data may be stored with the data of competitors.
Public cloud	• Made available to the general public or a large industry group • Owned by an organization selling cloud services	• Same as community cloud, plus: • Data may be stored in unknown locations and may not be easily retrievable.
Hybrid cloud	A composition of two or more clouds (private, community or public) that remain unique entities but are bound together by standardized or proprietary technology that enables data and application portability (e.g., cloud bursting for load balancing between clouds)	• Aggregate risk of merging different deployment models • Classification and labeling of data will be beneficial to the security manager to ensure that data are assigned to the correct cloud type.
ISACA, *Cloud Computing: Business Benefits With Security, Governance and Assurance Perspectives*, USA, 2009, fig. 2, p. 5, *www.isaca.org/Knowledge-Center/Research/ResearchDeliverables/Pages/Cloud-Computing-Business-Benefits-With-Security-Governance-and-Assurance-Perspective.aspx*		

backup equipment and storage, but also provides services for a BCP, if it is not yet available. The benefits of DRaaS include:

- The cost for an in-house DR infrastructure is reduced significantly. Because DR is often considered to be a necessity, rather than core business, the ROI in DR services can be significant.
- Offsite storage means that the DR environment is less likely to fail in the case of a major disaster.

Identity as a service (IDaaS) is a relatively new cloud service and currently has two interpretations:

- The management of identities in the cloud that is separated from the users and applications that use the identities. This can be either managed identity services, including provisioning, or management for both onsite or offsite services. Delivering a single sign-on (SSO) solution can also be part of the cloud service offering.

- The delivery of an identity and access management (IAM) solution. IDaaS is often a hybrid solution where access and roles are configured by the CSP and users are authorized by enterprise internal solutions. This is known as a federated model.

Data storage and data analytics as a service, or big data, is the next step in data analysis that makes it possible to analyze all types of data by taking away the constraints on volume, variety, velocity and veracity. These constraints are not taken away by new big data technologies; they are, rather, removed through a synergy between new technologies and the extended capabilities provided by cloud computing. Limitless volume availability and variety allow enterprises to reuse their "old" data for new purposes. Furthermore, big data technology facilitates the ability of enterprises to find patterns in their current data, which influences their way of doing business. In addition to the advice of experienced people, enterprises receive decision-making support from the information that results from big data analysis, such as real-time reporting and predictive analysis.

Information as a service (IaaS) builds on the big data concept—rather than providing the raw data or the algorithms that are used for trending, IaaS provides the required information. With this new service, the result of a query is more important than the query itself.

Integration platform as a service (IPaaS), also called "cloud integrator" by some, is defined by Gartner as "a suite of cloud services enabling development, execution and governance of integration flows connecting any combination of on premises and cloud-based processes, services, applications and data within individual or across multiple organizations" (Gartner IT Glossary, Integration Platform as a Service (IPaaS), *www.gartner.com/it-glossary/information-platform-as-a-service-ipaas)*.

Many enterprises are implementing a hybrid model in which some of their data, applications, services and infrastructure are maintained locally onsite, while others are provisioned by a cloud provider. Integrating all these business resources can be a complex mission. Cloud integrators can help a business cope with this complexity and facilitate the integration without the need to constantly modify and maintain diverse and often incompatible applications. The main advantage is that IPaaS enables efficient and cost-saving methods to ensure IT integration throughout the enterprise. Furthermore, IPaaS provides a more robust solution in the areas of data confidentiality, integrity and availability and data governance, risk and compliance.

Forensics as a service (FRaaS) is a relatively newer service that, according to Jon Rav Gagan Shende, "establishes a cloud forensic investigative process, which can be implemented within a cloud ecosystem, integrated with tolls that should ensure relevant information gathered, verified and stored in a manner that is forensically sound and legally defensible."

Cloud access security brokers (CASBs) are an emerging service to address cloud security issues. These are either on-premises or cloud-based policy control points. The control points are located between the consumer and service provider and primarily serve to enforce the consumer's policies. These may include authentication, authorization, SSO, tokenization, logging, notification and alerts, malware detection and prevention, and perhaps others depending on the particular vendor. It is likely that CASBs will see substantial adoption in the coming years because they offload a number of security issues and greatly simplify security for cloud-based services.

Advantages

Cloud computing is viewed as a significant change to the platform in which business services are translated, used and managed. Many consider it to be as large a shift in IT as the advent of the personal computer or Internet access. However, a major difference between the cloud and those technologies is that the introductions of those earlier technologies encompassed a slower development phase. With the cloud, the required pieces have come together more rapidly for implementation. Some of the drivers bringing the cloud to the attention of enterprise decision makers are:

- **Optimized resource utilization**—Enterprises typically use just 15 to 20 percent of server computing resources. This means that they have five times the computing capacity than is typically used. By using a pay-as-you-go cloud solution, resources become available when needed and are liberated when no longer needed; there is near-perfect alignment with actual demand.
- **Cost savings**—Increased server utilization plus the transition of computational capability from acquired and maintained computers to rented cloud services change the computing cost paradigm from a capital expenditure (CAPEX) to an operational expenditure (OPEX), with potentially significant up-front and total cost savings. Indeed, flexible, on-demand services enable solution testing without significant capital investments and provide transparency of usage charges to drive behavioral change within organizations.
- **Better responsiveness**—On-demand, agile, scalable and flexible services that can be implemented quickly provide organizations with the ability to respond to changing requirements and peak periods.
- **Faster cycle of innovation**—By using the cloud, innovation is handled a lot faster than when addressed within the enterprise. Patch management and upgrades to new versions become more flexible. For the cloud user, upgrading to a new software version is often nothing more than typing a different uniform resource locator (URL) into the web browser.
- **Reduced time for implementation**—Cloud computing provides processing power and data storage as needed and at the capacity needed, in near-real time, not requiring the weeks or months (or CAPEX) that accrue when a new business initiative is brought online in a traditional IT enterprise.
- **Resilience**—A large, highly resilient environment reduces the potential for system failure. The failure of one component of a cloud-based system has less impact on overall service availability and reduces the risk of downtime.

Depending on business needs, any or all of these benefits could be a sufficient reason to consider a cloud computing solution. The recent world economy has pushed many enterprises to be more fiscally conservative. In the IT space, cloud computing presents a potentially significant savings by enabling enterprises to maximize dynamic computing on a pay-per-use basis.

As with any evolving technology, cloud computing offers the possibility of high reward in terms of containment of costs and features such as agility and provisioning speed. However, it also brings the potential for unknown and potentially high risk. Cloud computing introduces a level of abstraction between the physical infrastructure and the owner of the information being stored and processed. Traditionally, the data owner has had direct or indirect control of the physical environment affecting his/her data. In the cloud, this is no longer the case. Due to this abstraction, there is already a widespread demand for greater transparency and a robust assurance approach of the cloud computing supplier's security and control environment.

After it has been determined that cloud services are a plausible solution for an enterprise, it is important to identify the business objectives and risk that accompany a cloud implementation. This will assist enterprises in determining what types of data should be trusted to the cloud, as well as which applications and services might deliver the greatest benefit. For many organizations, prudence dictates initially entrusting only relatively low-value, noncritical services to the cloud.

Security Considerations

For organizations in which security is not considered a high priority, security provided by a reputable cloud provider may be a significant improvement. However, for the information security manager, security considerations are an issue that must be carefully assessed. The loss of control over sensitive data must be considered. The location of data may be an issue as well. In organizations that store and transmit data across state or national boundaries, the information security manager may need to consider myriad laws, regulations and compliance requirements of various jurisdictions. Requirements for handling incidents may vary from one jurisdiction to another (e.g., breach notification laws). Availability of audit logs may also be limited or nonexistent from the cloud provider, and the actual level of security may be difficult to ascertain. The relationship of cloud services and risk is shown in **figure 3.13**.

Figure 3.13—Cloud Computing Risk Map

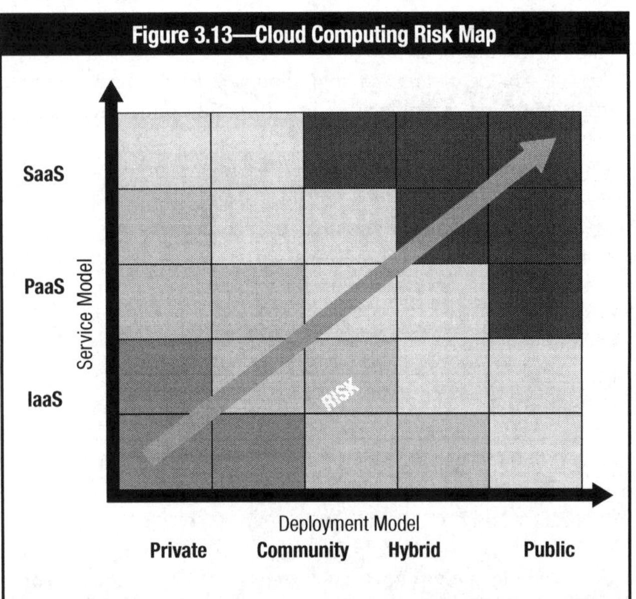

Source: ISACA, *Controls and Assurance in the Cloud Using COBIT 5*, USA, 2014, figure 3

Evaluation of Cloud Service Providers

As is the case with many outsourced services, the primary driver for cloud computing is usually cost. Depending on the criticality of the services provided, selection of a vendor based on cost alone could pose an excessive risk to the organization. Therefore, it is incumbent on the organization to conduct a thorough evaluation of the potential service providers based on the organization's security posture and risk appetite. It must be considered that the services provided are under the control of a number of other organizations, including the data center provider, backbone transport and Internet service provider, among others. Failure of any of these providers results in loss of possibly critical operations. Thorough evaluation of all providers can serve to quantify this risk to some extent and provide an informed basis for the decision to use these services.

There are a number of approaches that can identify the risk of cloud-based solutions and provide reasonable assurance of an acceptable level of performance.

The existing assurance frameworks can be classified into two broad categories:

- Existing, widely accepted frameworks customizable for the cloud (i.e., COBIT, ISO 2700x)
- Frameworks built for the cloud (i.e., CSA Cloud Control Matrix, Jericho Forum® Self-Assessment Scheme)

Figure 3.14 provides an overview of some of the various cloud standards, certifications and assurance frameworks available.

Knowledge Check: Outsourcing Contracts and Cloud Computing

1. What is the main purpose of developing a contract with a third-party service provider?
2. List five key points a contract with a third-party vendor should include, from the information security manager's perspective.
3. What are some additional security considerations that need to be taken into consideration when considering use of a cloud service provider?
4. To address the concerns in question 3, what provisions should the information security manager include in a contract with the cloud service provider?

Answers on page 205.

Figure 3.14—Cloud Standards, Certifications and Frameworks
AICPA/CICA Trust Services (SysTrust and WebTrust)—Intended to provide assurance that an enterprise's systems controls meet one or more of the Trust Services principles and related criteria. Areas addressed by the principles include security, online privacy, availability, confidentiality and processing integrity. SysTrust is similar to a SOC 1 report, but with predefined principles and criteria. However, these principles, while of the proper intent needed for cloud risk assurance, may lack the specificity required to be effective in a cloud environment. IT audit and assurance professionals could insert within these overarching controls specific risk control points, but the responsibility is on the user auditors to properly determine these more detailed control points. Also, effective in 2011, Trust Service reports can be issued as SOC 2 or SOC 3 reports under the SSAE standard noted previously.
AICPA Service Organization Control (SOC) Reports—A SOC report is an independent, third-party examination under the AICPA/Canadian Institute of Chartered Accountants (CICA) audit standards. Released under Statement of Standards for Attestation Engagements (SSAE) No. 16 and the International Standard on Assurance Engagements (ISAE) 3402, SOC reports replaced the previously used Statement on Auditing Standards No. 70 (SAS 70) third-party examination reports effective 15 June 2011. Under a SOC report, a CSP engages a CPA firm to perform an independent examination to provide the CSP clients and their internal and external auditor's assurance regarding the understanding and reliance on controls that support the CSP client's processes and systems. There are three SOC report forms available: • ***SOC 1*** reports apply to financial reporting processes and are most consistent with prior Statement on Auditing Standards 70 reports. • ***SOC 2*** and ***SOC 3*** reports are discussed in the previous row as AICPA Trust Services. SOC reports provide the client with an understanding of the nature and significance of the services provided and the relevant impact in identifying and assessing the risk and assurances by the CSP.
Background Intelligent Transfer Service (BITS)—The BITS Shared Assessment Program contains the Standardized Information Gathering (SIG) questionnaire and Agreed Upon Procedures (AUP). They are used primarily by financial operations evaluating the IT controls that their IT service providers have in place for security, privacy and business continuity. SIG is aligned with ISO/IEC 27002:2005, Payment Card Industry Data Security Standard (PCI DSS), COBIT and NIST and is also aligned with US Federal Financial Institutions Examination Council (FFIEC) guidance, the AICPA/CICA Privacy Framework and many other privacy regulatory guidance organizations. Like the other frameworks mentioned, BITS covers most, but not all, security elements of cloud computing, with a subset of the entire questionnaire. BITS has also mapped its control framework for CSPs. (In 2010, Shared Assessments published "Evaluating Cloud Risk for the Enterprise", a risk-based guide to evaluating cloud computing for the enterprise, *sharedassessments.org/media/pdf-EnterpriseCloud-SA.pdf.*)
Cloud Control Matrix—With a first version released by CSA in April 2010, this cloud security controls matrix is specifically designed to provide fundamental security principles to guide cloud vendors and assist prospective cloud clients in assessing the overall security risk of a CSP. The foundation of CSA's cloud control matrix is other industry-accepted security standards, regulations and controls frameworks. CSA's matrix is an amalgam of controls from Health Insurance Portability and Accountability Act (HIPAA), ISO/IEC 27001/27002, COBIT, PCI and NIST.
COBIT—Developed and maintained by ISACA, COBIT 5 provides management with a comprehensive framework for the management and governance of business-driven, IT-based projects and operations. Appendix A maps the COBIT 5 process practices to the cloud.
CSA STAR Certification—Closely related to the previous rating methodology is the recently released CSA STAR Certification (September 2013). Based on the perspective that enterprises that outsource services to CSPs have a number of concerns about the security of their data and information, achieving the STAR Certification will allow cloud providers of every size to give prospective customers a greater understanding of their levels of security controls. The CSA STAR certification is a technology-neutral certification based on a third-party independent assessment of the security of a CSP. It leverages the requirements of the above described ISO/IEC 27001:2005 management system standard with the CSA Cloud Control Matrix.
Federal Risk and Authorization Management Program—FedRAMP is a US government-wide program that provides a framework for security assessments and authorizing cloud computing services. FedRAMP is designed for federal agency use, but can be used for joint authorizations and continuous security monitoring services for both government and commercial cloud computing systems.
Jericho Forum® Self-Assessment Scheme (SAS)—A guideline for vendors to self-assess the security aspects of their cloud offering and for prospective cloud clients to include into their requests for proposal (RFPs). Jericho Forum's Self-Assessment Scheme is based on the organization's "11 Commandments," released in 2006, which are design principles for effective security in de-perimeterized environments. This Self-Assessment Scheme is designed to assess cloud security tools, either applications or devices.
ISO 20000—This was the first international standard on IT service management, established in 2005. It allows organizations to certify their "design, transition, delivery and improvement of services that fulfil service requirements and provide value for both the customer and the service provider." The ISO certification is not a one-off exercise: Maintaining the certificate requires reviewing and monitoring the Information Security Management System (ISMS) on an ongoing basis.
ISO 2700x—This is the specification against which an enterprise's information security management system (ISMS) is evaluated and by which certification is granted. The objective of the standard is to "provide a model for establishing, implementing, operating, monitoring, reviewing, maintaining and improving an ISMS." The ISO certification is not a one-time exercise: Maintaining the certificate requires reviewing and monitoring the ISMS on an ongoing basis. More than 1,000 certificates have been issued across the world. Additionally, CSA's Cloud Security Matrix has been identified as an appropriate ISO control subset for the cloud.[30]
NIST SP 800-53—The NIST IT security controls standards, much like ISO and COBIT, contain a controls framework required to address cloud security. Also similar to ISO and COBIT, the NIST IT security controls standards form an unspecified subset of the entire framework. Note: The current draft of NIST SP 800-146 contains additional guidance for using and implementing the various cloud deployment models.
Source: Adapted from ISACA, *Controls and Assurance in the Cloud Using COBIT 5*, USA, 2014, figure 29

3.11.11 INTEGRATION WITH IT PROCESSES

It is important to provide defined interfaces between the organization's security-related functions and other organizational assurance functions and ensure that there are clear channels of communication. For example, information security risk management activities should integrate well with the activities of an organizational risk manager to ensure continuity and efficiency of efforts. BCP is often a separate function that must integrate with incident response activities of the information security department.

Integration

The information security manager must ensure that the information security program interfaces effectively with other organizational assurance functions. This will help drive the organization to the strategic outcome of business assurance process integration, discussed in chapter 1. It will serve to reduce security gaps and duplication of efforts, situations where poor practices in one area serve to undermine information security efforts, or assurance process providers (e.g., compliance, QA, change and configuration management, legal, audit, physical security) working at cross-purposes.

These interfaces are often bidirectional; that is, security-related information is received from these departments and, in turn, information security provides relevant information to these units. The assurance functions provide input, requirements and feedback to the information security program, which, in turn, provides metrics and evaluation data for assurance evaluation. It is important that the organization's assurance units are part of the steering committee to ensure broader awareness of security issues. Broad ongoing involvement also helps prevent the unintegrated silo effect created when assurance functions operate in isolation.

To be effective, information security must be pervasive, affecting every aspect of the enterprise. As a consequence, the range of responsibilities for effective security management is broad and, in most cases, exceeds the direct authority of the information security manager. As a result, the information security manager is most likely to be successful by operating in a collaborative fashion, being a good communicator and developing a persuasive business case for security initiatives.

System Development Life Cycle Processes

Achieving effective information systems security is easiest when risk and protection considerations are included in the SDLC. Because these activities are usually the responsibility of other departments but have significant impact on security, the information security manager must develop approaches to integrating these functions with information security activities. It is also critical to create a process to inform the information security manager of proposed change so that he/she can ensure the associated risk is assessed and appropriate treatment is provided.

The traditional division of SDLC stages includes:
- Initiation
- Development or acquisition
- Implementation
- Operation or maintenance
- End of life/disposition

Some organizations have developed expanded versions of SDLC processes generally consisting of:
- Establishing requirements
- Feasibility
- Architecture and design
- Proof of concept
- Full development
- Integration testing
- Quality and acceptance testing
- Deployment
- Maintenance
- System end of life

Change Management

Most organizations employ some form of change management process, whether formal or informal. Security needs to be an integral part of the change management process because new vulnerabilities may be introduced as a result of system or process changes. The information security manager should identify all change management processes used by the organization and establish a notification process when changes are taking place that may impact security.

The information security manager needs to implement processes to ensure that security implications are considered as a standard practice. This is generally accomplished by a requirement that risk and potential impact of the proposed change is assessed prior to implementation and appropriately treated if risk exceeds the acceptable level. A part of the assessment is analysis of the extent to which existing controls may mitigate any change in risk. As changes are made to systems and processes over time, it is important to consider that there is often a tendency for existing security controls to become less effective. Therefore, it is critical for the security manager to ensure that information security controls and countermeasures are tested regularly and/or as a part of the change management process and, if needed, updated and adapted or modified to address changes in risk or impact.

Decentralized organizations can pose a special challenge to the security manager in terms of change management. Often, many of these divisions are highly autonomous, and it may be difficult to monitor and ensure compliance with corporate policies and procedures. It is important to understand the organizational structure during development and implementation of a security program to establish an effective change management approach.

To maintain accountability for policy compliance through inevitable system, environment, business and risk changes, an information security program must identify where in the organization IT changes are initiated, funded and deployed. The information security manager must negotiate hooks into these processes so those in job functions that specify, purchase and deploy new systems have policy compliance as part of their job functions. This includes providing timely notification so the information security manager can understand potential vulnerabilities in new systems or devices, identify new risk presented by the systems, and assist the implementation team in developing policy-compliant standards that can be handed to a release manager as preapproved for production deployment. This is one key element of effectively integrating the security program into the day-to-day work of the organization.

Configuration Management

Incorrect configuration is the major enabler of security breaches to information systems. As a consequence, it is essential that strong procedural and/or technical controls are implemented to effectively manage this risk by ensuring proper configuration prior to system or device activation on the network.

The typical underlying causes for failure to properly configure systems include a lack of clear standards or procedures for configuration and shorthanded staff failing to properly follow procedures or taking improper shortcuts. The information security manager must ensure that proper documentation exists on correct configuration, adequate oversight is provided and IT staff has sufficient training in performing these activities. If proper tested and validated procedural documentation does exist and adequate skills exist, then the workload of those responsible should be examined to see if time is the issue or if some form of compliance enforcement is required to ensure correct configuration.

Release Management

When properly implemented, release management reduces the chances of operational failure by ensuring adequate testing has been performed and required conditions exist for the correct operation of new software, devices or systems. The information security manager should ensure that proper standards and procedures exist so that products are not deployed to production prematurely. In addition, it is important to provide adequate monitoring and oversight to ensure that procedures are followed to avoid unexpected production system failure.

3.12 CONTROLS AND COUNTERMEASURES

In designing the information security program, planning must include both general and application-level controls as required to achieve control objectives. General controls are control activities that support the entire organization in a centralized fashion as a part of the security infrastructure. Because infrastructure is often shared among different departments in the same organization, the term "general controls" is often used to describe all controls in the infrastructure. These include control activities in support of an operating system, network and facility security. These controls typically include centralized user administration policies, standards and procedures, as well as technical elements such as access controls, firewalls and IDSs.

Where specific, noncentralized business information processing is protected by technology, the controls that are specific to providing security to an application are often referred to as application-level controls. In most organizations, general controls and application-level controls are managed by different groups (i.e., technical infrastructure is managed by IT while specific financial controls are managed by the finance department).

The information security manager must ensure the security-related roles and responsibilities are identified for these and all other groups. The roles and responsibilities should be documented in a matrix, such as the one in **figure 1.4**, to ensure a collective understanding across the organization.

A major part of security management is the design, implementation, monitoring, testing and maintenance of controls. Controls are defined as the policies, procedures, practices, technologies and organizational structures designed to provide reasonable assurance that business objectives are achieved and undesirable events are prevented or detected and corrected.

Controls are essentially any regulatory process, whether physical, technical or procedural. The choice of controls is based on a number of considerations, including effectiveness, cost, restrictions to business activities and optimal form.

Controls are one of the primary methods of managing information security risk and a major responsibility of information security management. It is important to understand that controls for physical elements, such as administrative processes and procedures, are just as critical as controls applied to technology. **Most security failures can ultimately be attributed to failures of management, and it must be remembered that management problems typically do not have technical solutions.** Inevitably, people and physical processes exist at each end of technical processes and constitute the greatest risk to information security. As a consequence, the information security manager must be careful not to place excessive focus and reliance on technology.

The information security manager must be aware that standards or procedures that are too restrictive or prevent the organization from meeting its business objectives are likely to circumvented. The objective is to balance the need for controls with the requirements of the business. Therefore, the information security manager must have a good business perspective, understand the risk to the organization's information resources, interpret the information security policies and implement security controls that consider all these aspects. An important perspective is to use the approach that is least restrictive and disruptive to the business that nevertheless meets the criteria for acceptable risk at an acceptable cost. The trade-off between the greatest security and least impact on business activities is the fine line that the effective information security manager must strive to achieve.

Information security controls must be developed for both IT- and non-IT-related information processes. This includes secure marking, handling, transport and storage requirements for physical information as well as considerations for handling and preventing social engineering. Environmental controls must also be taken into account, so that otherwise secure systems are not subject to simply being stolen, as has occurred in some well-publicized cases.

There are a number of standards and guides available for information security management that should be familiar to the information security manager. Two of the most accepted references for information security are COBIT 5 and ISO/IEC 27001 and 27002. Numerous other sources of guidance are available such as the *US Federal Information Processing Standards (FIPS) Publication 200*, NIST 800-53 and the *Standard of Good Practice for Information Security* published by the Information Security Forum.

3.12.1 CONTROL CATEGORIES

Control categories include:

- **Preventive**—Preventive controls inhibit attempts to violate security policy and include such controls as access control enforcement, encryption and authentication. Preventive controls directly address risk.
- **Detective**—Detective controls warn of violations or attempted violations of security policy and include such controls as audit trails, intrusion detection methods and checksums.
- **Corrective**—Corrective controls remediate impact. Backup restore procedures are a corrective measure as they enable a system to be recovered if harm is so extensive that processing cannot continue without recourse to corrective measures.
- **Compensating**—Compensating controls are internal controls that reduce the risk of an existing or potential control weakness resulting in errors and omissions; for example, adding a challenge response component to weak access controls can compensate for the deficiency and insurance can compensate for breach losses. Corrective and compensating controls address impact.
- **Deterrent**—Deterrent controls provide warnings that can deter potential compromise. Examples include warning banners on login screens or offering rewards for the arrest of hackers. Deterrent controls address threat.

Controls and their effect are shown in **figure 3.15**. Note that compensating and corrective controls are often combined because they both address impact.

Knowledge Check: Control Categories

Sort the controls listed below into the appropriate control category:

- Audit trail
- Business interruption insurance
- Data backup
- Data leak prevention
- Employee awareness training
- Error correction
- Segregation of duties
- Intrusion detection system
- Restricted Access sign
- User policy

Preventive	Detective	Corrective	Compensating	Deterrent

Answers on page 206.

3.12.2 CONTROL DESIGN CONSIDERATIONS

Based on the current regulatory environment, controls and countermeasures are most efficiently approached based on a top-down, risk-based approach. This is because control objectives are essentially determined by management's defined acceptable risk levels. The acceptable risk levels are the objectives the controls must be designed to achieve. The control objectives are consequently both the design goal and the subsequent control metric for effectiveness (i.e., the primary control effectiveness metric is the extent to which the control meets the objectives).

Figure 3.15—Control Types and Effect

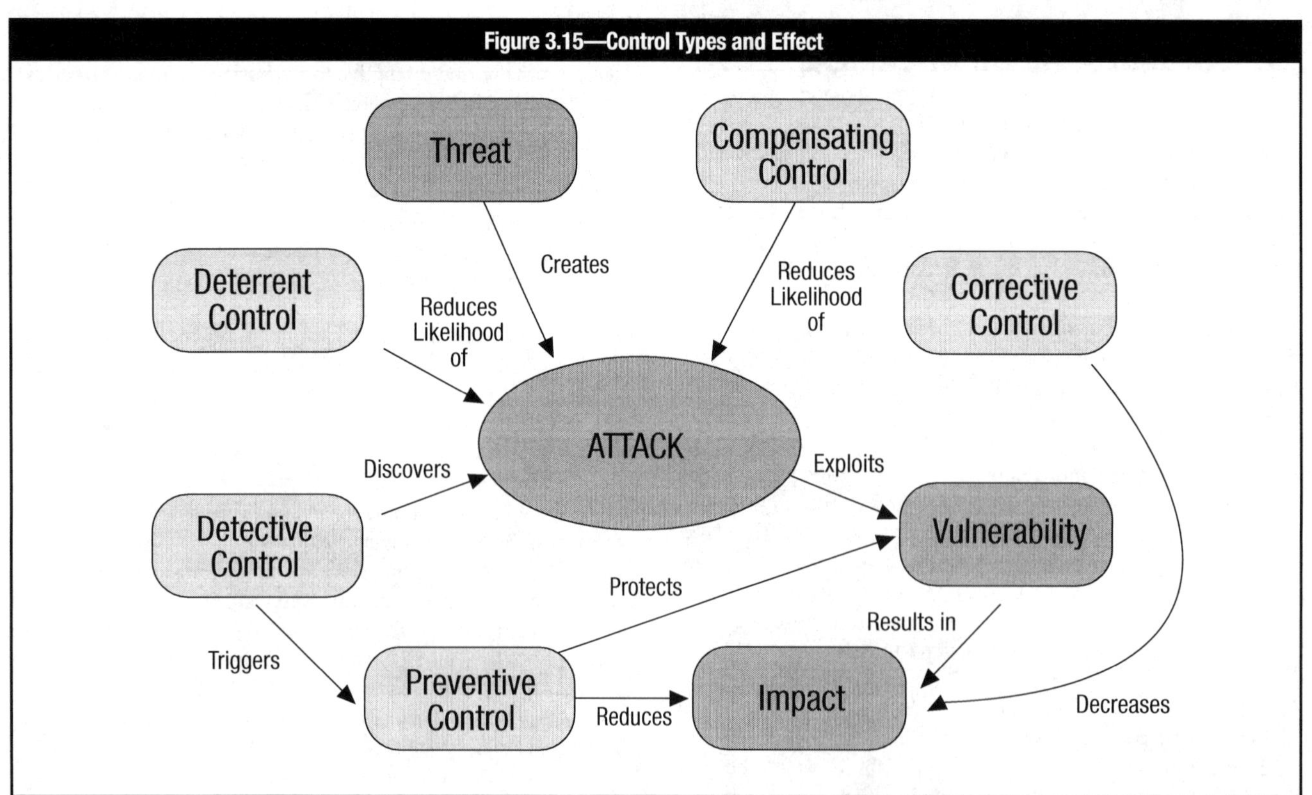

Defining the control objectives is essential in program development and is pertinent to physical, administrative and technical controls.

Achieving control objectives requires combinations of several control types (i.e., a technical control such as a firewall may require a physical protection control, a configuration procedural control and oversight via an administrative control).

As with other security activities, selection of the best options is usually best based on cost-effectiveness comparisons of available options, but this is only one of the criteria. Other selection elements that may need to be considered include:

- Impacts on productivity/inconvenience to users
- Acquisition and implementation costs
- Training costs
- Operational costs
- Maintenance and testing costs
- User acceptance
- Cultural and ethical acceptability
- Legal and regulatory requirements and restrictions
- Adaptability to changing risk
- Scalability
- Ability to monitor
- Provide notification
- Robustness
- Resilience
- Reliability
- Ease of testing
- Self-testing capability
- Acceptable failure mode
- Tamper resistance

Controls as Strategy Implementation Resources

Implementing the strategy largely consists of designing, developing, testing and implementing controls of various types in various combinations. Development of the strategy includes determining acceptable risk and risk tolerance, as discussed in chapters 1 and 2. Acceptable risk levels are used to determine control objectives, which, in turn, set the main requirements for the controls. Controls are also subject to some or all of the criteria described in the previous section.

Effective information security requires controls that will affect all aspects of an organization, including people, technology and processes. Often, a combination of controls is required to achieve the control objectives. The options for controls are essentially unlimited, which adds to the challenge.

For example, access control is a preventive control that prevents unauthorized access that may result in harm to systems. Intrusion detection is a detective control because it enables unauthorized access to be detected. Backup and restoration procedures are a corrective control that enables a system to be recovered if harm is so extensive that data are lost or irreparably damaged, thereby addressing impact. Compensating controls (e.g., insurance) are similar to corrective controls and also compensate for an impact as the result of a compromise.

Security products often provide various combinations of these different types of controls. A typical control is a firewall, which is a product that filters network traffic to limit what protocols (or ports) can be used to enter or exit an internal network, as well as what address or address range is allowed as a source and destination. This is a preventive control because it prevents access to specific network ports, protocols or destinations not specifically allowed. The same firewall may have more advanced features that allow it to examine inbound network traffic for malware and send alerts to an operations center if suspect traffic passes through the device. This is a detective control. The firewall may also have a feature that allows operations to divert incoming traffic to a backup site (although this would typically be accomplished by changes at the Internet router) if, upon responding to the virus alert, it is found that a virus has reduced the capacity at the primary site. That is a recovery, or corrective, control because it allows the systems to resume normal operations. The proxy service that runs on the firewall may be capable of displaying a warning banner as a deterrent control against unauthorized access.

Note that the deterrent, preventive, detective or corrective (or compensatory) control features of the firewall are fully describable, technically, without using the word "firewall." The point is that the information security manager must recognize the security value of technology product features independently of the label given to a product by the product vendor. It is the features and how they are used in achieving the control objectives and enabling control points to be established that are important, not the choice of product or its name.

Controls should be automated as much as possible so that it is technically infeasible to bypass them. (See section 3.12.6 Countermeasures.) Some common control practices that make it hard for users to bypass controls are mechanisms that embody these principles:

- **Access (logical) control**—Users of information should be identified, authenticated and authorized prior to accessing information. There are a variety of ways to implement access control. Most access control models fall into one of two types: mandatory access control (MAC) or discretionary access control (DAC). MAC refers to a means of restricting access to data based on security requirements for information contained in the data and the corresponding security clearance of users. MAC is typically used for military applications where, for example, a secret clearance is required to access data classified as "secret." There is generally also a second requirement of need-to-know for access authorization. DAC refers to means of restricting access to objects based on the identity of subjects and/or groups to which they belong. The controls are discretionary in the sense that rules may allow a subject with certain access permission to pass that permission on to another subject. The choice of appropriate access control mechanisms in a given situation depends on organizational requirements for data protection.
- **Secure failure**—This refers to a device designed to shut down and stop processing information whenever it detects a malfunction that may affect its access control mechanisms. Secure failure as a control policy must be carefully considered because it affects availability. It may also pose an obvious hazard

when electrically controlled physical access fails in a locked condition, preventing egress in the case of fire or other disaster.

- **Principle of least privilege**—This refers to a resource access design strategy that provides administrative capability to partition resource access so that those requiring fewer resources than others have the minimum system privileges that they require to accomplish their responsibilities.
- **Compartmentalize to minimize damage**—This refers to the capacity of system architecture to contain access to subsets of system resources by requiring a separate set of authorization controls per subset. For example, a system in which administrative privileges are not granted when they are requested on certain network interfaces can be combined with network port restrictions to lock Internet users out of administrative functions.
- **Segregation of duties (SoD)**—This refers to the ability of software to restrict a user from having two functions that are meant to provide supervisory or oversight features. For example, software should prevent a person who has the ability to print checks from being able to change the name of the check recipient before and after it is printed.
- **Transparency**—This refers to the ability of the average layperson to understand how system security is supposed to work so that all stakeholders can easily see what effect their activities have on systems security. Users, administrators, engineers and architects should be able to converse about system controls in a way that all can verify that they are working as expected. Transparency is often achieved by keeping the technology design as simple as possible to avoid confusion as to system functionality.
- **Trust**—This refers to a design strategy that includes the existence of a security mechanism whereby the identity of a user can be determined by its relationship to an identity provider that is trusted by a relying party. The relying party has some mechanism for determining the authenticity of a connection from the identity provider and relies on that information to allow the identity provider to pass to it the identity of the user. A typical application is the use of a trusted third party in PKI architecture known as the certificate authority (CA), which attests to the identity of an entity by issuing a certificate.
- **Trust no one**—This refers to a design strategy that includes oversight controls as part of the information system design rather than designating trusted individuals to administer the system and expecting them to follow procedure or relying on subsequent audit to verify if they did. A typical application is the use of closed-circuit television (CCTV) to monitor activities.

3.12.3 CONTROL STRENGTH

Strength of controls can be measured by the type of control being evaluated (preventive, detective, manual, automated, etc.) and its quantitative and qualitative compliance testing results. As such, although an automated control is typically preferable to a manual control, detailed analysis may reveal that a manual control is better. An automated control design may create alerts and generate automatic reports. However, careful evaluation of the process may determine that no evidence of review can be produced and subsequent response actions up to and including resolution cannot be measured. In this scenario, the control fails. On the other hand, if handwritten notes were recorded within IDS log reports on a daily basis with initials and dates, and the same notes contained analysis, action plans, ticket numbers and resolution, then the manual control is far more effective than the automated one. Of course, no final conclusion can be reached as to the strength of the control until it has been adequately tested.

The strength of a control can be measured in terms of its inherent, or design, strength and the likelihood of its effectiveness. An example of an inherently strong control is balancing the books to account for all cash and/or segregating accounting responsibilities among multiple employees. An example of an inherently strong control by design is requiring dual control to access sensitive areas or materials.

To demonstrate value and alignment with business objectives, risk mitigation must be tied to supported business functions. This ensures that information security and IT governance initiatives are inherently followed, and cost justification for the treatment process is readily available and self-explanatory.

3.12.4 CONTROL METHODS

Security controls encompass the use of technical and nontechnical methods. These include administrative, technical and physical controls (**figure 3.16**). Technical controls are safeguards that are incorporated into computer hardware, software or firmware (e.g., access control mechanisms, identification and authentication mechanisms, encryption methods, intrusion detection software). Nontechnical controls are management and operational controls such as security policies, standards, operational procedures and personnel, and physical and environmental security. Controls such as two-factor authentication required for high-security situations can include technical and manual processes (e.g., smart cards requiring a personal identification number). Adequate physical controls are an essential requirement for achieving acceptable overall risk levels.

Figure 3.16—Control Methods

Category	Description
Managerial (administrative)	Controls related to the oversight, reporting, procedures and operations of a process. These include policy, procedures, balancing, employee development and compliance reporting.
Technical	Controls also known as logical controls and are provided through the use of technology, piece of equipment or device. Examples include firewalls, network or host-based intrusion detection systems (IDSs), passwords, and antivirus software. A technical control requires proper managerial (administrative) controls to operate correctly.
Physical	Controls that are locks, fences, closed-circuit TV (CCTV), and devices that are installed to physically restrict access to a facility or hardware. Physical controls require maintenance, monitoring and the ability to assess and react to an alert should a problem be indicated.

Source: ISACA, *CISA Review Manual 26th Edition*, USA, 2015

3.12.5 CONTROL RECOMMENDATIONS

Elements of controls that should be considered when evaluating control strength include whether the controls are preventive or detective, manual or automated, and formal (documented in procedure manuals and evidence of their operation is maintained) or *ad hoc*. During this step of the process, controls that could mitigate or eliminate the identified risk (as appropriate to the organization's operations) are provided. The goal of the recommended controls is to reduce the level of risk to information resources to an acceptable level. The following factors should be considered in recommending controls and alternative solutions to reduce identified risk to acceptable levels:
- Effectiveness of recommended options
- Compatibility with other impacted systems, processes and controls
- Relevant legislation and regulation
- Organizational policy and standards
- Organizational structure and culture
- Operational impact
- Safety and reliability

The control recommendations are the results of the risk assessment and analysis process and provide input to the risk treatment process. During the risk treatment process, the recommended procedural and technical security controls are evaluated, prioritized and implemented. To determine which ones are required and appropriate for a specific organization, a cost-benefit analysis should be conducted for the proposed controls to demonstrate that the costs of implementing the controls can be justified by the reduction in the level of risk or impact.

3.12.6 COUNTERMEASURES

In addition to the general safeguards that standard controls provide, the information security manager may occasionally require a control against a specific threat. Such a control is called a **countermeasure**. Countermeasures often provide specific protection, making them more effective, but less efficient, than broader, more general safeguards—although not necessarily less cost-effective, depending on the original and residual annual loss expectancy (ALE) associated with the threat being countered.

Countermeasures are controls that are put into place in response to a specific threat that is known to exist. They may be preventive, detective or corrective, or any combination of the three. They are not recognized in ISO/IEC 27001:2013 and can be considered a form of targeted control. Countermeasures can be nontechnical as well, such as offering a reward for information leading to the arrest of hackers.

Countermeasures deployed to narrowly address specific threats or vulnerabilities are often expensive, both operationally and financially, and can become a distraction from core security operations. Their deployment should commence only with clear justification and due caution, and only when an existing or more general control cannot adequately mitigate the threat.

Countermeasures can be used just like any other control. Because they are implemented in response to a specific threat, they are applied as incremental enhancements to existing controls. For example, an organization that scans all email to block incoming viruses may encounter a threat to its email infrastructure due to an influx of spam. The organization may implement a countermeasure to the spam attacks by enhancing the virus scanner to block incoming mail from a list of known spammers.

An information security program must be flexible enough to be able to implement a countermeasure with very little warning. Emergency changes may need to bypass standard change control processes, but they must be used with caution and thoroughly documented, and they must still pass through the change management process, even if it is after the fact. Countermeasures are commonly deployed in response to specific threats. A technical countermeasure might be a cell phone jammer to prevent phone usage in sensitive areas.

3.12.7 PHYSICAL AND ENVIRONMENTAL CONTROLS

When implementing an information security program, it is critical to understand that all efforts to protect information have as their foundation a strong physical barrier protecting the physical media on which the information resides. In many organizations, physical security is a service provided as part of facilities management. The physical security organization may set requirements on a building-by-building basis and enforce these requirements using a combination of physical security technology measures and manual procedures. An information security manager should validate technology choices in support of physical security processes and must ensure that policies and standards are developed to ensure adequate physical security.

Physical and environmental controls are a specialized set of general controls on which all computing facilities and personnel depend. In addition, some technologies have features that allow physical mechanisms to override logical controls. For example, unauthorized physical access to technology devices may allow unauthorized access to information. Though the information security manager is often not responsible for physical access controls, it is important that roles and responsibilities with respect to these controls are assigned and the information security manager has an escalation path to ensure that requirements are met. It must be impossible for an unauthorized person to physically connect equipment to the network. In addition, equipment and removable media (including documents and discarded items) must be protected from theft.

Methods to keep unauthorized individuals from gaining access to tangible information resources include identification badges and authentication devices such as smart cards or access controls based on biometrics, security cameras, security guards, fences, lighting, locks and sensors. A variety of intrusion sensors are available, including vibration sensors, motion detectors and many others.

Physical controls are also intended to prevent or mitigate damage to facilities and other tangible resources that might be caused by natural or technological events (e.g., backup power sources can sustain operations if a hurricane damages the power lines serving the facility or if the power grid fails). Environmental controls include air conditioning, water drainage, fire suppression and other measures designed to ensure that the facilities in which systems are stored are designed with the physical limitations of computer system

operation as requirements. Without adequate environmental controls to prevent, detect and recover from physical damage to information systems, other general and application control activities could be rendered ineffective or useless.

In a large and geographically dispersed organization, an IT site operations manager may be assigned at each site to ensure that all equipment is inventoried and configured to policy-compliant standards. An information security manager may establish this role and responsibility as a resource to interface with local physical security organizations on behalf of the information security program.

3.12.8 CONTROL TECHNOLOGY CATEGORIES

When determining the types of control technologies that must be considered by the information security manager, it may be useful to consider operational authority and the types of controls available. Because the majority of technical controls are the direct responsibility of the IT department, consideration must be given to how security will be maintained. Technologies typically fall under one of three different categories in terms of the type of controls that are available: native, supplemental and support control technologies. In some cases, operational authority may be split between IT and the security department.

Native Control Technologies

Native control technologies are out-of-the-box security features that are integrated with business information systems. For example, most web servers include functions providing authentication capabilities, access logging and SSL transport encryption. All these controls would be considered native to the web server technology.

Although the policies and standards governing their use are established by the information security function, most native control technologies are generally configured and operated by IT. This is because native controls often directly impact production operations and providing information security staff with configuration rights to core production systems often violates the SoD principal and can create risk by complicating change control processes and system ownership issues.

Native technology controls exist on all information technology devices including:
- Servers
- Databases
- Routers
- Switches

Supplemental Control Technologies

Supplemental control technologies are usually components that are added on to an information systems environment. They often provide some function that is not available from native components (e.g., network intrusion detection) or that is more appropriate to implement outside of primary business application systems for architectural or performance reasons (e.g., a single network firewall vs. individual host-based network filtering).

Supplemental control technologies tend to be more specialized than native control technologies and, therefore, they are often operated by security specialists. Even when these technical security specialists are resources of the information security group, security technical operations can benefit by leveraging the support of information technology units. In some cases, it may be appropriate to share responsibility for a particular supplemental control technology, particularly if it is deeply embedded in both the security and business application domains. Federated identity and access management technologies are a common example of a supplemental technology for which responsibilities are shared across the security and technology organizations. Typical supplemental control technologies include:
- Federated identity management systems
- SSO
- IPSs
- Firewalls

Management Support Technologies

Management support technologies serve to automate a security-related procedure, provide management information processing, or otherwise increase management efficiency or capabilities. A few examples include security information management (SIM) tools, security event analysis systems and compliance monitoring scanners. These technologies are primarily used by the security organization and they do not directly impact the production environment. For these reasons, and to help enforce SoD, technologies that support security operations are commonly implemented and operated by the information security group in relative independence from the IT department.

Security-related procedures can frequently be automated to increase the information security program's efficiency or capabilities. The use of these supporting technologies to increase resource productivity is an important strategy in developing an efficient and effective program. Some of the most common supporting technologies include:
- SIM tools
- SIEM systems
- Compliance monitoring and management tools
- Access management workflow systems
- Vulnerability scanning tools
- Security configuration monitoring tools
- Policy management and distribution systems

3.12.9 TECHNICAL CONTROL COMPONENTS AND ARCHITECTURE

Information security management includes dealing with a wide range of technical components. Generally, these technical mechanisms have been previously categorized as native control technologies, supplementary control technologies and management support technologies.

Analysis of Controls

Control and support technologies collectively form the technical security architecture. This construct can be applied to individual business applications or the enterprise as a whole, with the objective of revealing how the interaction of individual technical components provides for overall enterprise or application security. This holistic view of technical component capabilities prevents a point-solution perspective that leads to poor overall security. Technical security architecture analysis should be coordinated closely with reviews and analysis of threat and risk factors. The information security manager must ensure that the components of the technical security architecture are aligned with the organization's risk and threat postures as well as business requirements. Provided that the technical architecture is in alignment with the higher levels of architecture (i.e., logical, conceptual), this alignment with business objectives should occur naturally.

When analyzing technical security architecture, the information security manager should use a clearly defined set of measurable criteria to enable tracking of performance metrics. A common approach to the development of metrics is to pose questions that need to be answered and then develop methods to provide answers. A few possible criteria for analyzing technical security architecture and components and developing suitable metrics and monitoring include:

- **Control placement**
  - Where are the controls located in the enterprise?
  - Are controls layered?
  - Is control redundancy needed?
  - Are controls on or near the perimeter efficient providers of broad access protection?
  - Are there uncontrolled access channels to processing services or data? (Consider physical, network, system-level, application and message access vectors.)
- **Control effectiveness**
  - Are controls reliable?
  - Are controls the minimum required?
  - Do controls inhibit productivity?
  - Are controls automated or manual?
  - Are key controls monitored? In real time?
  - Are control easily circumvented?
- **Control efficiency**
  - How broadly do the controls protect the environment?
  - Are controls specific to one resource or asset?
  - Can they and should they be more fully utilized?
  - Is any one control a single point of application failure?
  - Is any one control a single point of security failure?
  - Is there unnecessary redundancy in controls?
- **Control policy**
  - Do controls fail secure or fail open?
  - Do controls implement a restrictive policy (denial unless explicitly permitted) or a permissive policy (permission unless explicitly denied)?
  - Is the principle of least-needed functionality and access enforced?
  - Does the rationale for the control configuration align with policy, corporate expectations and other drivers?
- **Control implementation**
  - Is each control implemented in accordance with policies and standards?
  - Are controls self-protecting?
  - Will controls alert security personnel if they fail or detect an error condition?
  - Have controls been tested to verify that they implement the intended policy?
  - Are control activities logged, monitored and reviewed?
  - Do controls meet defined control objectives?
  - Are control objectives mapped to organizational goals?

Details of specific security technologies, technical control architecture, and analysis of technical control requirements are an important component of information security program development, as well as ongoing management and administration. The information security manager should be well versed in the technologies that are part of the technical security architecture and have access to technical specialists who will be responsible for installing, configuring and maintaining these technologies.

3.12.10 CONTROL TESTING AND MODIFICATION

Changes to the technical or operational environment can often modify the protective effect of controls or create new weaknesses that existing controls are not designed to mitigate. Periodic testing of controls is mandatory in most publicly traded organizations and it should be implemented as a regular practice in all organizations to ensure that procedural controls are being carried out consistently and effectively.

Changes to technical or operational controls must be made with caution. Changes to technical controls should be made under change control procedures and stakeholder approval. The information security manager should analyze the proposed control environment to determine if new or recurring vulnerabilities exist in the design and to ensure that the control is properly designed (i.e., self-protecting, contains a failure policy, can be monitored). Upon implementation, acceptance testing must be conducted to ensure that prescribed policies are enforced by the mechanisms.

Changes to operational procedures should also undergo review and approval by appropriate stakeholders. Requisite changes to process inputs, activity steps, approvals or reviews, and process results should be considered and modifications to related processes and technologies should be coordinated. Workload considerations should also be taken into consideration to ensure that changes to operational controls do not overload resources and impact operational quality. If additional staff training is required to implement changes, it should be coordinated and completed prior to implementation of the change. The operational control should be reviewed in the form of a walkthrough shortly after implementation to ensure that all elements are understood and appropriately implemented.

Extensive information is available on developing control objectives and implementing specific controls from COBIT 5 and other sources such as ISO/IEC 27001 and 27002.

3.12.11 BASELINE CONTROLS

Defined baseline security controls should be a standing requirement for all new systems development. Baseline security requirements should be defined and documented, usually in standards, as an essential part of the system documentation. Adequate traceability of the security requirements should be ensured and supported across the different phases of the life cycle. A few examples include authentication functions, logging, role-based access control and data transmission confidentiality mechanisms. The information security manager must understand the organization's risk appetite and should refer to industry and

regional sources to determine a baseline set of security functions appropriate to organizational policies and acceptable risk levels. Supplemental controls may be warranted based on vulnerability, threat and risk analysis, and these controls should be included in the requirements-gathering process.

The information security team may be consulted during the design and development phases to evaluate the ability of solution options to fulfill acceptable risk requirements. Rarely is a perfect solution found, and there will always be trade-offs among security requirements, performance, costs and other demands. It is important that the information security manager exercise diligence in identifying and communicating solution deficiencies and developing mitigating or compensating controls as required to achieve control objectives. The information security manager should also employ internal or external resources to review coding practices and security logic during development to ensure that adequate practices are being employed.

During the quality and acceptance phases, the information security manager should coordinate testing of originally established functional security requirements in addition to testing system interfaces for vulnerabilities. This testing should verify that the system's security mechanisms meet control objectives and provide the information security team with needed administrative and feedback functions. If functional security shortcomings, coding vulnerabilities or exploitable logic flaws are identified, the information security manager should work with the project team to prioritize and resolve issues. If issues cannot be repaired or mitigated prior to planned rollout, the senior management team should review security issues and associated risk to decide if the system should be deployed prior to resolution of identified vulnerabilities. If the system is to be deployed with unresolved security issues, the security manager should ensure that the issues are documented and there is an agreed-on timetable to resolve issues. If there is no viable resolution available at the time, the information security manager must track and periodically reassess the issue and to determine whether a viable resolution has become available.

The information security manager should ensure that appropriate segregation of duties is considered throughout the SDLC. Testing and QA plans must also be subject to review by the information security manager to ensure that the security elements are properly tested and certified. In some cases, for software developed for critical operations, it may be necessary to perform a code review in addition to the QA testing process to ensure there are no unexpected vulnerabilities. Code reviews are often outsourced for an independent review.

3.13 SECURITY PROGRAM METRICS AND MONITORING

In the process of information security program management, several aspects of metrics must be considered. First, are the metrics necessary to track and guide program development itself at the project, management and strategic levels? Second, will the development of metrics be needed for ongoing management of the results of the program? Because one of the essential elements of controls selection is whether they can be effectively monitored and measured, it is critical to consider this aspect during program design and development. Key controls that cannot be monitored pose an unacceptable risk and should be avoided.

The security of an organization involves much more than specific technical controls like policy, firewalls, passwords, intrusion detection and disaster recovery plans. It also includes processes that surround technical controls and people issues that make it a complex system involving many components.

For any complex system, applying basic system engineering concepts will improve the performance of the system. The concepts of design, planned implementation, scheduled maintenance and management can significantly increase the effectiveness and performance of a security program. One of the fundamental principles of systems engineering is the ability to measure and quantify. Measurement enables proper design, accurate implementation to specifications, and effective management activities including goal setting, tracking progress, benchmarking and prioritizing. In essence, measurement is a fundamental requirement for security program success.

An effective security program involves design and planning, implementation, and ongoing management of the people, processes and technology that impact all aspects of security across an organization.

It may be useful to clarify the distinction between managing the technical IT security systems at the operational level and the overall management of an information security program. Technical metrics are obviously useful for the purely tactical operational management of the technical security infrastructure (e.g., antivirus servers, intrusion detection devices, firewalls). They can indicate that the infrastructure is operated in a sound fashion and technical vulnerabilities are identified and addressed. However, these metrics are of less value from a strategic or overall security program management standpoint. That is, they say nothing about strategic alignment with organizational objectives or how well risk is being managed; they provide few measures of policy compliance or whether objectives for acceptable levels of potential impact are being reached. They also provide no information on information security program direction, velocity or objective proximity.

From a management perspective, while there have been improvements in technical metrics, they are generally incapable of providing answers to such questions as:
- How secure is the organization?
- How much security is enough?
- How do we know when we have achieved adequate security?
- What are the most cost-effective solutions?
- How do we determine the degree of risk?
- How well can risk be predicted?
- Are we moving in the right direction?
- What impact is lack of security having on productivity?
- What impact would a catastrophic security breach have?
- What impact will proposed security solutions have on productivity?

3.13.1 METRICS DEVELOPMENT

The information security governance process described in section 1.1 Information Security Governance Overview should produce a set of goals for the information security program—goals that are tailored to the organization. As discussed in section 3.3.1 Concepts, these goals will generate control objectives and corresponding planning activity designed to achieve the objectives that result in the desired outcomes. Information security program metrics that directly correspond to these control objectives are essential for managing the program.

It should be evident that it will not be possible to develop meaningful security management metrics without the foundation of governance to set goals and create points of reference. That is to say, measurements without a reference in the form of objectives or goals are not metrics and not likely to be useful in program guidance.

Ultimately, metrics serve only one purpose: decision support. The organization measures to manage. It measures to provide the information upon which to base informed decisions relative to what it is trying to accomplish (i.e., the goals).

There are a number of considerations when developing useful and relevant metrics in the course of program development. Because the purpose of metrics is decision support, it is essential to know what decisions are made at various levels of the organization and, consequently, what metrics information is needed to make those decisions correctly. That means that roles and responsibilities have to be defined to know what information is required by whom. The primary parameter of metrics design can be summed up as:
- Who needs to know?
- What do they need to know?
- When do they need to know it?

Metrics need to provide information at one or more of the following three levels:
- Strategic
- Management
- Operational

Strategic

Strategic metrics are often a compilation of other management metrics designed to indicate that the security program is on track, on target and on budget to achieve the desired outcomes. At the strategic level, the information needed is essentially navigational (i.e., determining whether the security program is headed in the right direction to achieve the defined objectives leading to the desired outcomes). This information is needed by both the information security manager as well as senior management to provide appropriate oversight.

Management

Management (or tactical) metrics are those needed to manage the security program such as the level of policy and standards compliance, incident management and response effectiveness, and manpower and resource utilization. At the security management level, information on compliance, emerging risk, resource utilization, alignment with business goals and other topics is needed to make the decisions required for effective management.

The information security manager also requires a summary of technical metrics to ensure that the machinery is operating properly in acceptable ranges, much as the driver of an automobile wants to know that there is fuel in the tank and the oil pressure and water temperature are in an acceptable range. By the same token, while these metrics will not ensure that the car is heading in the right direction or that it will arrive at its destination, they can indicate that the destination will not be reached.

Operational

Operational metrics are the more common technical and procedural metrics such as open vulnerabilities and patch management status. Purely technical metrics are primarily useful for IT security managers and system administrators. These include the usual malware mitigation measures, firewall configuration data, syslog reviews and other operational matters.

There are a number of other considerations for developing metrics. The essential attributes that must be considered include:
- **Manageable**—This attribute suggests that the data should be readily collected, condensed, sorted, stored, correlated, reviewed and understood.
- **Meaningful**—This attribute suggests that the data should be understandable to the recipient and relevant to the objectives and provide a basis for the decisions needed to manage.
- **Actionable**—Just as a compass makes it clear to a pilot in which direction to head to stay on course, so should management metrics make it clear whether to turn left or right. Information that merely invites further investigation may just be clutter.
- **Unambiguous**—Information that is not clear is not likely to be useful.
- **Reliable**—It is essential to be able to rely on the various feedback mechanisms and on their consistency in providing the same result for the same conditions each time they are measured. For example, when the gas tank is actually empty, the gauge should indicate it every time it is measured. In addition, to be reliable, the metric must measure what you think it is measuring, not an unrelated event or spurious artifact.
- **Accurate**—Showing a heading of north when going south is worse than having no information at all. The degree of accuracy depends on how critical the measure is and varies considerably. Qualitative metrics may be approximations, but nevertheless may be adequate in many situations. Quantitative metrics need to be accurate to be of any use (e.g., if the gauge shows 10 gallons of fuel, but the actual amount is only five, the metric can be dangerous).
- **Timely**—To be useful, feedback must occur when needed. It is better to know that the barn door is open before the horse escapes.
- **Predictive**—To the extent possible, leading indicators are very valuable.
- **Genuine**—Metrics subject to manipulation are less reliable and may suffer from inaccuracy as well.

These attributes can be used as metametrics (i.e., a measure of metrics themselves), serving to rank the metrics and determine which are most useful. Any metric that substantially fails to meet these criteria is suspect. In many cases, the metrics in use by most organizations will not rate well on these criteria, but nevertheless

may be the best available. It must also be understood that even well-rated metrics can fail. A prudent approach is to strive to create a system of metrics that cross-reference each other for the purpose of validation. This can be accomplished by measuring two separate aspects of the same thing and ensuring agreement between them. For example, if an engine consumes five gallons of fuel each hour, a 10-gallon tank should last two hours. After one hour, the fuel gauge should indicate that half of the tank remains, assuming it was full at the start. In a similar manner, discrepancies between internal measures of compliance and the results of an audit must be investigated to determine which measure is faulty. If they both indicate approximately a similar level of compliance, it serves to validate the measures.

A number of other descriptive attributes can be developed, but the ones listed previously in this section are the most significant. The question then becomes whether these attributes can be prioritized or whether any metrics or combinations of metrics that do not include most or all of the characteristics listed should be discarded.

Security management metrics must be implemented to determine the ongoing effectiveness of security to meet the defined objectives at the strategic, management and operational levels. The information required to make decisions about security will be different for each of these levels, as will be the metrics to capture it.

Monitoring processes are required to ensure compliance with applicable laws and regulations to which the organization is subject. Monitoring of all relevant metrics is required to ensure that they are operating and the information they provide is properly distributed and handled.

In recent years, a number of industries have become subject to specific regulations to ensure the security and privacy of sensitive information, especially in financial and healthcare organizations, and to reduce operational risk in national critical infrastructure organizations. Compliance failure in these cases can have adverse legal implications, so adequate monitoring is a requirement.

To assess the effectiveness of an organization's security program(s), the information security manager must have a thorough understanding of how to monitor security programs and controls on an ongoing basis. Some monitoring is technical and quantitative in nature while other aspects are, by necessity, imprecise and qualitative. Technical metrics can be used to provide quantitative metrics for monitoring and can include elements such as the number or percentage of:

- Unremediated vulnerabilities
- Open or closed audit items
- User accounts in compliance with standards
- Perimeter penetrations
- Unresolved security variances

Qualitative metrics that should be monitored can be used to determine trends and can include such things as:

- CMMI levels at periodic intervals
- KGIs
- KPIs
- KRIs
- Business balanced scorecard (BSC)
- Six Sigma quality indicators
- ISO 9001:2015 quality indicators
- COBIT 5 PAM

Other relevant measures of significance can include the cost-effectiveness of controls and the extent of control failures.

Other monitoring activities relate to organizational compliance with security policies and procedures established by the organization as a security baseline. As information resources change over time, it is important to be aware that the security baseline and the resources must adapt to changing threats and new vulnerabilities. It is important that all stakeholders are aware of these changes and an appropriate consensus is reached.

3.13.2 MONITORING APPROACHES

It is important for the security manager to develop a consistent, reliable method to determine the overall ongoing effectiveness of the program. One way is to regularly conduct risk assessments and track improvements over time. Another standard tool is the use of external and internal scanning and penetration testing to determine system vulnerabilities, although this will indicate the effectiveness of only one facet of the overall program. Doing so on a regular basis and tracking the results can be a useful indicator of trends in technical security. Most organizations conduct regular vulnerability scans to determine if open vulnerabilities are addressed and new ones appear. Steady improvement is the hallmark of an effective program, although it is of limited value from a management perspective unless it is accompanied by information about viable threats and potential impacts to provide enough information to inform appropriate security decisions.

In addition to monitoring automated security activities, the organization's change management activities should also feed the information security's monitoring program. Metrics are important, but of little use if adverse trends are not dealt with in a timely manner. The information security manager should have a process in place whereby metrics are reviewed on a regular basis and any unusual activity is reported. An action plan to react to the unusual activity should be developed as well as a proactive plan to address trends in activity that may lead to a security breach or failure.

Monitoring Security Activities in Infrastructure and Business Applications

Because an organization's vulnerability to security breaches likely exists 24/7, continuous monitoring of security activities is a prudent business practice (and increasingly a regulatory requirement) that the information security manager should implement. A great deal of this monitoring will be technical and may be performed by IT personnel. In this case, the monitoring requirements must be defined in suitable operating standards along with severity criteria and escalation processes.

Continuous monitoring of IDSs and firewalls can provide real-time information on attempts to breach perimeter defenses. Training help desk personnel to escalate suspicious reports that may signal a breach or an attack can serve as an effective monitoring and early warning system. This information can be critical to taking corrective action in a timely manner.

Determining Success of Information Security Investments
It is important for the information security manager to have processes in place to determine the overall effectiveness of security investments and the extent to which objectives have been met. There is always competition for resources in organizations, and senior management will seek to obtain the best ROIs and justify costs.

During the design and implementation of the security program, the information security manager should ensure that KPIs are defined and agreed to, and a mechanism to measure progress against those indicators is implemented. This way, the information security manager can assess the success or failure of various components of the security program and whether they are cost-justifiable. This will be helpful when developing a business case for other elements of the security program.

Actual costs for various components of a security program need to be accurately calculated to determine cost-effectiveness. It is useful to use the concept of TCO for evaluating the various components of a security program. In addition to initial procurement and implementation costs, it is important to include:
- Costs to administer controls
- Training costs
- Maintenance costs
- Monitoring costs
- Update fees
- Consultant or help desk fees
- Fees associated with other interrelated systems that may have been modified to accommodate security objectives

3.13.3 MEASURING INFORMATION SECURITY MANAGEMENT PERFORMANCE

The information security manager should understand how to implement processes and mechanisms that provide the ability to assess the success and shortcomings of the information security program. Measuring success consists of defining measurable objectives, tracking the most appropriate metrics, and periodically analyzing results to determine areas of success and improvement opportunities. The specific objectives of the information security program will vary according to the scope and operating level of the security department, but they must be conceptually and chronologically aligned with business goals.

An information security program generally includes a core set of common objectives:
- Achieve acceptable levels of risk and loss related to information security issues.
- Support achievement of overall organizational objectives.
- Support organizational achievement of compliance.
- Maximize the program's operational productivity.
- Maximize security cost-effectiveness.
- Establish and maintain organizational security awareness.
- Facilitate effective logical, technical and operational security architectures.
- Maximize effectiveness of program framework and resources.
- Measure and manage operational performance.

While each organization must develop its specific goals, the following sections cover the most common objectives and suggested methodologies for measuring their successful delivery.

3.13.4 MEASURING INFORMATION SECURITY RISK AND LOSS

The primary objective of an information security program is to ensure that risk is managed appropriately and impacts from adverse events fall within acceptable limits. Attaining perfect security while retaining system usability is virtually impossible. Determining whether the security program is functioning at a suitable level—balancing operational efficiency against adequate safety—can be approached in a number of ways from different perspectives.

The following are possible approaches to periodically measuring the program's success against risk management and loss prevention objectives:
- The technical vulnerability management approach poses the following questions:
  - How many technical or operational vulnerabilities exist?
  - How many have been resolved?
  - What is the average time to resolve them?
  - How many recurred?
  - How many systems (critical or otherwise) are impacted by them?
  - How many have the potential for external exploit?
  - How many have the potential for gross compromise (e.g., remote privileged code execution, unauthorized administrative access, bulk exposure of sensitive printed information)?
- The risk management approach is concerned with the following questions:
  - How many high-, medium- and low-risk issues are unresolved? What is the aggregate ALE?
  - How many were resolved during the reporting period? If available, what is the aggregate ALE that has been eliminated?
  - How many were completely eliminated vs. partially mitigated vs. transferred?
  - How many were accepted because no mitigation or compensation method was tenable?
  - How many remain open because of inaction or lack of cooperation?
- The loss prevention approach is concerned with the following questions:
  - Were there loss events during the reporting period? What is the aggregate loss, including investigation, recovery, data reconstruction and customer relationship management?
  - How many events were preventable (i.e., risk or vulnerability identified prior to the loss event)?
  - What was the average amount of time taken to identify loss incidents? To initiate incident response procedures? To isolate incidents from other systems? To contain event losses?

In addition to these quantitative metrics, a number of qualitative measures can be applied to risk management success monitoring. Some of these include:
- Do risk management activities occur as scheduled?
- Have IRPs and BCPs been tested?
- Are asset inventories, custodianships, valuations and risk analyses up to date?

- Is there consensus among information security stakeholders as to acceptable levels of risk to the organization?
- Do executive management oversight and review activities occur as planned?

3.13.5 MEASURING SUPPORT OF ORGANIZATIONAL OBJECTIVES

The information security program must support core organizational objectives. Measuring this set of objectives is largely subjective and organizational objectives can change rapidly in the face of evolving operational pressures and market conditions. The following qualitative measures can be reviewed by the information security steering committee and/or executive management:

- Is there a documented correlation between key organizational milestones and the objectives of the information security program?
- How many information security objectives were successfully completed in support of organizational goals?
- Were there organizational goals that were not fulfilled because information security objectives were not met?
- How strong is consensus among business units, executive management and other information security stakeholders that program objectives are complete and appropriate?

The information security manager should recognize that much of a successful measure's value is in analysis of why an objective was or was not met. Qualitative measures such as those represented by support of organizational objectives should be handled as such. For missed objectives, the reasons why they were not accomplished should be analyzed and the feedback used to guide ongoing optimization of the information security program.

3.13.6 MEASURING COMPLIANCE

An essential, ongoing and primary concern for information security program management is policy and standards compliance. Given that most security failures are the result of personnel failing to follow procedures in compliance with standards, compliance is an essential element of security management. Criticality and sensitivity must be known, as they mandate the level of compliance that must be achieved to manage risk to acceptable levels. Anything less than 100 percent compliance is unacceptable when piloting passenger jets or operating nuclear power plants because impacts are likely to be catastrophic. For any activity that is not life- or organization-threatening, the cost and level of compliance efforts must be weighed against the benefits and potential impacts.

Measuring compliance with technical standards is often straightforward and can frequently be automated. Compliance with procedural or process standards is generally more difficult and may pose a challenge. Because this is frequently the area of failure leading to security compromise, it requires careful consideration. Some compliance requirements are sufficiently critical to warrant direct continuous monitoring such as access controlled by guards and sign-in procedures. In other cases, detective controls such as logging and checklists may suffice.

Compliance requirements may be statutory, contractual or internal. If the organization must comply with compulsory or voluntary standards involving information security, the information security manager must ensure that program goals are aligned with these requirements. Likewise, the policies, procedures and technologies implemented by the program must fulfill requirements of adopted standards. Measurements of compliance achievement are often tied to the results of internal or external audits. The information security manager may also wish to implement automated or manual compliance monitoring with higher frequency and/or broader scope than achievable with incremental audits. In addition to actual point-in-time compliance, the program should be measured on the effectiveness of resolving identified compliance issues.

3.13.7 MEASURING OPERATIONAL PRODUCTIVITY

No information security program has unlimited resources. Coupled with rapid growth of IT enterprises, it is crucial for the information security manger to maximize operational productivity.

Ways in which productivity can be improved include using automation technologies, outsourcing low-value operational tasks and leveraging the activities of other organizational units. Security management automation technologies can act as workforce multipliers, increasing the accomplishment of operational tasks many times over. Vulnerability scanning tools are an outstanding example of this effect: Manual vulnerability assessment that took several individuals a number of weeks can now be accomplished in hours or days. The personnel cost savings can often justify the expense of such tools.

Productivity measures are most useful when employed in a time-based comparison analysis. This approach provides a powerful demonstration of security automation value by showing productivity before and after a productivity-boosting technology was applied. Using this approach, the information security manager can demonstrate returns on security investments.

Productivity is a measure of work product generated per resource. For example, if event log entries analyzed is the work product and security analyst is the resource, the measure of productivity is event log entries analyzed per analyst. Productivity measures can be applied to technology resources in addition to personnel (e.g., network packets processed per IDS node).

The information security manager should set periodic goals for increasing the productivity of the information security program through specific initiatives. These goals should be reviewed to determine the productivity gains achieved. Where possible, the information security manager should analyze data such as hourly employee cost and effort expended per task to demonstrate the financial value of productivity improvement initiatives to senior management.

3.13.8 MEASURING SECURITY COST-EFFECTIVENESS

It is important for the information security program to be financially sustainable because financial constraints are a common reason for security lapses, including failure to plan for ongoing maintenance requirements. The information security manager must work to maximize the value of each security investment to control information security expenses and ensure sustainable achievement of objectives.

This process begins with accurate cost forecasting and budgeting. The success of this activity is generally established by monitoring budget utilization vs. original projections and can help identify issues with security cost planning. In addition to measuring budgeting effectiveness, the information security manager should implement procedures to measure the ongoing cost-effectiveness of security components, most often accomplished by tracking cost-result ratios. This approach establishes cost-efficiency goals for new technologies and improvement goals for existing technologies by measuring the total cost of generating a specific result.

To be accurate, costs must include maintenance, operations and administration costs for the period analyzed (e.g., month, quarter, year). Including all pertinent costs provides the information security manager with the TCO for the security investment being analyzed. Ratios of result-units per currency-unit (e.g., 7,400 network packets analyzed per US dollar annually) or vice versa (0.04 euros per thousand emails scanned annually) can be used to demonstrate cost-efficiency and cost of results. Other examples include:
- Costs of vulnerability assessment per application
- Costs for workstation security controls per user
- Costs for email spam and virus protection per mailbox

Technology purchase and deployment efforts represent only a fraction of full life cycle costs. The information security manager must regularly consider the total costs of maintaining, operating and administering technical security components. In addition, the personnel costs associated with ongoing operational and management activities should also be considered. These analyses should be shared with the security steering committee to help identify areas of opportunity for improving cost-effectiveness and assist with forecasting future resource needs.

3.13.9 MEASURING ORGANIZATIONAL AWARENESS

Even in a tightly controlled technical environment, personnel actions can present threats that can be mitigated only through education and awareness. It is important for the information security program to implement processes for tracking the ongoing effectiveness of awareness programs.

Tracking organizational awareness is most commonly achieved at the employee level. As such, the information security manager should work with the HR department to implement metrics for tracking organizational awareness success. Records of initial training, acceptance of policies and usage agreements, and ongoing awareness updates are useful metrics relative to the actual training program. In addition to identifying individuals in need of training, this helps identify organizational units that may not be fully engaged in the security awareness program.

Another measure of awareness program effectiveness is employee testing. The information security manager should develop tools such as short online or paper examinations that are administered immediately following training to determine the effectiveness of the training. In addition, conducting additional quizzing on a random sampling of employees several months after training will help determine the long-term effectiveness of awareness training and other efforts (e.g., information security newsletters).

3.13.10 MEASURING EFFECTIVENESS OF TECHNICAL SECURITY ARCHITECTURE

The technical security architecture is often one of the most tangible manifestations of an information security program. It is important for the information security manager to establish quantitative measures of the effectiveness of the technical control environment. The range of possible technical metrics is quite broad, and the information security manager should identify those metrics most meaningful to the identified recipients. Technical security metrics can be categorized for reporting and analysis purposes by protected resource and geographic location. Some examples of technical security effectiveness metrics include:
- Probe and attack attempts repelled by network access control devices; qualify by asset or resource targeted, source geography and attack type
- Probe and attack attempts detected by IDSs on internal networks; qualify by internal vs. external source, resource targeted and attack type
- Number and type of actual compromises; qualify by attack severity, attack type, impact severity and source of attack
- Statistics on viruses, worms, and other malware identified and neutralized; qualify by impact potential, severity of larger Internet outbreaks and malware vector
- Amount of downtime attributable to security flaws and unpatched systems
- Number of messages processed, sessions examined and kilobytes (KB) of data examined by IDSs

In addition to the quantitative success metrics above, there are a number of important qualitative measures that apply to the technical control environment. Some examples include the following:
- Individual technical mechanisms have been tested to verify control objectives and policy enforcement.
- The security architecture is constructed of appropriate controls in a layered fashion.
- Control mechanisms are properly configured and monitored in real time, self-protection is implemented, and information security personnel are alerted to faults.
- All critical systems stream events to information security personnel or to event analysis automation tools for real-time threat detection.

The information security manager should bear in mind that, although these measures and metrics are useful for IT and information security management and other individuals, most will have little meaning or interest to senior management. A composite summary indicating the security department is performing according to expectations will probably be much more useful to senior and executive management.

3.13.11 MEASURING EFFECTIVENESS OF MANAGEMENT FRAMEWORK AND RESOURCES

Efficient information security management maximizes the results produced by the components and processes it implements. Mechanisms for capturing process feedback, identifying issues

and opportunities, tracking consistency of implementation, and effectively communicating changes and knowledge help maximize program effectiveness. Methods of tracking the program's success in this area include:
- Tracking the frequency of issue recurrence
- Monitoring the level of operational knowledge capture and dissemination
- The degree to which process implementations are standardized
- Clarity and completeness of documented information security roles and responsibilities
- Incorporating information security requirements into every project plan
- Efforts and results in making the program more productive and cost-effective
- Overall security resource utilization and trends
- Ongoing alignment with, and support of, organizational objectives

The information security manager should implement such mechanisms with the goal of extracting additional latent value from the framework, procedures and resources that make up the program.

3.13.12 MEASURING OPERATIONAL PERFORMANCE

Measuring, monitoring and reporting on information security processes help the information security manager ensure that operational components of the program effectively support control objectives. Measures of security operational performance include:
- Time to detect, escalate, isolate and contain incidents
- Time between vulnerability detection and resolution
- Quantity, frequency and severity of incidents discovered after their occurrence
- Average time between vendor release of vulnerability patches and their application
- Percentage of systems that have been audited within a certain period
- Number of changes that are released without full change control approval

The information security manager should determine the most appropriate metrics for tracking security operations within all organizational units. These metrics should be compiled, analyzed and distributed to stakeholders and responsible management on a regular basis. Performance issues should be analyzed for root cause by the security steering committee and improvement solutions implemented.

3.13.13 MONITORING AND COMMUNICATION

There are a number of monitoring considerations in implementing or managing a security program, regardless of its scope. New or modified controls, in addition to numerous other design considerations, require methods to determine if they are operating as intended. Monitoring technical controls often consists of reviewing logs and various alerts, such as an IDS or firewalls, for potential security vulnerabilities or emerging threats.

Procedural and process controls are typically just as important but more difficult to implement. Technical monitoring of physical processes is likely to be most efficient and effective. Personnel usually interface with information systems at various points in typical processes and these will be the most promising control points to monitor. Monitoring earlier in processes, in addition to watching for suitable outcomes, provides earlier warning of impending problems.

Monitoring of information systems security is a critical operational component of any information security program. The information security manager should consider the development of a central monitoring environment that provides analysts visibility into all enterprise information resources. There is a broad range of security events that are logged and could be monitored, and each organization needs to determine which events are the most pertinent in terms of affected resource and event type. Some commonly monitored event types include:
- Failed access attempts to resources
- Processing faults that may indicate system tampering
- Outages, race conditions and faults related to design or other issues
- Changes to system configurations, particularly security controls
- Privileged system access and activities
- Technical security component fault detection

Procedures for analyzing events and taking appropriate responsive action must be developed. Security monitoring analysts should be trained on these procedures, and monitoring supervisors should have response procedures to address anomalies. Usually response procedures involve analyzing related events and system states, capturing additional event-related information, investigating suspicious activity, or escalating the issue to senior analysts or management. The escalation path for security events and incident initiation should be tested regularly.

In addition to real-time monitoring, the information security manager should periodically conduct analysis of trends in security-related events, such as attempted attack types or most frequently targeted resources. The longer view associated with this type of analysis can often reveal threat and risk patterns that would otherwise go unrecognized. Key controls should be monitored in real time whenever possible. Detective controls, such as log reviews, tend to be useful only to determine how something that has already happened might have unfolded. Even IDS alerts are reactive; depending on tuning, an IDS may not be effective as a means of triggering rapid response.

Results of ongoing monitoring may be rolled up to provide assurance to management that security is providing the appropriate levels of operational assurances and control objectives are being met.

3.14 COMMON INFORMATION SECURITY PROGRAM CHALLENGES

Initiating or expanding a security program will often result in a surprising array of unexpected impediments for the information security manager. These include:
- Organizational resistance due to changes in areas of responsibility introduced by the program
- A perception that increased security will reduce access required for job functions
- Overreliance on subjective metrics

- Failure of strategy
- Assumptions of procedural compliance without confirming oversight
- Ineffective project management, delaying security initiatives
- Previously undetected, broken or buggy security software

There are always cultural and organizational challenges in any job function and the path is not cleared for the information security manager simply by virtue of gaining senior management support. To implement a truly successful program, the cooperation of many others in the organization is important as well.

Figure 3.17 identifies some constraints inherent in road map development for three different organizations. An information security manager must be aware of the type of constraints inherent in the organization when designing control activities and strive to minimize their impact on the information security program objectives.

The last line in **figure 3.17** shows the information security program capabilities that result from the set of constraints associated with the corresponding organization. Organization 1 is the most resource-constrained with regard to the implementation of a successful information security program. In contrast, Organization 3 has abundant resources and strong management and technical support in place—all of which are conducive to developing an effective program. These examples are contrasted to demonstrate that the capability of an information security manager to produce an effective information security program is almost totally dependent on the environment in which he/she operates.

Many organizations still view information security as a low-level, technology-based cost center, forgoing security governance and strategic information security management. While universally recognized as necessary, information security is often viewed as obstructionist and an impediment to getting the job done. This situation can create a challenging environment for an information security manager. A few common manifestations of such an environment may include a lack of management support, poor resource levels for information security, a shortsighted approach to strategy and poor cooperation from other business units. Typically, the information security manager in this situation reports far down in the hierarchy and has the daunting task of trying to drive security up the organizational structure.

The situation is gradually changing for the better for a number of reasons, including legal and regulatory mandates that require improved security as a matter of national security and commercial necessity. Expectations of customers and business partners are an important consideration that has provided a positive push toward enhanced security. Another is the credit card industry's rigorous security requirements from the PCI DSS council for organizations that process credit card information. Since this impacts a very wide range of organizations, it is likely to significantly improve many aspects of information security globally. Whether the rate of adoption and improved security will match the pace of increasingly sophisticated and profitable criminal elements remains a challenge.

Future drivers will include increased litigation that is likely to result in substantial damage awards. Finally, the emergence of the necessity for cyber insurance, with the attendant requirements of

Figure 3.17—Constraints on Developing an Information Security Road Map

	Organization 1	Organization 2	Organization 3
Legal and regulatory requirements	Compliance requires major changes in application data flow.	Frequent spot checks interrupt progress on longer-term projects.	Restrictions exist on the sharing of data with service providers.
Physical and environmental factors	Computer room is located on easily accessible floor and subject to flooding.	Data center operation is outsourced.	Data center on fourth floor is a suitable environment control.
Ethics	Attitude is "if I see it on my screen, it is mine."	Attitude is "if I can use it to make money, it should be mine."	Attitude is "if I need it, my request should be approved."
Culture/regional variances	Culture promotes freedom of information sharing.	Turf wars forestall policy approval processes.	The tone at the top promotes information security goals.
Costs	Company is in bankruptcy and cannot spend money on IT.	All information security projects must be cost-justified.	Information security budget is approved and adequate.
Personnel	Former hackers are hired by departments seeking competitive information.	Background checks are done sporadically based on human resources risk assessments.	Personnel screening processes are uniformly implemented.
Logistics	The information security manager is located in a branch with limited network access.	The information security manager is located at headquarters and has no data center access.	The information security manager is well situated between headquarters and the data center.
Resources	No staff or equipment is dedicated to security.	The information security manager staff lacks technical skills.	The information security manager staff has IT experience and daily access to technology.
Capabilities	Documentation only	Process coordination	Control implementation

insurers for adequate risk management, will be a driver for better, more effective security. This is likely to come about in the same manner in which insurance has been instrumental in gradually improving automobile, product and fire safety.

Although the overall situation in terms of security is improving, the information security manager must deal with the situation as it exists. Regardless of organizational circumstances, a persuasive information security manager with a clear vision of the role of information security in the organization can often improve the overall security posture with an ongoing campaign to educate stakeholders in the role and relevance of information security. This includes defining and seeking agreement on information risk control objectives, determining the organization's risk appetite, and identifying mission-critical information assets.

It is also beneficial for information security activities to align with and support defined business objectives. This is most effective if it is an ongoing effort and suitably communicated to stakeholders. Developing meaningful KPIs and metrics is also useful in supporting information security objectives. Each information security manager must determine the appropriate breadth and depth of metrics for his/her own organization to provide the information needed for management. Information security managers should also implement some form of consistent reporting to promote awareness of the importance that information security management has in the achievement of organizational objectives.

Other issues the information security manager may need to deal with occur in situations where information security is a relatively new function within an organization. Even for mature information security programs, requirements and demands are rapidly changing, driven by technical and regulatory pressures. The challenges described in the following sections do not represent an exhaustive list, but they do illustrate methods for assessing and addressing several of the most common concerns. The information security manager should be cognizant of common challenges to effective information security management, the reasons behind those challenges and strategies for addressing them.

Management Support

Lack of management support is most common in smaller organizations or those of any size that are not in security-intensive industries. Such organizations often have no compulsory requirement to address information security and, therefore, often view it as a marginally important issue that adds cost with little value. These views often reflect misunderstanding of the organization's dependence on information systems, the threat and risk environment, or the impact the organization faces or may be unknowingly experiencing.

In such circumstances, the information security manager must use resources such as industry statistics, organizational impact and dependency analyses, and reviews of common threats to the organization's specific information processing systems. In addition, management may require guidance in what is expected of them and approaches that industry peers are taking to address information security. Even if initial education does not result in immediate strengthening of support, ongoing education should still be conducted to develop awareness of security needs.

Funding

Inadequate funding for information security initiatives is one of the most frustrating and challenging issues the information security manager must address. While this issue may be a symptom of an underlying lack of management support, there are often other factors the information security manager is able to influence. Some funding-related issues that may need to be addressed by the information security manager include:

- Management not recognizing the value of security investments
- Security being viewed as a low-value cost center
- Management not understanding where existing money is going
- The organizational need for a security investment not being understood
- The need for more awareness of industry trends in security investment

If additional funding to close financial gaps is not available, the information security manager must exercise strategies that minimize the impact of the financial shortfall on the organization's information risk posture. Some common strategies that can be applied include:

- Leveraging the budgets of other organizational units (e.g., product development, internal audit, information systems) to implement needed security program components
- Improving the efficiency of existing information security program components
- Working with the information security steering committee to reprioritize security resource assignments and providing senior management with analysis of what security components will become underresourced and the associated risk implications

It is important for the information security manager to pay close attention to funding issues and work on them on an ongoing basis. It is often too late for analyzing needs and educating management once the actual budgeting process begins, delaying needed investments by months or years.

It is also important for the information security manager to understand that financing an information security program is often difficult. If the program is working successfully, management may not see where the money is going and may wonder why they are spending so much. If the program is not successful, management may not understand why they are budgeting significant amounts of money for systems that are still being compromised. It is essential for the information security manager to find ways to show that security is important, functional and business-related.

Staffing

The root causes of funding issues extend to the challenge of inadequate staff levels to meet security program requirements. Obstacles to obtaining effective staffing levels might include:

- Poor understanding of what activities new resources will do
- Questioning the need or benefit of new resource activities
- Lack of awareness of existing staff utilization levels or activities
- Belief that existing staff are underutilized
- Desire to examine outsourcing alternatives

When presented with these issues, the information security manager should utilize workload management procedures to generate personnel workload analyses, utilization reports and other metrics that demonstrate the level of effort currently expended. In addition, charts that associate specific information security roles or teams with the protection they provide to enterprise information systems are helpful. Demonstrating high or growing levels of productivity also help demonstrate that the information security program is utilizing resources effectively and efficiently.

If the organization is unable to allocate additional human resources to the program, the information security manager may wish to consider implementing the following strategies to minimize the impact of understaffing on information security program effectiveness:

- Collaborate with other business units to determine if they can assume more information security responsibilities; delegate appropriate tasks with oversight.
- Analyze outsourcing possibilities, especially for high-volume operational activities; be prepared to demonstrate how freed resources would be immediately redeployed to higher-value activities.
- Work with the information security steering committee to reprioritize security personnel assignments; provide senior management with analysis of what security activities will not be addressed with current staff and communicate risk implications.

3.15 CASE STUDY

At the start of a multinational corporation's new financial year, a new information security manager was hired to oversee the firm's information security program. The information security manager requested and obtained policies and procedures, prior year internal audits, and results from the most recent external audits.

Upon reviewing information provided by internal audit, the information security manager focused on the change management component of the overall information security program. In assessing the control design effectiveness for the change management process, it was noted through interviews with key IT personnel that the firm uses a proprietary change control system (dubbed "GLASS") to enter change requests and then track them through the change management process—taking into account approvals and plans for both rolling out and backing out changes. GLASS also keeps a record of all changes made to the billing system, along with the associated date, using unique identifiers for each change and audit logs that capture other billing system activities.

To determine operational effectiveness for the control process, the information security manager conducted interviews with technology owners for the billing system and the firm's audit personnel, then used a random sample of changes taken from the proprietary change control system. The sampling methodology used concurred with the firm's internal audit policies. The resulting control assessment involved in the audit was also conducted, along the parameters required by the firms' policies and procedures. The change control system passed and no exceptions were found.

However, during the firm's external audit, it was determined that there were changes to the system that did not have the required approvals. Reviewing the policies and procedures, the information security manager could not find any obvious reason why the changes were missed during the internal audit. Through subsequent investigation with the audit team and personnel responsible for migrating changes into one of the production systems, the information security manager determined that GLASS was not the sole point of entry for change requests, but instead could be bypassed.

The firm was then tasked to develop a monitoring solution that looked at all billing system and database system activity, then compared it against approved functionality (called "VIEW"). This tool would then alert IT security of any actions that did not reflect those permitted by change management policies and procedures, such as a change that did not have approvals being migrated into production.

1. **When evaluating control design effectiveness, reviewing the audit program and making inquiries with staff, what are some of the mistakes the information security manager made?**

2. **What type of control does the implementation of the VIEW tool represent?**
   A. Compensating
   B. Corrective
   C. Detective
   D. Preventive

Answers on page 206.

CHAPTER 3 ANSWER KEY

IN PRACTICE EXAMPLE (PAGE 168)

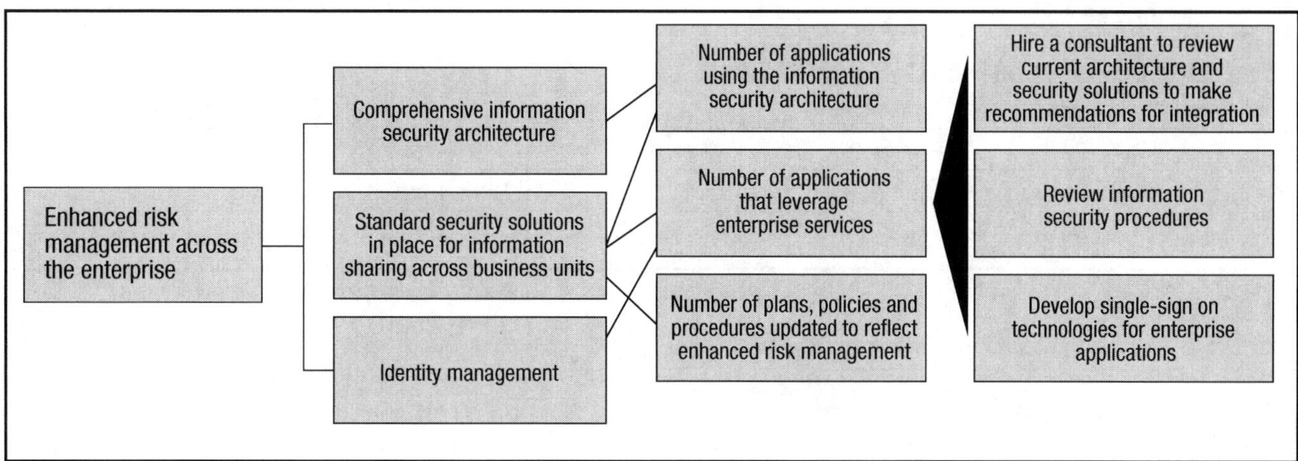

KNOWLEDGE CHECK: OUTSOURCING CONTRACTS AND CLOUD COMPUTING (PAGE 184)

Question 1: To ensure both parties (the outsourcer and the service provider) are aware of their responsibilities and rights within the relationship and provide the means to address disagreements once the contract is in force.

Question 2: Considerations include:

- Detailed specification of outsourced service
- Specific security requirements
- Restrictions on copying information and securing assets
- Prohibiting access without explicit authorization and maintaining a list of individuals who have access
- Right to audit and/or inspect
- Indemnity clauses to mitigate impacts caused by the service provider
- Requirements for IRPs and business continuity plans (BCPs)
- Level of service quality
- Integrity and confidentiality of business assets
- Nondisclosure agreements to be signed by the employees/agents of third parties
- Protection of intellectual property
- Ownership of information
- Requirement that applicable legal and regulatory requirements are met
- Return and/or destruction of information/assets at the end of the contract
- Duration up to which the confidentiality shall be maintained
- Employees or agents of the third party required to comply with security policies of the organization
- Escalation processes
- Predetermined reduction or increases in costs associated with changing service levels
- Response and implementation times for changing service requirements
- Conditions and circumstances for contract cancelation and associated costs

Question 3: Loss of control of sensitive data, location of data, laws, regulations and compliance requirements, processes for handling incidents between jurisdictions, availability of audit logs

Question 4: Considerations include:

- Specific security requirements
- Right to audit and/or inspect
- Indemnity clauses to mitigate impacts caused by the service provider
- Requirements for IRPs and BCPs
- Integrity and confidentiality of business assets
- Nondisclosure agreements to be signed by the employees/agents of third parties
- Protection of intellectual property
- Ownership of information
- Requirement that applicable legal and regulatory requirements are met

- Return and/or destruction of information/assets at the end of the contract
- Duration up to which the confidentiality shall be maintained
- Employees or agents of the third party required to comply with security policies of the organization
- Escalation processes

KNOWLEDGE CHECK: CONTROL CATEGORIES (PAGE 188)

Preventive	Detective	Corrective	Compensating	Deterrent
Employee awareness training	Intrusion detection system	Data backup	Data leak prevention	Restricted Access sign
Segregation of duties	Audit trail	Error correction	Business interruption insurance	User policy

CASE STUDY (PAGE 204)

Question 1: The information security manager's review and sampling methods were flawed. First, the information security manager would not be primarily involved in the audit function; however, he could consult the audit reports to gather additional information. Second, interviews should be supported with additional research because interviews can be unreliable; employees may omit information or not be forthcoming with information for a variety of reasons. Also, as seen in the case study, interviews did not uncover the more significant issues of changes occurring that bypassed the system.

When inquiring about a process that has a built-in version control and monitoring system, it is important to note all possible entry points, or completeness. Since it is the supplemental control technology, GLASS, that recorded activity, it was important to note whether or not there was a way around this system. The closer to the native control technology of a system the information security manager can get, the closer he/she will be to accessing core functionality. The information security manager missed this part of the investigation.

Question 2:

A. A compensating control is an internal control that reduces the risk of an existing or potential control weakness, resulting in errors and omissions. The VIEW reacts to attempted violations and logs the information but does not react based upon the actions of other controls and correct errors; therefore, it is detective.

B. A corrective control would have the power to remediate impact. In the case of the VIEW tool, it is a monitor against activity and sends out alerts when a breach of acceptable activity occurs.

C. A detective control warns of violations or attempted violations of security policy—in this case, logging of system activity and comparing it against authorized actions. The VIEW tool can react only once an action has occurred; therefore, it is detective.

D. The VIEW reacts based on the activity already implemented within the targeted systems; therefore, it cannot be a preventive tool. For it to be preventive, the tool would have to be empowered with preventing users from completing an action that was not authorized.

Chapter 4:

Information Security Incident Management

Section One: Overview

Section Two: Content

view

knowledge necessary to
spond to and subsequently
rganization's information
systems and infrastructure.

DOMAIN DEFINITION

Plan, establish and manage the capability to detect, investigate, respond to and recover from information security incidents to minimize business impact.

LEARNING OBJECTIVES

The objective of this domain is to ensure that the CISM candidate has the knowledge and understanding necessary to:

- Identify, analyze, manage and respond effectively to unexpected events that may adversely affect the organization's information assets and/or its ability to operate.
- Identify the components of an incident response plan.
- Evaluate the effectiveness of an incident response plan.
- Understand the relationship among an incident response plan, a disaster recovery plan and a business continuity plan.

CISM EXAM REFERENCE

This domain represents 19 percent of the CISM examination (approximately 28 questions).

TASK AND KNOWLEDGE STATEMENTS

TASK STATEMENTS

There are 10 tasks within this domain that a CISM candidate must know how to perform:

- T4.1 Establish and maintain an organizational definition of, and severity hierarchy for, information security incidents to allow accurate classification and categorization of and response to incidents.
- T4.2 Establish and maintain an incident response plan to ensure an effective and timely response to information security incidents.
- T4.3 Develop and implement processes to ensure the timely identification of information security incidents that could impact the business.
- T4.4 Establish and maintain processes to investigate and document information security incidents in order to determine the appropriate response and cause while adhering to legal, regulatory and or ganizational requirements.
- T4.5 Establish and maintain incident notification and escalation processes to ensure that the appropriate stakeholders are involved in incident response management.
- T4.6 Organize, train and equip incident response teams to respond to information security incidents in an effective and timely manner.
- T4.7 Test, review and revise (as applicable) the incident response plan periodically to ensure an effective response to information security incidents and improve response capabilities.
- T4.8 Establish and maintain communication plans and processes to manage communication with internal and external entities.
- T4.9 Conduct postincident reviews to determine the root cause of information security incidents, develop corrective actions, reassess risk, evaluate response effectiveness and take appropriate remedial actions.
- T4.10 Establish and maintain integration among the incident response plan, business continuity plan and disaster recovery plan.

KNOWLEDGE STATEMENTS

The CISM candidate must have a good understanding of each of the areas delineated by the knowledge statements. These statements are the basis for the exam.

There are 18 knowledge statements within the information security incident management area:

- K4.1 Knowledge of incident management concepts and practices
- K4.2 Knowledge of the components of an incident response plan
- K4.3 Knowledge of business continuity planning (BCP) and disaster recovery planning (DRP) and their relationship to the incident response plan
- K4.4 Knowledge of incident classification/categorization methods
- K4.5 Knowledge of incident containment methods to minimize adverse operational impact
- K4.6 Knowledge of notification and escalation processes
- K4.7 Knowledge of the roles and responsibilities in identifying and managing information security incidents
- K4.8 Knowledge of the types and sources of training, tools and equipment required to adequately equip incident response teams
- K4.9 Knowledge of forensic requirements and capabilities for collecting, preserving and presenting evidence (e.g., admissibility, quality and completeness of evidence, chain of custody)
- K4.10 Knowledge of internal and external incident reporting requirements and procedures
- K4.11 Knowledge of postincident review practices and investigative methods to identify root causes and determine corrective actions
- K4.12 Knowledge of techniques to quantify damages, costs and other business impacts arising from information security incidents
- K4.13 Knowledge of technologies and processes to detect, log, analyze and document information security events
- K4.14 Knowledge of internal and external resources available to investigate information security incidents

K4.15 Knowledge of methods to identify the potential impact of changes made to the operating environment during the incident response process
K4.16 Knowledge of techniques to test the incident response plan
K4.17 Knowledge of applicable regulatory, legal and organization requirements
K4.18 Knowledge of key indicators/metrics to evaluate the effectiveness of the incident response plan

RELATIONSHIP OF TASK TO KNOWLEDGE STATEMENTS

The task statements are what the CISM candidate is expected to know how to perform. The knowledge statements delineate each of the areas in which the CISM candidate must have a good understanding to perform the tasks. The task and knowledge statements are mapped, insofar as it is possible to do so. Note that although there is often an overlap, each task statement will generally map to several knowledge statements.

Task and Knowledge Statements Mapping

Task Statement		Knowledge Statements	
T4.1	Establish and maintain an organizational definition of, and severity hierarchy for, information security incidents to allow accurate classification and categorization of and response to incidents.	K4.1 K4.2 K4.3 K4.4 K4.5 K4.6 K4.16	Knowledge of incident management concepts and practices Knowledge of the components of an incident response plan Knowledge of business continuity planning (BCP) and disaster recovery planning (DRP) and their relationship to the incident response plan Knowledge of incident classification/categorization methods Knowledge of incident containment methods to minimize adverse operational impact Knowledge of notification and escalation processes Knowledge of techniques to test the incident response plan
T4.2	Establish and maintain an incident response plan to ensure an effective and timely response to information security incidents.	K4.2 K4.3 K4.4 K4.5 K4.6 K4.7 K4.9 K4.10	Knowledge of the components of an incident response plan Knowledge of business continuity planning (BCP) and disaster recovery planning (DRP) and their relationship to the incident response plan Knowledge of incident classification/categorization methods Knowledge of incident containment methods to minimize adverse operational impact Knowledge of notification and escalation processes Knowledge of the roles and responsibilities in identifying and managing information security incidents Knowledge of forensic requirements and capabilities for collecting, preserving and presenting evidence (e.g., admissibility, quality and completeness of evidence, chain of custody) Knowledge of internal and external incident reporting requirements and procedures
T4.3	Develop and implement processes to ensure the timely identification of information security incidents that could impact the business.	K4.2 K4.4 K4.7 K4.13	Knowledge of the components of an incident response plan Knowledge of incident classification/categorization methods Knowledge of the roles and responsibilities in identifying and managing information security incidents Knowledge of technologies and processes to detect, log, analyze and document information security events
T4.4	Establish and maintain processes to investigate and document information security incidents in order to determine the appropriate response and cause while adhering to legal, regulatory and organizational requirements.	K4.1 K4.3 K4.4 K4.8 K4.9 K4.11 K4.13 K4.14 K4.15 K4.17 K4.18	Knowledge of incident management concepts and practices Knowledge of business continuity planning (BCP) and disaster recovery planning (DRP) and their relationship to the incident response plan Knowledge of incident classification/categorization methods Knowledge of the types and sources of training, tools and equipment required to adequately equip incident response teams Knowledge of forensic requirements and capabilities for collecting, preserving and presenting evidence (e.g., admissibility, quality and completeness of evidence, chain of custody) Knowledge of postincident review practices and investigative methods to identify root causes and determine corrective actions Knowledge of technologies and processes to detect, log, analyze and document information security events Knowledge of internal and external resources available to investigate information security incidents Knowledge of methods to identify the potential impact of changes made to the operating environment during the incident response process Knowledge of applicable regulatory, legal and organization requirements Knowledge of key indicators/metrics to evaluate the effectiveness of the incident response plan

Task and Knowledge Statements Mapping *(cont.)*			
Task Statement		**Knowledge Statements**	
T4.5	Establish and maintain incident notification and escalation processes to ensure that the appropriate stakeholders are involved in incident response management.	K4.6 K4.7 K4.14	Knowledge of notification and escalation processes Knowledge of the roles and responsibilities in identifying and managing information security incidents Knowledge of internal and external resources available to investigate information security incidents
T4.6	Organize, train and equip incident response teams to respond to information security incidents in an effective and timely manner.	K4.7 K4.8 K4.9 K4.14 K4.16	Knowledge of the roles and responsibilities in identifying and managing information security incidents Knowledge of the types and sources of training, tools and equipment required to adequately equip incident response teams Knowledge of forensic requirements and capabilities for collecting, preserving and presenting evidence (e.g., admissibility, quality and completeness of evidence, chain of custody) Knowledge of internal and external resources available to investigate information security incidents Knowledge of techniques to test the incident response plan
T4.7	Test, review and revise (as applicable) the incident response plan periodically to ensure an effective response to information security incidents and improve response capabilities.	K4.7 K4.11 K4.16 K4.18	Knowledge of the roles and responsibilities in identifying and managing information security incidents Knowledge of postincident review practices and investigative methods to identify root causes and determine corrective actions Knowledge of techniques to test the incident response plan Knowledge of key indicators/metrics to evaluate the effectiveness of the incident response plan
T4.8	Establish and maintain communication plans and processes to manage communication with internal and external entities.	K4.6 K4.7 K4.10	Knowledge of notification and escalation processes Knowledge of the roles and responsibilities in identifying and managing information security incidents Knowledge of internal and external incident reporting requirements and procedures
T4.9	Conduct postincident reviews to determine the root cause of information security incidents, develop corrective actions, reassess risk, evaluate response effectiveness and take appropriate remedial actions.	K4.10 K4.11 K4.12 K4.17 K4.18	Knowledge of internal and external incident reporting requirements and procedures Knowledge of postincident review practices and investigative methods to identify root causes and determine corrective actions Knowledge of techniques to quantify damages, costs and other business impacts arising from information security incidents Knowledge of applicable regulatory, legal and organization requirements Knowledge of key indicators/metrics to evaluate the effectiveness of the incident response plan
T4.10	Establish and maintain integration among the incident response plan, business continuity plan and disaster recovery plan.	K4.2 K4.3	Knowledge of the components of an incident response plan Knowledge of business continuity planning (BCP) and disaster recovery planning (DRP) and their relationship to the incident response plan

TASK STATEMENT REFERENCE GUIDE

The following section contains the task statements a CISM candidate is expected to know how to accomplish mapped to the areas in the review manual with information that support the execution of the task. The references in the manual focus on the knowledge the information security manager must know to accomplish the tasks and successfully negotiate the exam.

Task Statement Reference Guide		
Task Statement		**Review Manual Reference**
T4.1	Establish and maintain an organizational definition of, and severity hierarchy for, information security incidents to allow accurate classification and categorization of and response to incidents.	4.1 Incident Management Overview 4.2.5 Incident Management Systems
T4.2	Establish and maintain an incident response plan to ensure an effective and timely response to information security incidents.	4.2 Incident Response Procedures 4.2.4 Incident Response Concepts 4.4.2 Incident Response Technology Concepts 4.9.9 Challenges in Developing an Incident Management Plan 4.10.7 Response and Recovery Strategy Implementation 4.10.8 Response and Recovery Plan
T4.3	Develop and implement processes to ensure the timely identification of information security incidents that could impact the business.	4.8.3 Vulnerabilities 4.9 Developing an Incident Response Plan 4.9.5 Help/Service Desk Processes for Identifying Security Incidents
T4.4	Establish and maintain processes to investigate and document information security incidents in order to determine the appropriate response and cause while adhering to legal, regulatory and organizational requirements.	4.2.2 Outcomes of Incident Management 4.2.4 Incident Response Concepts 4.4.4 Roles and Responsibilities 4.13.2 Documenting Events
T4.5	Establish and maintain incident notification and escalation processes to ensure that the appropriate stakeholders are involved in incident response management.	4.2.5 Incident Management Systems 4.4.5 Skills 4.6 Incident Management Metrics and Indicators 4.7.1 Detailed Plan of Action for Incident Management 4.9.4 Escalation Process for Effective Incident Management 4.9.8 Incident Notification Process 4.10.10 Notification Requirements
T4.6	Organize, train and equip incident response teams to respond to information security incidents in an effective and timely manner.	4.9.6 Incident Management and Response Teams 4.9.7 Organizing, Training and Equipping the Response Staff
T4.7	Test, review and revise (as applicable) the incident response plan periodically to ensure an effective response to information security incidents and improve response capabilities.	4.10.9 Integrating Incident Response With Business Continuity 4.10.16 Updating Recovery Plans 4.11.1 Periodic Testing of the Response and Recovery Plans 4.11.2 Testing for Infrastructure and Critical Business Applications 4.11.3 Types of Tests 4.11.4 Test Results 4.11.5 Recovery Test Metrics
T4.8	Establish and maintain communication plans and processes to manage communication with internal and external entities.	4.2.2 Outcomes of Incident Management 4.4.3 Personnel 4.5.3 Assurance Process Integration 4.9 Developing an Incident Response Plan 4.9.4 Escalation Process for Effective Incident Management 4.9.8 Incident Notification Process 4.10.10 Notification Requirements
T4.9	Conduct postincident reviews to determine the root cause of information security incidents, develop corrective actions, reassess risk, evaluate response effectiveness and take appropriate remedial actions.	4.13 Postincident Activities and Investigation 4.13.1 Identifying Causes and Corrective Actions 4.13.2 Documenting Events
T4.10	Establish and maintain integration among the incident response plan, business continuity plan and disaster recovery plan.	4.10.9 Integrating Incident Response With Business Continuity

SUGGESTED RESOURCES FOR FURTHER STUDY

Alberts, Christopher J.; Audrey J. Dorofee; Georgia Killcrece; Robin Ruefle; Mark Zajicek; *Defining Incident Management Processes for CSIRTs: A Work in Progress*, Carnegie Mellon University, USA, 2004

Burtles, Jim; *Principles and Practice of Business Continuity: Tools and Techniques,* Rothstein Associates Inc., USA, 2007

Carnegie Mellon University, Software Engineering Institute, *Create a CSIRT, www.cert.org/csirts/Creating-A-CSIRT.html*

Cloud Security Alliance (CSA), *The Notorious Nine: Cloud Computing Top Threats in 2013*, USA, 2013

CSA, *Top Threats to Cloud Computing V1.0*, USA, 2010

CSA, *Top Threats to Mobile Computing*, USA, 2012

Federal Emergency Management Agency, USA, *www.fema.gov*

Hiles, Andrew; *The Definitive Handbook of Business Continuity Management, 3rd Edition*, John Wiley & Sons Inc., USA, 2011

International Organization for Standardization (ISO), *ISO 22301:2012 Societal security—Business continuity management systems—Requirements*, Switzerland, 2012

Kabay, M.E.; *CSIRT Management,* USA, 2009, *www.mekabay.com/infosecmgmt/csirtm.pdf*

National Institute of Standards and Technology (NIST), *NIST Special Publication 800-61 Revision 2 Computer Security Incident Handling Guide*, USA, 2012

NIST, *NIST Special Publication 800-83 Revision 1 Guide to Malware Incident Prevention and Handling*, USA, 2013

NIST, *NIST Special Publication 800-86 Guide to Integrating Forensic Techniques into Incident Response*, USA, 2006

Snedaker, Susan; *Business Continuity & Disaster Recovery Planning for IT Professionals, Second Edition*, Syngress Publishing Inc., USA, 2014

Note: Publications in bold are stocked in the ISACA Bookstore.

SELF-ASSESSMENT QUESTIONS

CISM exam questions are developed with the intent of measuring and testing practical knowledge in information security management. All questions are multiple choice and are designed for one best answer. Every CISM question has a stem (question) and four options (answer choices). The candidate is asked to choose the correct or best answer from the options. The stem may be in the form of a question or incomplete statement. In some instances, a scenario or a description problem may also be included. These questions normally include a description of a situation and require the candidate to answer two or more questions based on the information provided. Many times a CISM examination question will require the candidate to choose the most likely or best answer.

In every case, the candidate is required to read the question carefully, eliminate known incorrect answers and then make the best choice possible. Knowing the format in which questions are asked, and how to study to gain knowledge of what can be tested, will go a long way toward answering them correctly.

4-1 The **PRIMARY** goal of a postincident review is to:

A. gather evidence for subsequent legal action.
B. identify individuals who failed to take appropriate action.
C. prepare a report on the incident for management.
D. derive ways to improve the response process.

4-2 Which of the following is the **MOST** appropriate quality that an incident handler should possess?

A. Presentation skills for management report
B. Ability to follow policy and procedures
C. Integrity
D. Ability to cope with stress

4-3 What is the **PRIMARY** reason for conducting triage?

A. Limited resources in incident handling
B. As a part of the mandatory process in incident handling
C. To mitigate an incident
D. To detect an incident

4-4 Which of the following is **MOST** important when deciding whether to build an alternate facility or subscribe to a hot site operated by a third party?

A. Cost to rebuild information processing facilities
B. Incremental daily cost of losing different systems
C. Location and cost of commercial recovery facilities
D. Estimated annual loss expectancy from key risk

4-5 Which of the following documents should be contained in a computer incident response team manual?

A. Risk assessment
B. Severity criteria
C. Employee phone directory
D. Table of all backup files

4-6 Which of the following types of insurance coverage would protect an organization against dishonest or fraudulent behavior by its own employees?

A. Fidelity
B. Business interruption
C. Valuable papers and records
D. Business continuity

4-7 Which of the following practices would **BEST** ensure the adequacy of a disaster recovery plan?

A. Regular reviews of recovery plan information
B. Tabletop walkthrough of disaster recovery plans
C. Regular recovery exercises, using expert personnel
D. Regular audits of disaster recovery facilities

4-8 Which of the following procedures would provide the **BEST** protection if an intruder or malicious program has gained superuser (e.g., root) access to a system?

A. Prevent the system administrator(s) from accessing the system until it can be shown that they were not the attackers.
B. Inspect the system and intrusion detection output to identify all changes and then undo them.
C. Rebuild the system using original media.
D. Change all passwords, then resume normal operations.

4-9 Which of the following is likely to be the **MOST** significant challenge when developing an incident management plan?

A. Misalignment between plan and organizational goals
B. Implementation of log centralization, correlation and event tracking
C. Development of incident metrics
D. Lack of management support and organizational consensus

4-10 If a forensics copy of a hard drive is needed, the copied data are **MOST** defensible from a legal standpoint if which of the following is used?

A. A compressed copy of all contents of the hard drive
B. A copy that includes all files and directories
C. A bit-by-bit copy of all data
D. An encrypted copy of all contents of the hard drive

ANSWERS TO SELF-ASSESSMENT QUESTIONS

4-1 A. Forensic evidence should have been gathered earlier in the process.
B. A postincident review should not be focused on finding and punishing those individuals who did not take appropriate action or learning the identity of the attacker.
C. Although a postincident review can be used to prepare a report/presentation to management, it is not the primary goal.
D. The primary goal of a postincident review is to derive ways in which the incident response process can be improved.

4-2 A. Presentation skills are useful for preparing management reports but not the most essential quality.
B. The ability to follow policy and procedures is important, but incidents are unanticipated and chaotic. It is likely that there are no specific policies or procedures to deal with them, and if the individual cannot cope with the stress of an incident, the ability is of little value.
C. Integrity is an essential quality, but if an employee lacks it, they probably should not be employed by the organization.
D. Incident handlers work in high-stress environments when dealing with incidents. Incorrect decisions are likely made if the person is unable to cope with stress; thus, the primary quality of incident handlers is to cope with stress.

4-3 **A. Triage is conducted primarily because incident handling resources are limited, and they must be used to the greatest benefit. With categorization, prioritization and assignment of incidents based on their criticality, resources can be allocated more efficiently.**
B. Triage is not generally considered a mandatory process in incident handling.
C. Triage does not mitigate an incident, but applies available resources most effectively to address the impact.
D. Triage does not serve to detect incidents.

4-4 A. The cost of rebuilding the primary processing facility is not a factor in choosing an alternate recovery site.
B. The daily cost of losing systems is the same whether the alternate site is built or rented.
C. The decision whether to build an alternate facility or rent hot site facilities from a third party should be based entirely on business decisions of cost and ensuring the location is not susceptible to the same environmental risk as the primary facility.
D. Annual loss expectancy is not a factor in choosing to build or rent an alternate site.

4-5 A. Risk assessments would be available to the response team. However, they typically change every year or more often, so it would not make sense to include them in the manual.
B. Severity criteria will remain relatively static and is the only one of the choices that is appropriate for the manual. The other choices will change frequently, and it would not make sense to reprint the manual every time phone numbers or backup files change.
C. A phone directory will change frequently and would not be included in the manual.
D. A table of backup files would typically be very large and change frequently and would not be included in the manual.

4-6 **A. Fidelity coverage means insurance coverage against loss from dishonesty or fraud by employees.**
B. Business interruption insurance protects against losses from events that prevent the business from operating.
C. Valuable papers and records insurance protects against the costs associated with the destruction of business records due to fire, flood or other incident.
D. Business continuity insurance is similar to business interruption coverage, but generally provides broader protection.

4-7 **A. The most common failure of disaster recovery plans is a lack of maintaining the current essential operational information.**
B. Table top walkthroughs are useful only if the information about systems and versions is current and up to date.
C. Recovery exercises are critical for testing plans and procedures, but expert personnel have the knowledge to recover systems without using the plans and written procedures, which makes the recovery test less useful because there is no assurance that in a real disaster these individuals would be available.
D. Audits can be helpful, but they are typically infrequent and use sampling; therefore, they provide limited and only occasional assurance that information in recovery plans is current and up to date.

4-8 A. Preventing access by system administrator(s) provides no protection and does nothing to restore the system.
B. Root access makes it possible to initiate changes that are difficult or impossible to locate and is not an acceptable choice to resolve the issue.
C. If someone, or a malicious program, gains superuser privileges on a system without authorization, the organization never really knows what the perpetrator or program has done to the system. The only way to assure the integrity of the system is to wipe the system clean by either performing a low-level format on the hard disk or replacing it with a new one (usually after making a bit copy backup for the purpose of further analysis and to prevent the destruction of data that may not exist elsewhere) and starting over again by reinstalling the operating system and applications using original media.
D. Changing passwords provides no protections against any malicious changes made to the system.

4-9 A. The incident management plan is a subset of the security strategy, which already aligns to organizational goals and, therefore, does not represent a major challenge.
B. Implementation of log centralization, correlation and event tracking is required, but it is not the most significant challenge.
C. Incident metrics must be developed, but they are straightforward and not a significant challenge.
D. Getting senior management buy-in is often difficult, but it is the necessary first step to move forward with any incident management plan.

4-10 A. Whether a copy is compressed is irrelevant, and a straight copy operation will not include everything on the hard disk that is not identified by the operating system as a standard file.
B. A copy of all files and directories will not be an image of the hard disk and will fail to copy a variety of data, including data between the end of a file and the end of the disk sector ("slack space") and deleted files that have not been overwritten.
C. There is no alternative to making a bit-by-bit copy. For legally sufficient evidence, only a bit copy will result in a true image of the hard drive.
D. Whether the data are encrypted is not relevant, and copying all files and folders will miss certain data such as data between the end of a file and the end of the disk sector ("slack space").

Section Two: Content

4.0 INTRODUCTION

Incident management is defined as the capability to effectively manage unexpected operationally disruptive events to the organization with the objective of minimizing impacts and maintaining or restoring normal operations within defined time limits.

Incident response is the operational capability of incident management that identifies, prepares for and responds to incidents to control and limit damage; provide forensic and investigative capabilities; and maintain, recover and restore normal operations as defined in service level agreements (SLAs).

In most organizations, incident response for information- and information-system-related events is the responsibility of the information security manager. Typically, such events require technical expertise coupled with information security competence. The information security manager is usually responsible for developing and testing the incident management and response plans and also ensuring it correlates with the business continuity and disaster recovery plans in case incident response is insufficient to resolve an event.

Just as asset classification is important due to the need to protect high-value information assets, the information security manager should use a similar approach to incidents. The organization must have an agreed-upon definition for what constitutes an incident and subsequently apply categorization levels, which are typically based on a severity level or impact pervasiveness throughout the organization. This allows the information security manager to prioritize incidents and enables an appropriate response from the incident management resources within the organization.

The information security manager must ensure that a formal (approved) incident response plan (IRP) is established and maintained for multiple objectives, such as:
- Demonstrating that incident response efforts have senior management support
- Ensuring that the IRP is in a communicable form to allow for organizational distribution, review and revision based on incident-handling experience and organizational changes
- Outlining the goals for a consistent and systematic approach in addressing and remediating incidents in a timely manner that is consistent with business objectives

Timely identification of information security incidents refers to the overall effectiveness of the incident identification process. Effectiveness, in this instance, refers to both the timeliness and accuracy with which incidents are identified. An overzealous system that identifies a large number of false positives increases the resource and cost burden attributed to the incident-handling strategy, potentially reducing the organization's ability to successfully identify and remediate valid incidents. Timeliness, as a component of effectiveness when identifying information security incidents, is the time that passes between incident identification and acceptance as a valid incident. This must be kept within organizationally determined acceptable tolerances. If excessive time passes, the impact of the incident may spread, making initial classification appear to be incorrect and remediation efforts more difficult. As a result, more resources would be required to reclassify and reprioritize the incident.

The information security manager must ensure that identified incidents are properly investigated and documented and a response commensurate with incident severity is determined. Effective documentation during the incident discovery process is important to ensure participants in incident response activities consistently understand the incident so they can work together in responding appropriately. In addition, documentation includes data to help forensics examiners (such as a legal investigation). When incident investigations involve individuals, it is important to ensure compliance with legal and regulatory requirements in the jurisdiction and internal policies. It is prudent to involve both legal and human resources (HR) when investigating individuals.

Once an incident is identified and classified, the next step for the information security manager is to notify necessary parties as described in the IRP. These can include the incident response team (IRT) and other support teams, IT and business owners. Operational agreements should be set up, including specifications on timeliness of notifications. Distribution lists of stakeholders must be available and up to date. Escalation procedures must be set up that include alternates in the event some stakeholders cannot be reached or the incident remediation effort encounters complications.

The information security manager must ensure that the organization has dedicated resources that specialize in aspects of incident response to assure that the handling of an incident has minimal impact on the organization. Without properly trained and equipped IRTs, the impact of an incident can quickly absorb operational resources and potentially result in a disaster. For example, in a distributed denial of service (DDoS) attack, without proper IRTs to quickly identify the incident, contact the appropriate resources and initiate containment, the attack is likely to result in major operational disruption and downtime.

The business environment changes over time. Large predetermined events, such as organizational restructurings or changes to business objectives, require the information security manager to test and revise the IRP because new risk is introduced and the IRP is no longer aligned with the organization. Alignment risk may also be present during regular business-as-usual activities. For example, employee turnover means that distribution lists in the communication/escalation component of the IRP may no longer be valid. The information security manager must maintain a periodic and purposeful testing schedule for the IRP to ensure that the plan continues to function properly and is in alignment with the organization's business objectives.

Depending on the nature and extent of a particular incident, the information security manager may involve internal and external resources such as a public relations (PR) representative, audit and legal counsel. Communication plans should be formally documented and resource listings maintained. The incident handlers should be equipped with a methodology to determine whether or not the incident requires communication to various stakeholders and at which point that contact should be established.

To properly address root causes, form lessons learned and maintain an accurate archive of incident events, the information security manager must establish a formal postincident review process. Having a consistent postincident review process allows the organization to review all events, identify trends and pervasive causes, and support management buy-in for constructive solutions and policy changes as needed.

An incident constitutes an unplanned interruption of business activities. This interruption varies in its nature, extent and organizational impact. At a certain point based on the organization's disaster declaration criteria, the impact of an incident results in the declaration of a disaster by the individuals identified as having that authority and should trigger the organization's business continuity plan (BCP) and disaster recovery plan (DRP). For example, an incident that leads to a major interruption of the organization's business activities will usually require the organization to run its operations from an alternate data center. Integrated plans make it more likely that an effective transition to the disaster recovery and business continuity teams will be achieved.

4.1 INCIDENT MANAGEMENT OVERVIEW

Incident management and response can be considered the emergency operations component of risk management. Included are activities that result from unanticipated attacks, losses, theft, accidents or any other unexpected adverse events that occur as a result of the failure or lack of controls.

There is no single best approach that will satisfy incident management requirements for every organization; rather, the approach depends on a variety of factors, including:
- Constituency to be served
- Mission, goals and objectives
- Services provided
- Organizational model and the relationship with the parent organization or customer base
- Funding for start-up costs and ongoing operations
- Resources needed by the computer security incident response team (CSIRT)

Numerous organizations—such as the Carnegie Mellon University Software Engineering Institute (CMU SEI), the US National Institute of Standards and Technology (NIST), the SANS Institute and ISACA—provide information, approaches and methodologies to develop incident management capabilities.

Incident management involves all of the actions taken prior to (including testing and planning), during and after an information security incident occurs. The actions taken should be designed to mitigate the impact of an incident with the following goals in mind:
- Provide an effective means of addressing the situation in such a way that it minimizes the impact to the enterprise.
- Provide management with sufficient information to decide on appropriate courses of action.
- Maintain or restore continuity of enterprise services.
- Provide a defense against subsequent attacks.
- Provide additional deterrence through the use of technology, investigation and prosecution.

The purpose of incident management is to identify and respond to unexpected disruptive events with the objective of controlling impacts within acceptable levels. These events can be technical, such as attacks mounted on the network via viruses, denial of service (DoS) or system intrusion, or they can be the result of mistakes, accidents, or system or process failure. Disruptions can also be caused by a variety of physical events such as theft of proprietary information, social engineering, or lost or stolen backup tapes or laptops, and environmental conditions such as floods, fires or earthquakes. Any type of incident that can significantly affect the organization's ability to operate, or that may cause damage, must be considered by the information security manager and will normally be a part of incident management and response capabilities.

Incident management can include activities that serve to minimize the possibility of occurrences, lessen impacts or both, although this is usually one of the functions of risk management. An example would be securing laptops physically to lessen the possibility of theft and encrypting hard disks to reduce the impact of theft or loss.

As with other aspects of risk management, risk assessments and business impact analyses (BIAs) form the basis for determining the priority of resource protection and response activities.

Incident management, problem management and disaster recovery planning are essential parts of business continuity planning although they are separate, but complementary, processes. As a first responder to adverse events related to information security, the objective of incident management is to prevent incidents from becoming problems and problems from becoming disasters.

The extent of incident management and response capabilities must be carefully balanced with baseline security, business continuity (BC) and disaster recovery (DR). For example, if there is little or no response capability, it may be prudent to raise baseline security levels. The level of incident management capability must also be considered in the context of business continuity planning and disaster recovery planning. There is some point where it will be more cost effective to resort to alternate processing options than to maintain a high level of incident management and response capability.

The goals of incident management and response activities can be summarized as the following:
- Detect incidents quickly.
- Diagnose incidents accurately.
- Contain and minimize damage.
- Restore affected services.
- Determine root causes.
- Implement improvements to prevent recurrence.
- Document and report.

The steps to effectively handle incidents are shown in **figure 4.1**.

Figure 4.1—Incident Management Life Cycle Phases	
Phase	**Activities**
Planning and preparation	• Creating policies, acquiring management support, developing user awareness, building a response capability • Conducting research and development • Building checklists and acquiring necessary tools • Developing a communication plan and awareness training
Detection, triage and investigation	• Defining events vs. incidents and notification process • Detecting and validating incidents • Prioritizing and rating incidents • Implementing intrusion detection systems (IDSs), intrusion prevention systems (IPSs) and security information events monitoring (SIEM) • Utilizing anti-malware and vulnerability management systems • Conducting and participating in global incident awareness, e.g., CERT • Conducting log and audit analysis
Containment, analysis, tracking and recovery	• Executing containment strategy for various incidents • Performing forensic analysis according to evidence-handling processes • Executing recovery procedures in line with the enterprise business continuity plans (BCPs) and disaster recovery plans (DRPs) • Determining the source of the incident
Postincident assessment	• Conducting postmortem: – Exactly what happened, and at what times? – How well did staff and management perform in dealing with the incident? Were the documented procedures followed? Were they adequate? – What corrective actions can prevent similar incidents in the future? • Reporting on incident management related metrics, e.g., mean-time-to-incident-discovery, cost of recovery • Providing feedback of lessons learned
Incident closure	• Conducting incident response postmortem analysis • Submitting reports to management and stakeholders
Source: ISACA, *Incident Management and Response*, USA, 2012	

By definition, incidents are unexpected and often confusing. The ability to detect and assess the situation, determine the causes, and quickly arrive at solutions can mean the difference between an inconvenience and a disaster. An important consideration is knowing the point at which an incident becomes a problem and, subsequently, when the inability to adequately address a problem calls for the declaration of a disaster. The time to make those determinations is not in the middle of a crisis.

Rigorous planning and commitment of resources are necessary to adequately plan for such events. As with other aspects of security, it is critical to achieve stakeholder consensus and senior management support for an effective incident management capability. Support can be gained as a result of unacceptable impacts of prior incidents to the organization and/or from developing a persuasive business case detailing examples of similar events in other organizations that show the consequences of inadequate incident management. Incident management and response can result in lower overall security costs by setting baselines to address common events by providing protection against relatively rare occurrences with an adequate response, containment and recovery capability. This is similar to the common approach for protecting against fire by a combination of reasonable preventive controls, coupled with an effective fire detection system, evacuation plans and triggers for fire department response, complemented with insurance policies in the rare event that incident management and response measures fall short. While a totally fireproof structure could be built, optimal cost-effectiveness uses a combination of risk management approaches developed through trial and error.

Many organizations have a separate department responsible for business continuity planning and disaster recovery and the extent of information security involvement and authority will vary widely. However these departments are structured, it is essential that they work together closely and their plans are complementary and well integrated. The individuals in charge of various types of events must be clearly defined and there must be clear severity and declaration criteria.

Severity criteria should be consistent, concisely described and easy to understand so severity levels of similar events are uniformly determined. Declaration criteria must also be established so it is clear who has the authority to determine the response level, activate the teams, declare a disaster and mobilize the recovery process. Severity criteria and their use must be widely published. Personnel must be trained to recognize potential incidents and provide proper classification, and they must be trained in notification, reporting and escalation requirements. The information security manager should be aware that, regardless of how well incidents are planned for and handled, there is always the possibility that events will escalate to a disaster.

4.2 INCIDENT RESPONSE PROCEDURES

There is no guarantee that even the best controls will prevent disruptive or catastrophic incidents from occurring. Adverse events such as security breaches, power outages, fires and natural disasters can bring IT and business operations to a halt. Response management enables a business to respond effectively when an incident occurs, to continue operations in the event of disruption and survive interruptions or security breaches in information systems.

The following section covers the necessity for incident response capabilities and the typical responsibilities of an information security manager. These responsibilities can vary significantly in different organizations and, in some cases, may include some DR and BC activities.

4.2.1 IMPORTANCE OF INCIDENT MANAGEMENT

As organizations increasingly rely on information processes and systems, and significant disruption to those activities result in unacceptably severe impacts, the criticality of effective incident management and response has grown. Some of the factors that compound the necessity of effective incident management include:

- The trend of both increased occurrences and escalating losses resulting from information security incidents
- The increase of vulnerabilities in software or systems affecting large parts of an organization's infrastructure and impact operations
- Failure of security controls to prevent incidents
- Legal and regulatory mandates requiring the development of an incident management capability
- The growing sophistication and capabilities of profit-oriented and nation-state attackers
- Advanced persistent threats (APT)
- Increasing zero-day attacks

4.2.2 OUTCOMES OF INCIDENT MANAGEMENT

Outcomes of good incident management and response include the following:

- The organization can deal effectively with unanticipated events that might threaten to disrupt the business.
- The organization will have sufficient detection and monitoring capabilities to ensure that incidents are detected in a timely manner.
- Well-defined severity and declaration criteria will be in place, as will defined escalation and notification processes.
- Personnel will be trained in the recognition of incidents, the application of severity criteria, and proper reporting and escalation procedures.
- The organization will have response capabilities that demonstrably support the business strategy by being responsive to the criticality and sensitivity of the resources protected.
- The organization will serve to proactively manage the risk of incidents appropriately in a cost-effective manner and will provide integration of security-related organizational functions to maximize effectiveness.
- The organization will provide monitoring and metrics to gauge performance of incident management and response capabilities, and it will periodically test its capabilities and ensure that information and plans are updated regularly, current and accessible when needed.

These monitoring and metrics activities will ensure that:

- Information assets are adequately protected and the risk level is within acceptable limits.
- Properly trained and equipped incident management and response teams are in place.
- Effective IRPs are in place and understood by relevant stakeholders (e.g., management, IT departments, end users, incident handlers).
- Incidents are quickly identified, categorized correctly and contained, and the root cause is addressed to allow recovery within an acceptable interruption window (AIW).
- Communication flows to different stakeholders and external parties are well controlled, as documented in the communication plan.
- Lessons learned are documented and shared with stakeholders to increase the level of security awareness and serve as a basis for improvement.
- Assurance is provided to internal and external stakeholders (e.g., customers, suppliers, business partners) that the organization has adequate control and is prepared to ensure business survivability in the long term.

4.2.3 THE ROLE OF THE INFORMATION SECURITY MANAGER IN INCIDENT MANAGEMENT

Depending on the organization, the extent of the information security manager's involvement in business continuity/disaster recovery planning and incident response will vary considerably. The typical situation is that the information security manager has, at a minimum, responsibility as first responder to information-security-related incidents, regardless of the causes.

To deal effectively with security incidents, it is important for the information security manager to have a good conceptual and practical understanding of what is required to adequately address those responsibilities. In addition, there must be a good understanding of the BC and DR processes. This is to ensure that incident management and response plans and activities integrate well with the overall BCP and DRP in the event that an incident escalates to a disaster.

4.2.4 INCIDENT RESPONSE CONCEPTS

There are a few concepts important to incident response. The following are some key terms as defined by CMU SEI.

Incident handling is one service that involves all the processes or tasks associated with handling events and incidents. It involves multiple functions:

- Detection and reporting—The ability to receive and review event information, incident reports and alerts
- Triage—The action taken to categorize, prioritize and assign events and incidents to maximize the effectiveness of limited resources
- Analysis—The attempt to determine what has happened, the impact and threat, the damage that has resulted, and the recovery or mitigation steps that should be followed
- Incident response—The action taken to resolve or mitigate an incident, coordinate and disseminate information, and implement follow-up strategies to prevent recurring incidents

Effective incident management ensures that incidents are detected, recorded and managed to limit impacts. Recording incidents is required so incident response activities can be tracked, information can be provided to aid planning activities and no aspect of an incident is inadvertently overlooked. Recording is also required to properly document information that potentially includes forensics data that can be used to pursue disciplinary or legal options. Incidents must be classified to ensure that they are

correctly prioritized and routed to the correct resources. Incident management includes initial support processes that allow new incidents to be checked against known errors and problems so that any previously identified workarounds can be quickly identified.

Incident management provides a structure by which incidents can be investigated, diagnosed, resolved and then closed. The process ensures that the incidents are owned, tracked and monitored throughout their life cycle. There may be occasions when major incidents occur that require a response above and beyond that provided by the normal incident process and may require activating BC/DR capabilities.

Incident management often includes other functions such as vulnerability management and security awareness training. It may include proactive activities intended to help prevent incidents.

Incident response is the last step in an incident-handling process that encompasses the planning, coordination and execution of appropriate mitigation, containment, and recovery strategies and actions.

The information security manager also must be aware of the possibility of nontechnical incidents that must be planned for and addressed. These incidents can include social engineering, lost or stolen backup tapes or laptop computers, physical theft of sensitive materials, and natural disasters.

4.2.5 INCIDENT MANAGEMENT SYSTEMS

The sheer amount of information and activities in increasingly complex systems has driven the development of automated incident management systems in recent years. These systems automate many manual processes that provide filtered information that can identify possible technical incidents and alert the incident management team (IMT).

> **Note:** Some organizations may have a designated IMT that determines how to manage incidents and is separate from the IRT. In other organizations, the IMT and IRT may be the same group of people. For the purposes of this manual, IMT will be differentiated when necessary, but in many cases, the IRT will perform these duties.

An example of a distributed incident management system is one that contains multiple specific incident detection capabilities (e.g., network intrusion detection systems [NIDSs], host-based intrusion detection systems [HIDSs] and server/appliance logs).

An example of a centralized incident management system is a security information and event manager (SIEM). This tool is an automated log reader that combines critical events and logs from many different systems and devices and correlates them into more meaningful incident information. This is accomplished by collecting data from many system and device logs that are normalized in a single database where it is data-mined using correlation and inference tools to create near-real-time notifications and alerts; it can also identify policy violations. Further processing can be done from there (e.g., prioritizing incidents based on their potential severity or business impacts or performing specific notifications/escalations based on the impact ratings).

Another significant feature of incident management systems is their ability to track an incident during its life cycle. Tracking is a powerful feature that ensures that incidents are not overlooked but instead receive necessary attention based on criticality. Tracking enables users to provide more information and receive status updates along the event life cycle until it is closed. The system is most often offered in a web-based format for easy accessibility.

An effective SIEM will:
- Consolidate and correlate inputs from multiple systems
- Identify incidents or potential incidents
- Notify staff
- Prioritize incidents based on business impact
- Track incidents until they are closed
- Provide status tracking and notifications
- Integrate with major IT management systems
- Implement good practices guidelines

There are potential efficiencies and cost savings that can be realized using automated incident management systems once they are properly configured. However, that is usually a complex and time-consuming activity that often takes several years. Some considerations for the information security manager can include:
- **Operating costs**—In the absence of an automated and centralized incident management system, information security staff may be required to monitor different security devices, correlate events and process the information manually. With this approach, there are additional costs for training and maintaining the staff over the long term. There also is a higher probability of human error.
- **Recovery costs**—An automated system, when configured properly, is able to detect and escalate incidents significantly faster than when a manual process is used. The amount of damage can be controlled and further damage prevented when the recovery actions are initiated earlier rather than later. In the case of a manual management system, a longer analysis process may contribute to further damage before incidents are contained.

4.3 INCIDENT MANAGEMENT ORGANIZATION

The incident management capability in an organization acts as the first responder for a variety of incidents, including information processing and processes. It responds to and manages incidents to contain and minimize damage, limit disruptions to business processes, and restore operations as quickly as possible. Poorly managed incidents have the capacity to become disasters. An example is a financial institution being infected with Trojan malware. The absence of a tested IRP and the lack of appropriate response while the Trojan captures data unhindered could result in the viability of the entire organization being at stake due to substantial financial losses and reputational damage.

Just as there are requirements for addressing fire and medical emergencies using trained individuals and appropriate equipment, the information security manager must plan for the inevitable range of incidents likely to disrupt the organization's business operations to an unacceptable extent.

Incident management is, nominally, a component of risk management and can be considered the operational and reactive element. That is, if managing risk is insufficient to prevent a threat from materializing and causing an impact, incident management and response capabilities should be available to react appropriately to limit the damage and restore operations.

The information security manager should understand the various activities involved in a response and recovery program. This includes meeting with emergency management officials (e.g., federal, state/provincial, municipal/local) to understand what governmental capabilities are available. These officials are likely to have information concerning the nature of the risk to which the location or area is susceptible. Most countries and governments have civil defense and/or emergency management agencies that are tasked with advising and assisting the population in dealing with a wide range of natural and human-initiated threats. This will provide insight into historical events that will inform the organization on the sorts of events it must plan to manage.

Emergency management activities typically focus around the activities immediately after an incident. This includes activities during or after a physical disaster, fire, electrical failure or security-related incident. These events may require prompt action to recover operational status. Actions may necessitate restoration of hardware, software and/or data files. Emergency management activities also include measures to assure the safety of personnel, such as evacuation plans and creation of a command center from which emergency procedures can be executed. It also is important that information about an incident be communicated only on a need-to-know basis to control exposure of potentially sensitive information.

4.3.1 RESPONSIBILITIES

There are a number of incident management responsibilities that the information security manager must undertake, including:
- Developing information security incident management and response plans
- Handling and coordinating information security incident response activities effectively and efficiently
- Validating, verifying and reporting protective or countermeasure solutions, both technical and administrative
- Planning, budgeting and program development for all matters related to information security incident management and response

The approach to incident response varies depending on the situation and types of events that may occur, but the goals are constant and include:
- Maintaining incident response readiness
- Containing and minimizing the effects of the incident so damage and losses do not escalate out of control
- Notifying the appropriate people for the purpose of recovery or to provide needed information
- Recovering quickly and efficiently from security incidents
- Responding systematically and decreasing the likelihood of recurrence
- Balancing operational and security processes
- Dealing with legal and law-enforcement-related issues

The information security manager also needs to define what constitutes a security-related incident. Typically, security incidents include:
- Malicious code attacks
- Unauthorized access to IT or information resources
- Unauthorized use of services
- Unauthorized changes to systems, network devices or information
- DOS attacks
- DDoS attacks
- Misuse
- Surveillance and espionage
- Hoaxes/social engineering

It should be noted that many incidents that initially appear to be malicious instead turn out to be the result of human error. Typically, organizations experience nearly double the number of incidents due to human error than are caused by externally initiated security breaches.

4.3.2 SENIOR MANAGEMENT COMMITMENT

As is the case with other aspects of information security, senior management commitment is critical to the success of incident management and response activities.

A business case can show that, under many circumstances, effective incident management and response may be a less costly option than attempting to implement controls for all possible conditions. Tested incident management and response may also allow the organization greater business opportunities by allowing higher levels of acceptable risk based on a demonstrated capacity to handle security incidents. Adequate incident response, in combination with effective information security, is likely to offer the most cost-effective risk management approach and may be the most prudent resource management decision. These points should be included in the business case, which can be used to gain senior management commitment needed to ensure the program's success.

4.4 INCIDENT MANAGEMENT RESOURCES

A number of internal and external resources are used in the development of an incident management and response plan. In a typical organization, these resources include:
- IT department
- Internal audit
- HR department
- Legal department
- Physical security
- Risk management
- Insurance department
- PR department
- Sales and marketing
- Senior management
- Compliance office
- Privacy officer

As with other aspects of information security, it is essential to develop a clear scope, objectives and implementation strategy. The strategy must consider the elements needed to move from the current state of incident management to the desired state. This will clarify what resources are required and how they must be deployed.

4.4.1 POLICIES AND STANDARDS

The IRP must be backed by well-defined policies, standards and procedures. A documented set of policies, standards and procedures is important to:

- Ensure that incident management activities are aligned with the IMT mission
- Set correct expectations
- Provide guidance for operational needs
- Maintain consistency and reliability of services
- Clearly understand roles and responsibilities
- Set requirements for identified alternate personnel for all important functions

4.4.2 INCIDENT RESPONSE TECHNOLOGY CONCEPTS

IRTs must be familiar with the following security concepts:

- **Security principles**—Knowledge of security principles is important to understand potential problems that can arise if appropriate security measures have not been implemented correctly or fail and to appreciate the potential impacts to the organization's systems. These security principles include:
  - Confidentiality
  - Availability
  - Authentication
  - Integrity
  - Access control
  - Privacy
  - Nonrepudiation
  - Compliance
- **Security vulnerabilities/weaknesses**—This entails understanding how any specific attack is manifested in a given software or hardware technology. The most common types of vulnerabilities and associated attacks involve:
  - Physical security issues
  - Phishing resulting from lack of user awareness
  - Protocol design flaws (e.g., man-in-the-middle attacks, spoofing)
  - Malicious code (e.g., viruses, worms, Trojan horses)
  - Implementation flaws (e.g., buffer overflow, timing windows/race conditions)
  - Configuration weaknesses
  - User errors or lack of awareness
- **The Internet**—There must be security in the underlying protocols and services used on the Internet. A good understanding is important to anticipate the threats that might occur in the future. The following technologies that enable the Internet should be addressed in the incident response program:
  - Network protocols—Common (or core) network protocols such as Internet Protocol (IPv4 vs IPv6), Transmission Control Protocol (TCP), User Datagram Protocol (UDP), Internet Control Message Protocol (ICMP), Address Resolution Protocol (ARP) and Reverse Address Resolution Protocol (RARP); how they are used; common types of threats or attacks against the protocol; strategies to mitigate or eliminate such attacks; and Internet technologies
  - Network applications and services—For example, domain name system (DNS), network file system (NFS) and secure shell (SSH); how they work; common usage; secure configurations; and common types of threats or attacks against the application or service and mitigation strategies
  - Network security issues—To recognize vulnerable points in network configurations. Basic perimeter security, network firewalls (design, packet filtering, proxy systems, demilitarized zone [DMZ], bastion hosts, etc.) and router security are relevant to recognize the potential for information disclosure of data traveling across the network (e.g., packet monitoring or "sniffers") or threats relating to accepting untrustworthy information.
- **Operating systems**—Knowledge of operating systems such as UNIX, Windows, MAC, Linux, Android and any other operating systems used by the team or constituency is important, specifically how to:
  - Configure (harden) the system
  - Review configuration files for security weaknesses
  - Identify common attack methods
  - Determine whether a compromise attempt occurred
  - Determine whether an attempted system compromise was successful
  - Review log files for anomalies
  - Analyze the results of attacks
  - Manage system privileges
  - Recover from a compromise
- **Malicious code (viruses, worms, Trojan horse programs, APT)**—These can have different types of payloads that can cause a DoS attack or web defacement, or the code can contain more dynamic payloads that can be configured to result in multifaceted attack vectors. It is important to understand how malicious code is propagated through some of the obvious methods (e.g., CDs, universal serial bus [USB] drives, email, programs, malicious web sites) and how it can propagate through other means such as macros, multipurpose Internet mail extension (MIME), peer-to-peer file sharing, or viruses that affect operating systems running on PC and Macintosh platforms
- **Programming skills**—This includes the range of programming languages used by the organization's operating systems, concepts and techniques for secure programming, and how vulnerabilities can be introduced into code (e.g., through poor programming and design practices, lack of input validation, susceptibility to structured query language [SQL] injection, cross-site scripting, broken authentication, unvalidated redirects).

4.4.3 PERSONNEL

An IMT usually consists of an information security manager, steering committee or advisory board, permanent or dedicated team members, and virtual or temporary team members.

The information security manager usually leads the team. In larger organizations, it may be more effective to appoint a separate IRT leader/manager who focuses on responding to incidents. Above the information security manager, there is a set of senior management executives in a group, typically a security steering group (SSG), security advisory board or perhaps an executive committee. Whatever it is called, the SSG function is responsible for approving the charter and serves as an escalation point for the IMT. The SSG also approves deviations and exceptions to normal practice.

Permanent/dedicated team members have full-time work inside the IMT. They perform the primary tasks in the IMT/IRT. As information systems are broad and complex, it is inefficient and expensive to have all individuals covering all disciplines in performing IMT/IRT. Virtual/temporary team members may have specialized skills and are recruited when necessary to fill gaps within the internal skills portfolio.

Incident Response Team Organization

Incident handlers analyze incident data, determine the impact of the incident and act appropriately to limit the damage to the organization and restore normal services. Often, the team will depend on the participation and cooperation of complementary groups and of general users. IRT models that have proven to work in many organizations include:

- **Central IRT**—A single IRT handles all incidents for the organization, typically used in a small organization or one that is centrally located.
- **Distributed IRT**—Each of several teams is responsible for a logical or physical segment of the infrastructure, usually of a large organization or one that is geographically dispersed.
- **Coordinating IRT**—The central team may provide guidance to distributed IRTs, develop policies and standards, provide training, conduct exercises, and coordinate or support response to specific incidents. Distributed teams manage and implement incident response.
- **Outsourced IRT**—Successful IRTs may be comprised entirely of employees of the organization or may be fully or partially outsourced.

Permanent team members may include incident handlers, investigators and forensics experts, and IT and physical security specialists. Virtual team members normally consist of business representatives (e.g., middle management), legal staff, communications staff (e.g., PR), HR staff, other security groups (e.g., physical security), risk management and IT specialists.

The composition of incident response staff varies from team to team and depends on a number of factors such as:

- Type (i.e., centralized/decentralized) and size of the organization
- Mission and goals of the incident response program
- Nature and range of services offered
- Available staff expertise
- Constituency size and technology base
- Anticipated incident load
- Severity or complexity of incident reports
- Funding

In Practice: Consider your organization. What type of IRT model does it follow? Is this the appropriate model or would another model be more suitable for the needs of the organization?

4.4.4 ROLES AND RESPONSIBILITIES

Figure 4.2 provides common roles and responsibilities of IRT personnel. Note that each position should have an alternate in case the designee is incapacitated or unavailable.

Figure 4.2—Roles and Responsibilities

Position	Roles	Responsibilities
Security steering group	Highest structure of an organization's functions related to information security	1. Takes responsibility for overall incident management and response concept 2. Approves IMT charter 3. Approves exceptions/deviations 4. Makes final decisions
Information security manager	IMT leader and main interface to SSG	1. Develops and maintains incident management and response capability 2. Effectively manages risk and incidents 3. Performs proactive and reactive measures to control information risk level
Incident response manager	IRT leader	1. Supervises incident response tasks 2. Coordinates resources to effectively perform incident response tasks 3. Takes responsibility for successful execution of IRP 4. Presents incident response report and lessons learned to SSG members
Incident handler	IMT/IRT team member	1. Performs incident response tasks to contain exposures from an incident 2. Documents steps taken when executing the IRP 3. Maintains chain of custody and observes incident handling procedures for court purposes 4. Writes incident response report and lessons learned
Investigator	IMT/IRT team member	1. Performs investigative tasks for a specific incident 2. Finds root cause of an incident 3. Writes report of investigation findings
IT security specialist	IMT/IRT team member; subject matter expert in IT security	1. Performs complex and in-depth IT security-related tasks as part of the IRP 2. Performs IT security assessment/audit as proactive measure and part of vulnerability management (i.e., performing routine vulnerability scans and associated remediation)

Figure 4.2—Roles and Responsibilities *(cont.)*		
Position	**Roles**	**Responsibilities**
Business managers	Business function owners; information assets/system owners	1. Make decisions on matters related to information assets/systems when an incident happens, based on IMT/IRT recommendations 2. Provide clear understanding of business impact in BIA process or in IRP
IT specialists/representatives	Subject matter experts in IT services	1. Provide support to IMT/IRT when resolving an incident 2. Maintain information systems in a good condition per company policy and good practices
Legal representative	Subject matter expert in legal	1. Ensures that incident response actions and procedures comply with legal and regulatory requirements. 2. Acts as the liaison to law enforcement and outside agencies.
HR	Subject matter expert in HR area	1. Provides assistance in incident management/response when there is a need to investigate an employee suspected of causing an incident 2. Integrates HR policy to support incident management/response (sanctions to employees found to violate acceptable use of policy or involved in an incident)
PR representative	Subject matter expert in PR area	1. Provides controlled communication to internal and external stakeholders to minimize any adverse impact to ongoing incident response activities and protect an organization's brand and reputation 2. Provides assistance to IMT/IRT in communication issues, thus relieving the team to work on critical issues on resolving an incident
Risk management specialist	Subject matter expert in risk management	1. Works closely with business managers and senior management to determine and manage risk 2. Provides input (e.g., BIA, risk management strategy) to incident management
Physical security/ facilities manager	Knowledgeable about physical plant and emergency capabilities	1. Responsible for physical plant and facilities 2. Ensures physical security during incidents

4.4.5 SKILLS

To build an IRT with capable incident handlers, organizations need people with certain skill sets and technical expertise—abilities that enable them to respond to incidents, perform analysis tasks, and communicate effectively with the constituency and external contacts. They must also be competent problem solvers, easily adapt to change and be effective in their daily activities.

The set of basic skills that IRT members need can be separated into two broad groups: personal and technical skills.

Personal skills are the major parts of the incident handler's daily activity. They include:

- Communication—The ability to communicate effectively is a critical component of the skills needed by IRTs. They need to be effective communicators to ensure that they obtain and supply the information necessary to be helpful. They need to be good listeners, understanding what is said (or not said), to enable them to gain details about an incident that is being reported. They also remain in control of these communications to most effectively determine what is happening, what facts are important and what assistance is necessary.
  - Members of an IRT need to communicate with:
    - Team members
    - IT staff
    - Application owners
    - Users of the systems
    - Technical experts
    - Management and other administrative staff
    - Human resources
    - Law enforcement
    - Media/public relations staff
    - Vendors
  - Communication can take many forms, including:
    - Email responses concerning incidents
    - Documentation of event or incident reports, vulnerabilities, and other technical information
    - Notifications and/or guidelines that are provided to the constituency
    - Internal development policies and procedures
    - Other external communications to staff, management or other relevant parties
- Leadership skills—Members of an IRT are often faced with directing and getting support of other members of the organization, so leadership is an important attribute.
- Presentation skills—An IRT's skills are needed for technical presentations, management or sponsor briefings, a panel discussion at a conference, or some other form of public speaking engagement. The specialist member's skills might extend to providing expert testimony in legal or other proceedings on behalf of the team or users.
- Ability to follow policies and procedures—Team members need the ability to follow and support the established policies and procedures for incident response management.
- Team skills—IRT members must have the ability to work in a team environment, as productive and cordial team players; be aware of responsibilities; contribute to the goals of the team; and work together to share information, workload and experiences.

Members must be flexible and willing to adapt to change. There may be a need for one or more team members to act in a leadership role to support the smaller groups or technical teams.
- Integrity—Team members often deal with information that is sensitive and, occasionally, they may have access to information that is newsworthy. Members must be trustworthy, discreet and able to handle information in confidence.
- Self-understanding—Team members must be able to recognize their limitations and actively seek support from their team members, other experts or management.
- Coping with stress—The IRT is likely to face stressful situations. The members need to be able to recognize when they are becoming stressed, be willing to make their fellow team members aware of the situation, and take the necessary steps to control and maintain their composure.
- Problem solving—Without good problem-solving skills, team members could become overwhelmed with the volumes of data related to incidents and other tasks that need to be handled. Problem-solving skills also include an ability to think outside the box or look at issues from multiple perspectives to identify relevant information or data. This includes knowing who else to contact or approach for additional information, creative ideas or added technical insight.
- Time management—Team members might be confronted with a multitude of tasks, ranging from analyzing, coordinating and responding to incidents, to performing duties such as prioritizing their workload, attending and/or preparing for meetings, completing time sheets, collecting statistics, conducting research, giving briefings and presentations, traveling to conferences, providing onsite technical support, and prioritizing tasks. Team members must be able to balance efforts between completing the tasks and recognizing when to seek help or guidance.

Technical skills are the basic skills required by IRT members and are of two types:
- Technical foundation skills—These require a basic understanding of the underlying technologies used by the organization.
- Incident-handling skills—These require an understanding of the techniques, decision points and supporting tools (software or applications) required in daily activities.

4.4.6 AWARENESS AND EDUCATION

A lack of user awareness is often the cause of security breaches, and it is essential for the information security manager to ensure that an ongoing awareness campaign reduces susceptibility to activities that may increase the risk of a security breach (e.g., phishing attacks, malicious web sites).

Periodically, a skills assessment is useful to determine whether the required expertise is available in the organization for the IRT. In some cases, relevant training or education may serve to provide the necessary skills. If an organization is unable to find internal experts or hire/train staff to provide the necessary incident response specialist skills, the organization may be able to develop relationships with experts in the field to provide the necessary skills. When a situation arises where in-house knowledge is insufficient, these technical specialists can be called on to fill the gap in expertise.

4.4.7 AUDITS

Internal and external audits are performed to verify compliance with policies, standards and procedures defined for an organization. Internal audits are conducted by specialists within the organization and are usually intended to support compliance requirements or improve risk and incident management. External audits involve a third party, who performs the tasks. While most external audits are exercised as part of mandatory requirements, they are also often imposed as part of business collaboration.

Both types of audits can be useful in reviewing incident management and response plans and capabilities. Periodic audits of the processes and procedures specified in the plans can provide validation that security will not be compromised in the event of an incident and policy compliance and legal requirements are addressed appropriately. Audits can also provide an objective view of the overall completeness and functionality of the incident management and response plans and provide assurance that major gaps in the processes do not exist.

4.4.8 OUTSOURCED SECURITY PROVIDERS

Outsourcing incident management capability may be a cost-effective option especially for smaller organizations (see section 3.11.9 Outsourcing and Service Providers). These organizations simply may not have the internal resources to adequately provide the necessary IMT/IRT skills.

Organizations that outsource their IT operations may benefit from close integration if incident management is outsourced to the same vendor. These organizations still require an IRP overseen by an IRT, even if components of incident management are outsourced. It will be essential to clearly understand the outsourcer's capabilities, response times, etc., and develop proper SLAs containing appropriate indemnity clauses.

The information security manager should consider the following when security functions are fully or partially outsourced:
- **Matching the organization's incident reference numbers with the vendor's for each applicable incident:** This ensures a common understanding of incident details between organizations. This also helps to identify the actual organizational recovery time. For example, the vendor may report that service was restored at 4:00 p.m.; however, the outsourcing organization may note that the incident ended at 6:00 p.m. This means that it took the organization an additional two hours to recover internally from the incident. This can be determined by matching the organization's incident reference numbers with the vendor's.
- **Integration of the organization's change management functions with the vendor's (to the extent possible):** Depending on the nature of the service provided, the organization's change management functions may be linked via leaders on the change advisory board or have a platform where the security group can view the vendor's changes and follow-up with key points of contact from the vendor. Also, this contributes to awareness of system changes by both organizations.
- **Requirement from the vendor for periodic review of incidents that occurs on a regular basis (e.g., monthly,**

annually): Operational leadership from both the organization and the vendor attend these meetings, and all incidents/events are reviewed. Follow-up items are generally taken from these meetings to help prevent incidents from recurring.

4.5 INCIDENT MANAGEMENT OBJECTIVES

Incident management exists to address the inevitable events that threaten the operation of any organization. It serves as the next-to-last safety net after controls have failed to prevent or contain an event. Its purpose is to respond to and contain security incidents or quickly restore normal operations in the event of compromise. Failing to do so will result in the declaration of a disaster, and recovery operations will move to an alternate site to restore operations according to a BCP/DRP.

The objectives of incident management are to:
- Handle incidents when they occur so the exposure can be contained or eradicated to enable recovery within the recovery time objectives (RTOs)
- Restore systems to normal operations
- Prevent previous incidents from recurring by documenting and learning from past incidents
- Deploy proactive countermeasures to prevent/minimize the probability of incidents from taking place

Ideally, incident management and response must effectively address a wide range of possible unexpected events. It needs to have well-developed monitoring capabilities for key controls to provide early detection of potential problems. Personnel should be trained in assessing the situation, providing triage, and managing effective responses that maximize operational continuity and minimize impacts. Incident managers will have made provisions to capture all relevant information and apply previously learned lessons. They will be prepared for a disaster through well-defined criteria, experience, knowledge, and authority to invoke the disaster recovery processes necessary to maintain or recover operational status.

4.5.1 STRATEGIC ALIGNMENT

Similar to many other support functions, incident management must be aligned with an organization's strategic plan. The following components may help to accomplish this alignment:
- **Constituency**—To whom does the IMT provide services? It is important to know who the stakeholders are for this function and identify their expectations and information needs. For example, senior management in financial institutions may be bound by Basel III or another regulation. Thus, the IMT should meet certain performance and reporting requirements.
- **Mission**—The mission defines the purpose of the team and the primary objectives and goals that are provided by IMT. An example of a possible IMT mission statement follows: "The mission of the incident management team is to develop, maintain and deliver incident management capabilities and services to safeguard the organization's information assets against computer incidents. We strive to provide assurance to our stakeholders that risk and cyber- or IT-related incidents are dealt with in an efficient and effective manner and that we will prevent/minimize losses resulting from such incidents."
- **Services**—Services provided by IMT should be clearly defined to manage stakeholder expectations. Services offered in organizations may differ significantly, and normally have a positive correlation with the size of the organization and the extent of senior management buy-in.
- **Organizational structure**—The structure of the IMT should effectively support the organization's structure. For multinational companies, a geography-based structure may be the best. For organizations with multiple subsidiaries, one IMT may be developed for each major subsidiary. The best structure would provide the business with the maximum availability of IMT services on the most cost-effective basis.
- **Resources**—Sufficient staffing is needed to be effective. Because incident management covers a wide range of services, most of the time it is not possible to have all the resources available within one IMT. One way to solve this issue is to establish virtual team members and/or complement the team with external resources.
- **Funding**—The IMT usually consists of highly specialized members. The equipment they use in the course of providing services may also be specialized, requiring greater capital expenditures. In view of this, sufficient funding is required to ensure the continuity of critical incident response services.
- **Management buy-in**—Senior management buy-in is essential for establishing and supporting the incident management function. The lack of buy-in normally results in suboptimal IMT performance because there may be significant limitation in budgets or the availability of suitable personnel.

4.5.2 RISK MANAGEMENT

Successful outcomes of risk management include effective incident management and response capabilities. Any risk that materializes that is not prevented by the organization's internal controls constitutes an incident that must be managed and responded to with the intent that it does not escalate into a disaster.

4.5.3 ASSURANCE PROCESS INTEGRATION

The type and nature of incidents the information security manager may deal with often require the involvement of a number of other organizational functions, such physical security, HR and legal. As a consequence, it is important to ensure incident management and recovery plans actively incorporate and integrate those functions where required. An effective outcome is a set of plans that defines which departments are involved in various incident management and response activities and specifies that those linkages have been tested under realistic conditions.

4.5.4 VALUE DELIVERY

Incident management capabilities must be closely integrated with business functions and provide the last line of defense of cost-efficient risk management. Incident management does not consist only of technology to prevent or respond to incidents; it includes a set of process that can provide an optimal balance among prevention, containment and restoration.

To deliver value, incident management should:
- Integrate with business processes and structures as seamlessly as possible
- Improve the capability of businesses to manage risk and provide assurance to stakeholders

- Integrate with BCP
- Become part of an organization's overall strategy and effort to protect and secure critical business function and assets
- Provide the backstop and optimize risk management efforts

4.5.5 RESOURCE MANAGEMENT

Resource management encompasses time, people, budget and other factors to achieve objectives efficiently under given resource constraints. Incident management and response activities consume resources that must be managed to achieve optimal effectiveness. This is accomplished by ensuring appropriate oversight, monitoring of resources and regular reporting. When it is not possible to achieve all objectives, effective resource management ensures that the most important priorities are addressed first. In terms of incident response, effective triage capabilities ensure that limited resources are applied most effectively to contain and limit damage. This is based on quickly identifying the assets that are too compromised to address quickly, the assets that are unaffected and can wait, and the ones that can most efficiently be restored with the available resources.

4.6 INCIDENT MANAGEMENT METRICS AND INDICATORS

Incident management metrics, measures and indicators are the criteria used to measure the effectiveness and efficiency of the incident management function. Metrics based on key performance indicators (KPIs) and program goals (KGIs) established for incident management should be presented to senior management as a basis of justification for continuous support and funding. It enables senior management to understand the incident management capability of the organization and areas of risk that need to be addressed.

KPIs are a quantifiable activity measure (e.g., the number of incidents per year resolved within two minutes of occurrence). KGIs are either quantitative or qualitative, depending on the situation, and are intended to show progress toward or relating to a predefined goal. For example, a business goal for the year is to have 1,000 incidents resolved within two minutes of incident occurrence. At the point of measurement, the KPI of incidents currently resolved in the target time frame is 100; therefore, the KGI at the point of measurement is 10 percent (100 incidents resolved out of 1,000 target incidents).

Incident management reports and measures are useful for the IMT for self-assessment and to understand what has been done satisfactorily and where improvements need to be made. Common criteria that are used as part of incident management metrics may include:

- Total number of reported incidents
- Total number of detected incidents
- Number of days without incident
- Average time to respond to an incident relative to the RTO
- Average time to resolve an incident
- Total number of incidents successfully resolved
- Incidents not resolved successfully
- Proactive and preventive measures taken
- Total number of employees receiving security awareness training
- Total damage from reported and detected incidents if incident response was not effective or not performed
- Total savings from potential damages from incidents resolved
- Total labor responding to incidents
- Detection and notification times

4.6.1 PERFORMANCE MEASUREMENT

The performance measurements for incident management and response focus on achieving the defined objectives and optimizing cost-effectiveness. KGIs and KPIs for the activity should be defined and agreed on by stakeholders and ratified by senior management. The typical range of KGIs encompasses the successful handling of incidents whether by live testing or under actual conditions. Key performance measures can be identified by successfully handling incidents that threaten business operations within the RTOs.

4.7 DEFINING INCIDENT MANAGEMENT PROCEDURES

There is no single, fixed, one-size-fits-all set of incident management procedures for every organization. However, there are a number of good practices that most organizations adopt and customize to meet their own specific needs. Commonly adopted approaches are available from CMU SEI and the SANS Institute, among others.

4.7.1 DETAILED PLAN OF ACTION FOR INCIDENT MANAGEMENT

The incident management action plan is also known as the IRP. There are a number of approaches to developing the IRP.

Figure 4.3—Incident Response Plan Process Flow

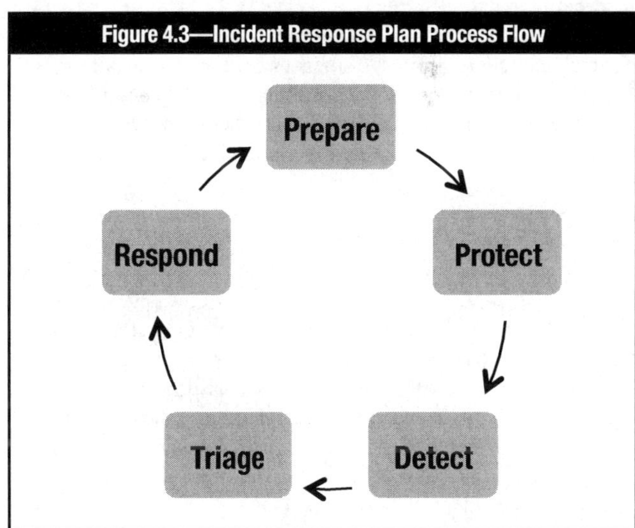

CMU/SEI's *Defining Incident Management Processes* describes the following approach (**figure 4.3**):

- **Prepare/improve/sustain (prepare)**—This process defines all preparation work that has to be completed prior to having any capability to respond to incidents. It contains subprocesses to evaluate incident-handling capability and postmortem review of incidents for improvements. Subprocesses include:
  - Coordinate planning and design:
    - Identify incident management requirements.
    - Establish vision and mission.

- Obtain funding and sponsorship.
- Develop implementation plan.
- Coordinate implementation:
  - Develop policies, processes and plans.
  - Establish incident-handling criteria.
  - Implement defined resources.
  - Evaluate incident management capability.
  - Conduct postmortem review.
  - Determine incident management process changes.
  - Implement incident management process changes.
- **Protect infrastructure (protect)**—The protect process aims to protect and secure critical data and computing infrastructure and its constituency when responding to incidents. It also proposes improvement on a predetermined schedule while keeping the appropriate security context in consideration. Subprocesses include:
  - Implement changes to computing infrastructure to mitigate ongoing or potential incident.
  - Implement infrastructure protection improvements from postmortem reviews or other process improvement mechanisms.
  - Evaluate computing infrastructure by performing proactive security assessment and evaluation.
  - Provide input to detect process on incidents/potential incidents.
- **Detect events (detect)**—The detect process identifies unusual/suspicious activity that might compromise critical business functions or infrastructure. Subprocesses include:
  - Proactive detection—The detect process is conducted regularly prior to an incident. The IMT monitors various information from online/periodic vulnerability scanning, network monitoring, antivirus and personal firewall alerts, commercial vulnerability alert services, risk analysis, and security audit/assessment.
  - Reactive detection—The detect process is conducted when there are reports from system users or other organizations. Users may notice unusual or suspicious activity and report it to the IMT. It is also possible that another organization's IMT will provide advisories when its system has received malicious activity from your organization.

For the IMT to receive the report promptly, there should be multiple communication channels from end users to the IMT. This can be in the form of phone calls, faxes, email messages, web-form reporting and automated intrusion detection systems (IDSs).

- **Triage events (triage)**—Triage is a process of sorting, categorizing, correlating, prioritizing and assigning incoming reports/events into (typically) three categories: the problems that cannot be readily resolved, those that can wait and those that can be efficiently addressed with the resources available. It is an essential element of any incident management capability to ensure maximum effectiveness of limited resources. When there are multiple incident reports coming into the IMT, triage allows events to be prioritized appropriately, thus maximizing response effectiveness. It can also serve as a single point of entry for any IMT communication and information.
  - Properly organized, triage can provide a snapshot of the current status of all incident activity reported, a central location for incident status reporting and an initial assessment of incoming reports for further handling. Triage prioritization can be done at two levels:
    - Tactical—Based on a set of criteria
    - Strategic—Based on the impact of business
  - Subprocesses include:
    - Categorization—This is the use of predetermined criteria to classify all incoming reports/events. There are many ways to categorize; for example, *NIST Special Publication 800-61 Revision 2 Computer Security Incident Handling Guide* uses:
      - DoS
      - Malicious code
      - Unauthorized access
      - Inappropriate usage
      - Multiple components
    - Correlation—This subprocess correlates a report/event with other relevant information. A higher correlation of a report/event provides more information that is useful for the IMT to decide on the appropriate response. For example, a specific event may be the cause of an incident whereas another may be a consequence and dealing with it would not resolve the issue.
    - Prioritization—In an ideal world, every undesirable event is followed up as soon as possible. However, resources are limited and it may not always be possible. To ensure minimal impact to critical business functions or information assets, incidents are usually prioritized based on their potential impact rather than a specific set of criteria. Remediation efforts obviously start with highest priority and then move down to lowest priority.
    - Assignment—When an incident or potential incident has been identified, it is assigned to the IMT to initiate the response effort. Assignment may be based on:
      - Workload of IMT members
      - IMT members who have handled similar incidents
      - Category or priority of the event
      - Relevant business unit
- **Respond**—The response process includes steps taken to address, resolve or mitigate an incident. CMU SEI defines three types of response activities:
  - Technical response—It is appropriate for technical IMT members, such as incident handlers and IT representatives, to analyze and resolve an incident. Technical response forms include the following:
    - Collecting data for further analysis
    - Analyzing incident-supporting information such as log files
    - Researching corresponding technical mitigation strategies and recovery options
    - Consulting telephone or email technical assistance
    - Securing onsite assistance
    - Analyzing logs
    - Developing and deploying patches and workarounds
  - Management response—The management response includes activities that require supervisory or management intervention, notification, interaction, escalation or approval as part of response effort. This response activity is normally executed by business managers, senior management and members of the affected business units.
  - Legal response—The legal response is associated with activity that relates to investigation, prosecution, liability, copyright and privacy issues, laws, regulations, and nondisclosure agreements. Because this response may require in-depth knowledge on legal matters, it is usually referred to the corporate legal team.

A detailed flow diagram of the incident management process is shown in **figure 4.4.**

4.8 CURRENT STATE OF INCIDENT RESPONSE CAPABILITY

Most organizations have some sort of incident response capability, either ad hoc or formal. The information security manager must identify what is already in place as a basis for understanding the current state. There are many ways to do this, including:

- **Survey of senior management, business managers and IT representatives**—Using a collection of senior management, business line managers and technology representatives, employ surveys and focus groups to gather information to help determine the past performance and perception of the IMT and its process capabilities.
- **Self-assessment**—Self-assessment is conducted by the IMT against a set of criteria to develop an understanding of current capabilities. This is the easiest method because it does not require participation from many stakeholders. The disadvantage of this method is that it may provide only a limited view on current capability and aspects that stakeholders may consider important.
- **External assessment or audit**—This is the most comprehensive option, and it combines interviews, surveys, simulation and other assessment techniques in the assessment. This option is normally used for an organization that already has an adequate incident management capability but is further improving it or reengineering the processes.

4.8.1 HISTORY OF INCIDENTS

Past incidents (both internal and external) can provide valuable information on trends, types of events and business impacts. This information is used as an input to the assessment of the types of incidents that must be planned for and considered.

A sample of historical information maintained associated with an incident is shown in **figure 4.5**. Please note that this represents a few sample items to track and is not meant to be a complete list.

4.8.2 THREATS

Threats are any event that may cause harm to an organization's assets, operations or personnel. There are a number of threats that must be considered, including:

- **Environmental**—Environmental threats include natural disasters. While natural disasters vary among locations, some natural disasters may occur for prolonged periods of time and others may occur annually. Some potentially catastrophic threats are rare enough and impractical to address effectively so they are either generally disregarded (e.g., comet strike) or might be addressed with some types of business interruption insurance.

Figure 4.4 – Incident Handling Process Flow

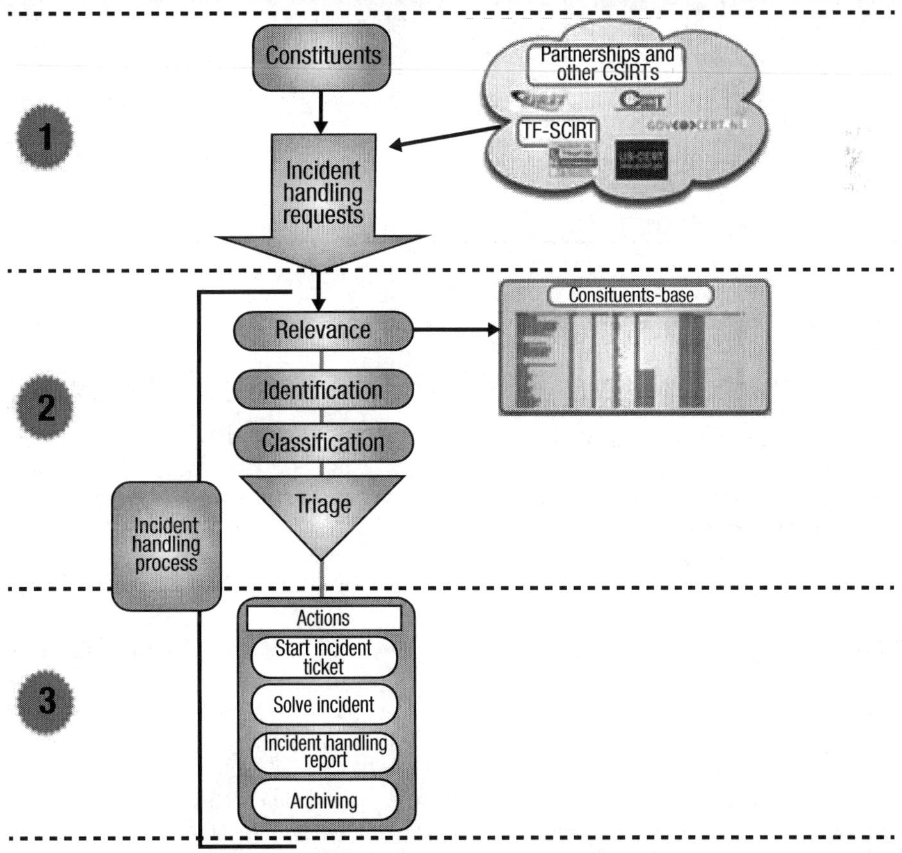

Source: European Union Agency for Network and Information Security, *http://www.enisa.europa.eu/activities/cert/support/incident-management/browsable/workflows/proposed-workflows/proposed-workflows-i*

Figure 4.5—Sample Historical Information to Track Regarding Security Incidents

Organization Incident #	Vendor Incident #	Start Date and Time	End Date and Time	Outage Duration	Description	Root Cause	Solution	Business Impact
12345	Z12111	1/14/2015 4:00 p.m. ET*	1/14/2015 5:00 p.m. ET	1 hr	Describes a specific event in an order entry system that caused a loss of functionality. Details depending on the business.	Describes the root cause of a functionality lost from the order entry system. This represents the key reason(s) behind why the function(s) became unavailable.	What was done specifically to solve the problem? Includes the group that performed the action and the system to which the action was directed.	Includes specific function lost, affected geographical or business areas, number of affected users (if applicable)
12346	Z12156	2/23/2015 7:00 a.m. ET	2/23/2015 7:31 a.m. ET	31 min	Describes a specific event in the organization's billing system that caused a loss of functionality. Details depending on the business.	Describes the root cause of a functionality lost from the billing system. This represents the key reason(s) behind why the function(s) became unavailable.	What was done specifically to solve the problem? Includes the group that performed the action and the system to which the action was directed.	Includes specific function lost, affected geographical or business areas, number of affected users (if applicable)
12347	None	4/16/2015 4:30 a.m. ET	4/17/2015 5:00 a.m. ET	24 hrs 30 min	Describes the DoS attack and the related system(s) affected. The key is detailing the functionality lost by this event.	Describes the root cause of a functionality lost from the system affected by the DoS attack. This represents the key reason(s) behind why the function(s) became unavailable.	What was done specifically to solve the problem? Includes the group that performed the action and the system to which the action was directed.	Includes specific function lost, affected geographical or business areas, number of affected users (if applicable)

* ET = Eastern Time Zone, United States of America

- **Technical**—Technical threats include fire; electrical failure; heating, ventilating and air conditioning (HVAC) failure; information system and software issues; telecommunication failure; and gas/water leakage. Technical threats are normally found in every organization and are quite common. With sufficient planning, most technical threats can be managed adequately. The exceptions may be APTs and zero-day attacks.
- **Man-made**—Threats that arise from man-made actions may include damage by disgruntled employees, corporate sabotage/espionage and political instability that impacts business functions. Many of these threats are normally easy to identify while some, such as espionage and embezzlement, will be far more difficult.

4.8.3 VULNERABILITIES

A weakness in a system, technology, process, people or control that can be exploited and result in compromise is a vulnerability. A vulnerability that can be exploited by threats results in risk. One aspect of risk management is managing vulnerabilities to maintain risk within acceptable limits as determined by the organization's risk appetite and tolerance. Vulnerability management is part of the incident management capability; it is the proactive identification, monitoring and repair of any weaknesses.

4.9 DEVELOPING AN INCIDENT RESPONSE PLAN

The IRP is the operational component of incident management. The plan details the actions, personnel and activities that take place in case adverse events result in the loss of information systems or processes. The plan should include the formation, management and maintenance of the IRT.

4.9.1 ELEMENTS OF AN INCIDENT RESPONSE PLAN

A common approach to developing an IRP is to base it on a six-phase model of incident response that covers preparation, identification, containment, eradication, restoration and follow-up. Those stages are defined as follows:

- **Preparation**—This phase prepares an organization to develop an IRP prior to an incident. Sufficient preparation facilitates smooth execution. Activities in this phase include:
  - Establishing an approach to handle incidents

  - Establishing policy and warning banners in information systems to deter intruders and allow information collection
  - Establishing a communication plan to stakeholders
  - Developing criteria on when to report an incident to authorities
  - Developing a process to activate the IMT
  - Establishing a secure location to execute the IRP
  - Ensuring equipment needed is available
- **Identification**—This phase aims to verify if an incident has happened and find out more details about it. Reports on possible incidents may come from information systems, end users or other organizations. Not all reports are valid incidents; they may be false alarms or may not qualify as an incident. Activities in this phase include:
  - Assigning ownership of an incident or potential incident to an incident handler
  - Verifying that reports or events qualify as an incident
  - Establishing chain of custody during identification when handling potential evidence
  - Determining the severity of an incident and escalating it as necessary
- **Containment**—After an incident has been identified and confirmed, the IMT is activated and information from the incident handler is shared. The team will conduct a detailed assessment and contact the system owner or business manager of the affected information systems/assets to coordinate further action. The action taken in this phase is to limit the exposure. Activities in this phase include:
  - Activating the IMT/IRT to contain the incident
  - Notifying appropriate stakeholders affected by the incident
  - Obtaining agreement on actions taken that may affect availability of a service or risk of the containment process
  - Getting the IT representative and relevant virtual team members involved to implement containment procedures
  - Obtaining and preserving evidence
  - Documenting and taking backups of actions from this phase onward
  - Controlling and managing communication to the public by the PR team
- **Eradication**—When containment measures have been deployed, it is time to determine the root cause of the incident and eradicate it. Eradication can be done in a number of ways: restoring backups to achieve a clean state of the system, removing the root cause, improving defenses and performing vulnerability analysis to find further potential damage from the same root cause. Activities in this phase include:
  - Determining the signs and cause of incidents
  - Locating the most recent version of backups or alternative solutions
  - Removing the root cause. In the event of worm or virus infection, it can be removed by deploying appropriate patches and updated antivirus software.
  - Improving defenses by implementing protection techniques
  - Performing vulnerability analysis to find new vulnerabilities introduced by the root cause
- **Recovery**—This phase ensures that affected systems or services are restored to a condition specified in the service delivery objectives (SDO) or BCP. The time constraint up to this phase is documented in the RTO. Activities in this phase include:
  - Restoring operations as defined in the SDO
  - Validating that actions taken on restored systems were successful
  - Getting involvement of system owners to test the system
  - Facilitating system owners to declare normal operation
- **Lessons learned**—At the end of the incident response process, a report should always be developed to share what has happened, what measures were taken and the results after the plan was executed. Part of the report should contain lessons learned that provide the IMT and other stakeholders valuable learning points of what could have been done better. These lessons should be developed into a plan to enhance the incident management capability and the documentation of the IRP. Activities in this phase include:
  - Writing the incident report
  - Analyzing issues encountered during incident response efforts
  - Proposing improvement based on issues encountered
  - Presenting the report to relevant stakeholders

4.9.2 GAP ANALYSIS—BASIS FOR AN INCIDENT RESPONSE PLAN

Gap analysis provides information on the gap between current incident response capabilities and the desired level. By comparing the two levels, improvements in capabilities, skills and technology can be identified, including:

- Processes that need to be improved to be more efficient and effective
- Resources needed to achieve the objectives for the incident response capability

The resulting gap analysis report can be used for planning purposes to determine the steps needed to resolve the gap between the current state and the desired state. It can also be useful in determining the most effective strategy to achieve the objectives and prioritize efforts. The priorities should be based on the areas of greatest potential impact and the best cost benefit.

4.9.3 BUSINESS IMPACT ANALYSIS

The next step in the incident response management process, after identifying all reasonably possible events, is to consider the potential impact of each type of incident that may occur. The argument is that one cannot properly plan for and prioritize response to an undesirable event if there is little idea of the likely impacts of different possible incident scenarios on the business/organization. The analysis of potential incidents and related business impacts is accomplished by conducting a BIA—a systematic activity designed to assess the impact of disruption or total loss of availability of the support of any critical information resource (system, network device, application, personnel and/or data) to an organization.

A BIA must:

- Determine the loss to the organization resulting from a function being unavailable
- Establish the escalation of that loss over time
- Identify the minimum resources needed for recovery
- Prioritize the recovery of processes and supporting systems

The purpose of a BIA is to create a report that helps stakeholders understand what impact an incident could have on the business.

The impact may be in the form of a qualitative (rating) or quantitative (monetary) value.

To perform this phase successfully, it is essential to understand the structure and culture of the organization, key business processes, acceptable risk and risk tolerance, and critical IT and physical resources. A successful BIA requires participation from business process owners, senior management, IT, physical security and end users.

BIAs have three primary goals:
- **Criticality prioritization**—Every critical business unit process must be identified and prioritized. The impact of an incident must be evaluated—the higher the impact, the higher the priority.
- **Downtime estimation**—The assessment is also used to estimate the maximum tolerable downtime (MTD) or maximum tolerable outage (MTO) that the business can endure and still remain viable. This can also mean the longest period of unavailability of critical processes/services/information assets before the company may cease to operate (i.e., AIW).
- **Resource requirement**—Resource requirements for critical processes are also identified at this time, with the most time-sensitive and highest-impact processes receiving the highest priority for resource allocation.

An assessment includes the following activities:
- **Gathering assessment material**—The initial step of the BIA is to identify which business units are critical to an organization. This step can be drilled down to the critical tasks that must be performed and the resources needed to ensure business survival.
- **Analyzing the information compiled**—During this phase, several activities take place, such as documenting required processes and resources, identifying interdependencies, and determining the acceptable period of interruption. Tasks in this phase include:
  - Identify interdependencies among the functions and departments classified as critical or high-impact.
  - Discover all possible disruptions that could affect the mechanism necessary to allow these departments to function together.
  - Identify and document potential threats that could disrupt interdepartmental communication.
  - Gather quantitative and qualitative information pertaining to those threats.
  - Provide alternative methods of restoring functionality and communication.
  - Provide a brief statement of rationale for each threat and corresponding information.
- **Documenting the result and presenting recommendations**—The last step of an assessment is documenting assessment results from the previous activities and creating a report for business units and senior management.

Another way of viewing a BIA is that it is an exercise designed to identify the resources that are most important to an organization and the impact resulting from disruption or availability loss. It should also result in an understanding of process flows in the organization needed to optimize recovery priorities based on system dependencies and interdependencies. If conducted properly, a BIA facilitates understanding the amount of potential loss (and various other unwanted effects) that could occur from

Knowledge Check: Gap Analysis

A small organization has an IT team of four people, including the manager. Currently, there is no documented IRP. The current plan consists of the IT team reporting suspicious events to the manager, who briefly confirms the finding and then reports the event to the chief operating officer (COO), typically by email. The COO then decides whether or not to take an action. The COO has requested that all activity that potentially could be an incident be reported to him, as senior leadership varies in what they consider an incident. The typical time it takes the COO to respond to these reports can be anywhere between two and six hours, depending on his schedule, which means it can take up to six hours before the IT team can respond to an event.

Comparable benchmarks among similar organizations show that typical response time is two hours. Good practice also recommends several guidelines for what constitutes an incident. The IT manager would like for his team to be able to follow these guidelines to help them to be more nimble in their response to events and reduce the time to respond.

Questions:
1. Briefly describe the current state of incident response for this IT team.

2. Briefly describe the desired state of incident response for this IT team.

3. List some gaps between the current and desired states of incident response for this IT team.

4. List some potential ways the organization can close the gaps between the current and desired states of incident response for this IT team.

Answers on page 253.

certain types of events. Potential loss includes direct financial loss and less tangible types of loss, such as reputational damage or failure to achieve regulatory compliance.

Despite the high level of importance of understanding the business impact of incidents on the business process, many organizations fail to undertake this assessment. Another common problem is the failure to update BIAs when systems and business functions are added or changed. A BIA is a part of and related to risk assessment (both qualitative and quantitative risk assessment) insofar as the degree of risk exposure will correlate to the potential for an event that causes an impact. The BIA is also used to determine the consequences of compromise, whereas the risk determination calculates the probability of compromise.

Elements of a Business Impact Analysis

The way in which BIAs are conducted varies from organization to organization. However, there are commonalities. In general, BIAs:

- Describe the business mission of each particular business/cost center
- Identify the functions that characterize each business function
- Determine dependencies such as required inputs from other operations
- Determine subsequent operations dependent on the function
- Identify critical processing cycles (in terms of time intervals) for each function
- Estimate the impact of each type of incident on business operations
- Determine required recovery time (i.e., RTO)
- Identify the resources and activities required to restore an acceptable level of operation
- Determine the amount of data that can be lost and must be re-created to determine RPOs
- Determine work-around possibilities such as manual or PC-based operation, or workload shifting
- Estimate the amount of time that recovering from each type of incident is likely to take in relation to the RTO. If the estimated time is greater than the RTO, additional resources may be needed for recovery. This includes restoration of all required dependencies as well.

A typical BIA includes obtaining the following information from the organization's functional units:

- **Function description:** What is the function of the business unit?
- **Dependencies:** What is dependent upon this function? What has to happen or needs to be available before the function can be performed?
- **Impact profile:** Is there a specific time, day, week or month that the function would be more vulnerable to risk/exposure or the impact to the business would be greater if the function is not performed?
- **Operational impacts:** When would operational impact to the business be realized if the function was not performed?
- **Financial impacts:** When would financial impact to the business be realized if the function was not performed?
- **Work backlog:** At what point will the backlog of work start to impact the business?
- **Recovery resources:** What kind of resources are needed to support the function, how many are needed and how soon are they needed after a disruption (e.g., phones, desks, PCs)?
- **Technology resources:** What software and/or applications are needed to support the function?
- **Stand-alone PCs or workstations:** Does the function require a stand-alone PC or workstation?
- **Local area networks (LANs):** Does the function require access to the LAN?
- **Workaround procedures:** Are there currently manual workaround procedures in place that would enable the function to be performed in the event that IT is unavailable? If so, how long could these workarounds be used to continue the function?
- **Work-at-home:** Can the function be performed from home?
- **Workload shifting:** Is it possible to shift workloads to another part of the business that might not be impacted by the disruption?
- **Business records:** Are there business records needed to perform the function and, if so, are they backed up? How? With what frequency?
- **Regulatory reporting:** Are regulatory documents created as a result of the function?
- **Work inflows:** What input is received, either internally or externally, that is needed to perform the function?
- **Work outflows:** Where does the output go after it leaves the functional area or, in other words, who would be impacted if the function were not performed?
- **Business disruption experience:** Has there ever been a disruption of the function? (If so, provide a brief description.)
- **Competitive analysis:** Would there be a competitive impact if the function were not performed? If so, when would the impact occur and when would a potential loss of the customer occur?
- **Other issues and concerns:** What are other issues or concerns relevant to the success of performing the function?

Collecting and analyzing this information for all significant functions across an organization provides a detailed picture of workflows, dependencies and interdependencies that determine options for the required order of restoration efforts (i.e., resources dependent on other functions require the restoration of those functions before they can be operational). It provides information on restoration and emergency operation options; critical timing; options for workarounds or other operational choices in the event of function failure; and other essential information needed to develop effective IRPs, DRPs and BCPs.

Benefits of Conducting a Business Impact Analysis

Conducting BIAs produces several important major benefits, including:

- Increasing the understanding of the amount of potential loss and various other undesirable effects that could occur from certain types of incidents resulting from the loss of a particular function, including catastrophic events that can place survival of the business at risk
- Prioritizing restoration activities and understanding recovery options
- Understanding dependencies of various functions and their interdependencies
- Raising the level of awareness for response management within an organization

4.9.4 ESCALATION PROCESS FOR EFFECTIVE INCIDENT MANAGEMENT

The information security manager should implement an escalation process to establish the events to be managed (i.e., in the event of a telecommunications shutdown). Events that appear routine can be related to a security compromise and constitute an incident (e.g., data corruption may not be due to an application problem, but rather to a virus or worm infection). As part of the emergency management and incident management policies and procedures, a detailed description of the escalation process and who has to authorize various recovery actions or disaster declaration must be clear and documented.

For every event, a list of actions should be described in the sequence to be performed. Every action listed should identify the responsible person, alternates in case of unavailability and an estimated time for execution.

When all the actions have been completed successfully, the process should continue in the section dedicated to "end of emergency." If any action cannot be executed or if the estimated time is reached, the process should continue in the next action. Each unsuccessful action wastes time. If the accumulated elapsed time reaches a predetermined limit, the emergency status may change to an alert condition (i.e., low, medium, high). An alert situation prompts notification of individuals and organizations with executive responsibilities. Entities and personnel that should receive an alert notification include:

- Senior management
- Response and recovery teams
- HR
- Insurance companies
- Backup facilities
- Vendors
- Customers

The process should continue until the emergency is resolved or the last alert has occurred. At this point, the emergency management team meets to evaluate the damages and mitigation alternatives, decide whether to declare a disaster and/or launch the response and recovery plan, and determine the appropriate strategy. The information security manager should develop a communication plan in consultation with PR, legal counsel and appropriate senior management to ensure the appropriateness of any information disclosures.

After the escalation process, tasks such as notifying personnel, activating backup facilities, containing security threats to information resources, making transportation arrangements and carrying them out, retrieving and unloading data, and testing must be executed. The total elapsed time should be in accordance with the established RTO as discussed in chapter 2.

The escalation process includes prioritizing event information and the decision process for determining when to alert various groups, including senior management, the public, shareholders and stakeholders, legal counsel, HR staff, vendors, and customers.

The information security manager should develop these escalation processes and decision authority through consultation with PR, legal counsel and appropriate senior management. This process should also include vendors and utility services.

Many organizations define the level of events and the escalation procedures differently for each level. These levels can be based on the severity of the event as well as the number of organizations that may be affected by the event and their specific need to be notified. The information security manager should also have mechanisms to communicate crisis or event information. These mechanisms may include using email (if computing systems and networks are operational), cellular phones, fax machines, electronic pagers, web sites or an emergency telephone number at which a message can be placed. Note that some types of communication, such as email messages, are by default in cleartext, making them subject to potential interception. The information security manager should also develop methods to encrypt email and other communications used in relaying crisis or event-related information to ensure that information is released only as prudent or according to plans.

4.9.5 HELP/SERVICE DESK PROCESSES FOR IDENTIFYING SECURITY INCIDENTS

The information security manager should have processes defined for help/service desk personnel to distinguish a typical request from a possible security incident. The help/service desk is likely to receive the first reports indicating a security-related problem. Prompt recognition of an incident in progress and quick referral to appropriate parties are critical to minimizing the damage resulting from such incidents.

By defining appropriate criteria and improving the awareness of help/service desk personnel, the information security manager develops another important method to detect a security incident. Proper training also helps to reduce the risk that the help/service desk could be successfully targeted in a social engineering attack designed to obtain access to accounts, such as a perpetrator pretending to be a user who has been locked out and requires immediate access to the system. In addition to identifying a possible security incident, help/service desk personnel should be aware of the proper procedures to report and escalate a potential issue.

4.9.6 INCIDENT MANAGEMENT AND RESPONSE TEAMS

The plan must identify teams and define their assigned responsibilities in the event of an incident. To implement the strategies that have been developed for business recovery, key decision-making, technical and end-user personnel to lead teams need to be designated and trained. Depending on the size of the business operation, the team may consist of a single person. The involvement of these teams depends on the level of the disruption of service and the types of assets lost, compromised, damaged or endangered. A matrix should be developed that indicates the correlations among the functions of the different teams. This will facilitate estimating the magnitude of the effort and activating the appropriate combination of teams. Examples of the kinds of teams usually needed include:

- **Emergency action team**—Designated first responders whose function is to deal with fires or other emergency response scenarios

- **Damage assessment team**—Qualified individuals who assess the extent of damage to physical assets and make an initial determination regarding what is a complete loss vs. what is restorable or salvageable
- **Emergency management team**—Responsible for coordinating the activities of all other recovery teams and handling key decision making
- **Relocation team**—Responsible for coordinating the process of moving from the affected location to an alternate site or to the restored original location
- **Security team**—Often called a CSIRT; responsible for monitoring the security of systems and communication links, containing any ongoing security threats, resolving any security issues that impede the expeditious recovery of the system(s), and assuring the proper installation and functioning of every security software package

A number of key decisions must be agreed on during the planning, implementation and evaluation phases of the response and recovery plan. These include:

- Goals/requirements/products for each phase
- KGIs and KPIs
- Reporting criteria
- Critical success factors and critical path aspects of implementation
- Alternate facilities in which tasks and operations can be performed
- Critical information resources to deploy (e.g., data and systems)
- Decision authority and persons responsible for completion
- Available resources—including financial, personnel and technical—to aid in deployment
- Scheduling of activities with established priorities

4.9.7 ORGANIZING, TRAINING AND EQUIPPING THE RESPONSE STAFF

Training the response teams is essential; the information security manager should develop event scenarios and test the response and recovery plans to ensure that team participants are familiar with their tasks and responsibilities. Through this process, the teams will also identify the resources they require for response and recovery, providing the basis for equipping the teams with needed resources. An added value of training is detecting and modifying ambiguous procedures to achieve clarity and determining recovery resources that may not be adequate or effective.

IMT members should undergo the following training program:

- **Induction to the IMT**—The induction should provide the essential information required to be an effective IMT member.
- **Mentoring team members regarding roles, responsibilities and procedures**—Existing IMT members can provide valuable knowledge to aid new members after induction. To facilitate effective mentoring, the buddy system can be used, pairing new members with experienced members.
- **On-the-job training**—This may serve to provide an understanding of company policies, standards, procedures, available tools and applications, acceptable code of conduct, etc.
- **Formal training**—Team members may require formal training to attain an adequate level of competence necessary to support the overall incident management capability.

4.9.8 INCIDENT NOTIFICATION PROCESS

Having an effective and timely security incident notification process is a critical component of any security program. The information security manager should understand how obtaining timely and relevant information can help the organization respond quickly and efficiently and will ultimately limit the potential loss and damage that may occur as the result of an incident. Notification mechanisms that enable an automated detection system or monitor to send email or phone messages should be used whenever possible. The following functions are most likely to need information concerning incidents when they occur:

- Risk management
- HR (when a security compromise appears to involve insiders)
- Legal
- PR
- Network operations
- Physical and information security
- Business process owners
- Senior management

Notification activities are effective only if knowledgeable personnel understand their responsibilities and perform them in an efficient and timely manner. The information security manager therefore needs to define the responsibilities and communicate them to key personnel. It can also be effective to work with HR to determine how these responsibilities can be documented in employees' job descriptions.

4.9.9 CHALLENGES IN DEVELOPING AN INCIDENT MANAGEMENT PLAN

When developing and maintaining an incident management plan, there may be unanticipated challenges as the result of:

- **Lack of management buy-in and organizational consensus**—Most challenges result from a lack of management buy-in and consensus among the business units. When an incident occurs, management response may not be provided as expected, thus hindering incident management efforts. This may happen when senior management and other stakeholders or constituents are not involved in incident management planning and implementation. Challenges can also be caused by the lack of regular meetings between the IMT and constituents. A sense of ownership among constituents in incident management helps to ensure that sufficient resources and support are available for the IMT.
- **Mismatch to organizational goals and structure**—Business operates at an accelerated rate and may change significantly over a short period of time. Incident management may not be able to cope with the speed or nature of changes happening within the organization. Consider a situation where the business expands to emerging countries. The presence in new countries may grow at a significant rate; however, it may be difficult to expand incident management capabilities. A round of discussions may be needed to identify critical business functions and local stakeholders and understand local regulations. Senior management is usually occupied with business matters at this stage and may not be able to invest time in incident management. It is the responsibility of the IMT to identify any critical issues and make these known to executive management.
- **IMT member turnover**—Developing an incident management plan may take a significant amount of time and involve frequent interaction with various stakeholders. The champion of

incident management, who is normally either a member of senior management or the information security manager, may leave the company unexpectedly, causing any planning or development efforts to come to a halt. The lack of a champion is likely to reduce the focus and resources devoted to implementing the plan.

- **Lack of communication process**—Ineffective communication processes may result either in undercommunication or overcommunication. In the case of undercommunication, relevant stakeholders may not receive the information they need. This may result in different understandings about the need for incident management planning; the benefits that can be obtained; or their role in developing, implementing and maintaining an incident management capability. Overcommunication may turn stakeholders against incident management because they may feel that the plan is too much to handle or it competes with priorities they have established.
- **Complex and broad plan**—The proposed plan may be good and cover many issues, but it is too complex and too broad. Constituents may not be prepared to participate and commit to plans that appear overreaching.

In Practice: Consider the challenges listed in this section. Which of these challenges exist within your organization? Looking at the guidance provided earlier in this section, list some potential solutions to address the challenges you face in your organization.

4.10 BUSINESS CONTINUITY AND DISASTER RECOVERY PROCEDURES

There are a number of considerations when developing response and recovery plans, including available resources; expected services; and the types, kinds and severity of threats faced by the organization. The state of monitoring and detection capabilities must also be known, and the level of risk the organization is willing to accept must be determined, as discussed in section 1.2.2 Determining Risk Capacity and Acceptable Risk (Risk Appetite). An effective strategy for recovery plans strikes the most cost-effective balance among risk management efforts, incident management and response, and business continuity/disaster recovery planning.

According to COBIT 5, business continuity is defined as "preventing, mitigating and recovering from disruption. The terms 'business resumption planning,' 'disaster recovery planning' and 'contingency planning' also may be used in this context; they focus on recovery aspects of continuity."

Therefore, the relationship between BC and DR is such that the DRP is a subset of the BCP. Specifically, while BCP goals include incident prevention and mitigation, the DRP is focused on what must be done to restore operations *after* an incident has already taken place. A BCP may be seen as a continuous process that is actively implemented in business-as-usual scenarios, while the DRP is reactive in nature and is implemented only upon satisfaction of a specific set of conditions (i.e., the business has incurred an incident).

4.10.1 RECOVERY PLANNING AND BUSINESS RECOVERY PROCESSES

The information security manager must understand the basic processes required to recover operations from information system incidents resulting in security breaches and system failures, natural disasters, and other probable events that would cause potential disruption of business operations.

Disaster recovery has traditionally been defined as the recovery of IT systems after disruptive events that cause major system failures, such as hurricanes and floods resulting in inaccessibility or prolonged power or communication outages. Business recovery is defined as the recovery of all the critical business processes necessary to continue or resume operations. Business recovery includes disaster recovery and all other required operational aspects.

Planning includes documenting the requirements for declaring a disaster (i.e., determining when an incident cannot be resolved by the available recovery processes). The declaration of a disaster pursuant to defined declaration criteria generally requires moving operations to the alternate processing site.

Not all events, incidents or critical disruptions should be classified as security incidents. For example, disruption in service can be caused by system malfunctions or accident. Regardless of the cause of operational disruption, satisfactory resolution requires prompt action to maintain or recover operational status. Actions may, among other things, necessitate restoration of hardware, software and/or data files.

Each of these planning processes typically includes several main phases, including:

- Conducting a risk assessment and BIA
- Defining a response and recovery strategy
- Documenting response and recovery plans
- Training that covers response and recovery procedures
- Updating response and recovery plans
- Testing response and recovery plans
- Auditing response and recovery plans

Prior to creating a detailed BCP, it is important to perform BIAs to determine the incremental daily cost of losing different systems. This provides the basis for deciding on appropriate RTOs and associated costs. This, in turn, affects the location and cost of offsite recovery facilities and the composition and mission of individual response and recovery teams.

4.10.2 RECOVERY OPERATIONS

Once the organization is up and running in recovery mode (which is usually from a disaster recovery site in the case of damage or inaccessibility of the primary facility), the business continuity teams should monitor the restoration progress at the primary site. This is done to assess when it is safe to return and perform tests to evaluate whether the primary data center and facilities are accessible, operational, and capable of functioning at normal capacities and processing load.

The teams that were responsible for relocating to the alternate site and making it operational perform a similar operation to return to the primary site. On complete restoration of the primary facility and data processing capabilities, the recovery teams update the business continuity leader, who will then declare normalcy in consultation with the crisis management team and migrate operations back to the primary site.

In the event that the primary site is completely destroyed or severely damaged, the organization may make a strategic decision to transform the alternate recovery site to the primary operations site or identify, acquire and set up another site where operations will eventually be restored and will function as the primary site. This is especially true in cases where the organization subscribes to a third-party disaster recovery site because the costs of operating from such a site for an extended period of time may prove to be prohibitively high.

To ensure effective and comprehensive contingency planning, organizations setting up a BCP should address the processes, roles and responsibilities of identifying an incident, declaring a disaster and managing operations in a disaster mode, but it should also define processes to restore operations at the primary site and announce the return to normalcy. A cold site may be identified during the strategy phase to be upgraded to a primary operating facility in the event that the incident renders the primary facility useless.

It is important to remember that information resources must still be protected, even during the potentially chaotic environment of a business interruption or disaster. The information security manager must ensure that information security is incorporated into all response and recovery plans. The security manager should carefully review these plans to ensure that their execution does not compromise information security standards and requirements. In cases where such a compromise is unavoidable, other mitigation options should be explored. Minimally, a focused risk assessment should be conducted to make management aware of the extent and potential impacts of the security risk introduced by execution of the plan.

In the likely event that the crisis is not catastrophic and the organization has switched to disaster recovery mode for only a limited time, it is important that the disaster recovery site be restored to an acceptable state of readiness after operations are reinstated at the primary facility.

Lessons learned and gaps identified in the plan when either switching to the disaster recovery site or reverting to the primary facility should be recorded and recommendations should be implemented to enhance the effectiveness of the plan. A realistic plan should cover all aspects of reestablishing the operations at the primary site, including people, facilities and technology areas.

4.10.3 RECOVERY STRATEGIES

Various strategies exist for recovering critical information resources. The most appropriate strategy is likely to be one that demonstrably addresses probable events with acceptable recovery times at a reasonable cost.

The total cost of a recovery capability is the cost of preparing for possible disruptions (e.g., purchasing, maintaining and regularly testing redundant computers, and maintaining alternate network routing, training and personnel costs) and the cost of putting these into effect in the event of an incident. Impacts of disruptions can, to some extent, be mitigated by various forms of business interruption insurance, which should be considered as a strategy option.

Depending on the size and complexity of the organization and the state of recovery planning, the information security manager should understand that the development of an incident management and response plan is likely to be a difficult and expensive process that may take considerable time. It may require the development of several alternative strategies, encompassing different capabilities and costs, to be presented to management for a final decision. Each alternative must be sufficiently developed to provide an understanding of the trade-offs among scope, capabilities and cost. It may be prudent to consider outsourcing some or all of the needed capabilities and determine associated costs for the purpose of comparisons. Once the decision is made for which strategy best meets management's objectives, that strategy provides the basis for the development of detailed incident management and response plans.

4.10.4 ADDRESSING THREATS

In the case of threats, some possible proactive strategies that may be considered as a part of incident management may include:

- **Eliminate or neutralize a threat**—Although removing or neutralizing a threat might seem like the best alternative, doing so when the threat is external is generally an unrealistic goal. If the threat is internal and specific, it may be possible to eliminate it. For example, the threat of a particular activity creating a security incident might be addressed by ceasing the activity.
- **Minimize the likelihood of a threat's occurrence**—The best alternative is often to minimize the likelihood of a threat's occurrence by reducing or eliminating vulnerabilities or exposure. This goal can be achieved by implementing the appropriate set of physical, environmental and/or security controls. For example, deploying firewalls, IDSs and strong authentication methods might substantially reduce the risk of a successful attack. Reducing exposure may be achieved by compartmentalization, such as network segmentation.
- **Minimize the effects of a threat if an incident occurs**—There are usually a number of ways to minimize impact if an incident occurs, such as effective incident management and response, insurance, redundant systems with automatic failover, or other compensating or corrective controls.

Each critical information processing system requires an approach to restoring operations in the event of disruption. There are many alternative strategies that can be considered both in terms of the incident management and response capability and from a disaster recovery perspective. These can range from redundant and mirrored systems to ensuring a high degree of system or process resilience and robustness.

4.10.5 RECOVERY SITES

The most appropriate alternatives for a recovery site must be based on probability of major outages occurring, the nature and extent of impact on the organization's ability to continue operations, and overall cost. Lengthier and more costly outages

or disasters that impair the primary physical facility are likely to require offsite backup alternatives. The types of offsite backup facilities that can be considered include:

- **Hot sites**—Hot sites are configured fully and ready to operate within several hours. The equipment, network and systems software must be compatible with the primary installation being backed up. The only additional needs are staff, programs, data files and documentation.
- **Warm sites**—Warm sites are complete infrastructures, but are partially configured in terms of IT, usually with network connections and essential peripheral equipment such as disk drives, tape drives and controllers. Sometimes a warm site is equipped with a less powerful central processing unit (CPU) than the primary site.
- **Cold sites**—Cold sites are a viable option only when organizations can afford relatively long downtime. Cold sites have only the basic environment (electrical wiring, air conditioning, flooring, etc.) to operate an information processing facility (IPF). The cold site is ready to receive equipment but does not offer any components at the site in advance of the need. Activation of the site may take several weeks. Because data and software are required for these strategies, special arrangements need to be considered for their backup to removable media and their safe, secure storage offsite. Several options for equipping a cold site exist:
  - Vendor or third party—Hardware vendors are usually the best source for replacement equipment. However, this may often involve a waiting period that is not acceptable for critical operations. It is unlikely that any vendor will guarantee a specific reaction to a crisis. Vendor arrangements are used best when an organization plans to move from a hot site to a warm or cold site, so advance planning is critical. Another source of equipment replacement is the used hardware market. This market can supply critical components or entire systems on relatively short notice, often at a substantially reduced cost. Establishing relationships with dealers well in advance of any actual emergency is critical.
  - Off-the-shelf—Such components are often available from the inventory of suppliers on short notice but may require special arrangements. To make use of this approach, several strategies must be used, including:
    - Avoiding the use of unusual and hard-to-get equipment
    - Regularly updating equipment to keep current
    - Maintaining software compatibility to permit the operation of newer equipment
- **Mobile sites**—Mobile sites are specially designed trailers that can be quickly transported to a business location or an alternate site to provide a ready-conditioned IPF. These mobile sites can be attached to form larger work areas and can be preconfigured with servers, desktop computers, communications equipment, and microwave and satellite data links. They are a useful alternative when there are no recovery facilities in the immediate geographic area. They are also useful in case of a widespread disaster and may be a cost-effective alternative for duplicate IPFs for a multisite organization.
- **Duplicate sites**—These facilities are dedicated recovery sites that are functionally similar or identical to the primary site that can quickly take over for the primary site. They range from a standby hot site to facilities available through a reciprocal agreement with another company. The assumption is that there are fewer problems in coordinating compatibility and availability in the case of duplicate sites. Large organizations with multiple data facilities can often develop failover capabilities among their own geographically dispersed data centers provided the following principles are followed:
  - The site chosen should be located so it is not subject to the same disaster event as the primary site. If, for example, the primary site is in an area subject to hurricanes, the recovery site should not be subject to the same hurricanes.
  - Coordination of hardware/software strategies is necessary. A reasonable degree of hardware and software compatibility must exist to serve as a basis for backup.
  - Resource availability must be assured. The workloads of the sites must be monitored to ensure that sufficient availability for emergency backup use exists.
  - There must be agreement concerning the priority of adding applications (workloads) until all the recovery resources are fully used.
  - Regular testing is necessary. Whether duplicate sites are under common ownership or under the same management, testing of the backup operation on a regular basis is necessary to ensure that it will work in the event of a disaster.
- **Mirror sites**—If continuous uptime and availability are required, a mirror site may be the best option. By definition, a mirror site is very similar or identical to the primary site. The mirror site is operational in concert with the primary site on a load-sharing basis. Typically, applications are launched by an automatic scheduler that balances the loads between the sites based on available operational capacity and applications can be executed in either one. Provided that sufficient reserve capacity exists, applications are seamlessly switched between the sites without interruption. Once again, organizations with multiple data facilities can often develop this capability among their own geographically dispersed data centers.
- **Reciprocal agreements**—(Note that although reciprocal agreements were once common, they are now seldom used.) Under the typical agreement, participants promise to provide computing time and network operations to each other when an emergency arises. If the recovery strategy is to use reciprocal agreements with one or more internal or external entities, it is essential that similar equipment and applications are available. It is important to consider that IT resources are generally used in a manner that approaches their maximum capacity and may not be able to support the requirements for recovery operations; organizations do not generally have sufficient reserve resources to satisfy even fairly small recovery requirements related to CPU, network bandwidth or storage capacity. The availability of competent personnel at an alternate facility must also be considered. It is likely that in the event of a serious incident or disaster, the availability of staff resources will be adversely affected. Extra expenses, integrating external personnel into operations and additional threats to physical security are some of the difficult and immediate problems inherent in this type of solution.
  - Additionally, creating a contract that provides adequate protection can be a difficult task, and the cost of dealing with changes over time is likely to be significant. It is important that contractual provisions for the use of third-party sites should cover important issues such as configuration of third-party hardware and software, speed of availability, reliability,

duration of usage, nature of intersite communications, and period of usage.

The type of site most suitable will largely be based on operational requirements determined by the BIA, the costs and benefits, and the risk appetite of management. It may be determined that some critical operational aspects must be mirrored to meet service level requirements while others can be adequately supported by hot, warm or cold site capabilities at the same or other facilities.

Additionally, part of the recovery of IT facilities involves telecommunications, for which the strategies usually considered include elements of network disaster prevention:
- Alternative routing
- Diverse routing
- Long-haul network diversity
- Protection of local resources
- Voice recovery
- Availability of appropriate circuits and adequate bandwidth
- Availability of out-of-band communications in case of failure of primary communications methods

After a strategy for the recovery of sufficient IT facilities to support critical business processes has been developed, it is critical that the strategies work for the entire period of recovery until all facilities are restored. The strategies may include:
- Doing nothing until recovery facilities are ready
- Using manual procedures
- Focusing on the most important customers, suppliers, products, systems, etc., with the resources that are still available
- Using PC-based systems to capture data for later processing or performing simple local processing

4.10.6 BASIS FOR RECOVERY SITE SELECTIONS

The type of site selected for a response and recovery strategy should be based on the following considerations:
- **AIW**—The total time that the organization can wait from the point of failure to the restoration of critical services/applications. After this time, the cumulative losses caused by the interruption may threaten the existence of the organization.
- **RTO**—The length of time from the interruption to the time that the process must be functioning at a service level sufficient to limit financial and operational impacts to an acceptable level. Some organizations express it as partial moments (i.e., from point of failure to technical recovery, or point of disaster declaration to full operations).
- **RPO**—The age of the data the organization needs to be able to restore in the event of a disaster (i.e., the amount of data that can be lost and need to be recreated). Sometimes this is expressed as the point of last known good data. This will be the starting point for operations at the recovery site. If full backups are infrequent, it may take too much time to re-create the amount of data lost and the result would be RTOs not being met.
- **SDO**—Level of services to be supported during the alternate process mode until the normal situation is restored. This must be directly related to business needs.
- **MTO**—The maximum time the organization can support processing in the alternate mode. Various factors will determine the MTO, including increasing backlogs of deferred processing. This, in turn, is affected by the SDO if it is less than that required during normal operations.
- **Proximity factors**—The distance from potential hazards, which can include flooding risk from nearby waterways, hazardous material manufacturing or storage, or other situations that may pose a risk to the operation of a recovery site
- **Location**—Sufficient distance needed to minimize the likelihood of both the primary and recovery facilities being subject to the same occurrence of an environmental event. When planning, consideration should be given to the typical impact area of the types of events that have a higher likelihood of occurrence for a given location. For example, in the case of hurricanes vs. tornados, the impact area of a hurricane is typically much larger than the impact area of a tornado, requiring greater distance between sites for a location where hurricanes frequently occur.
- **Nature of probable disruptions**—This must be considered in terms of the MTO. For example, a major earthquake is likely to render a primary site inoperable for a number of months. Clearly, the MTO in an area subject to this disruption must be greater than the probable duration of such an event.

To prepare a suitable recovery strategy, the information security manager must balance all of these parameters with the capabilities of different types of recovery sites, their costs and locations.

The complexity and cost of the response and recovery plans, as well as the type and cost of the recovery site, are proportionally inverse to these time objectives. An interruption window of two hours, for instance, dictates a hot or mirrored solution that is generally very expensive, but the corresponding recovery process is likely to be simple and inexpensive. In contrast, an interruption window of a month may permit the use of a cold site that is inexpensive, but the associated recovery process is likely to be complex and expensive.

4.10.7 RESPONSE AND RECOVERY STRATEGY IMPLEMENTATION

Based on the response and recovery strategy selected by management, a detailed response and recovery plan should be developed. It should address all issues involved in recovering from a disaster. Various factors should be considered while developing the plan, including:
- Preincident readiness
- Evacuation procedures
- How to declare a disaster
- Transition steps to disaster recovery if incident response fails
- Identification of the business processes and IT resources that should be recovered
- Identification of individuals with decision authority and responsibilities in the plan
- Identification of the people (and alternates) responsible for each function in the plan
- Identification of contact information
- The step-by-step explanation of the recovery options
- Identification of the various resources required for recovery and continued operations
- Ensuring that other logistics such as personnel relocation and temporary housing are considered

The response and recovery plan should be documented and written in simple language that is clear and easy to read. It is also common to identify teams of personnel who are responsible for specific tasks in case of incidents or disasters. Most BCPs are created as a set of procedures that accommodate system, user and network recovery strategies. Copies of the plan must be kept offsite to ensure that it is available when needed; this includes at the recovery facility, the media storage facility and the homes of key decision-making personnel.

4.10.8 RESPONSE AND RECOVERY PLAN

Organizations should have a formal, focused and coordinated approach to responding to incidents, including an IRP that provides the road map for implementing the incident response capability. Each organization needs a plan that meets its unique requirements, which relate to the organization's mission, size, structure and functions. The plan should lay out the necessary resources and management support. The IRP should include the following elements:
- Mission
- Strategies and goals
- Senior management approval
- Organizational approach to incident response
- Key decision-making personnel and responsibilities
- Communication with the rest of the organization and with other organizations
- Metrics for measuring the incident response capability and its effectiveness
- Road map for maturing the incident response capability
- How the program fits into the overall organization

The organization's mission, strategies and goals for incident response should help in determining the structure of its incident response capability. The incident response program structure should also be discussed within the plan. Once an organization develops a plan and gains management approval, the organization should implement the plan and review it at least annually to ensure the organization is following the road map for maturing the capability and fulfilling the goals for incident response.

4.10.9 INTEGRATING INCIDENT RESPONSE WITH BUSINESS CONTINUITY

Effective integration of incident response and business continuity/disaster recovery planning requires the relationships among RTO, RPO, SDO and MTO to be carefully considered. Because the transition from incident response to disaster recovery operations for any but mirrored or duplicate processing sites will require some time, RTO and AIW will be affected.

DR generally comprises the plan to recover an IT processing facility or the plan by business units to recover an operational facility. The incident management and recovery plan must be consistent with and support the overall IT plan of the organization. Business continuity planning, disaster recovery and incident response do not necessarily have to be combined into a single plan; however, each must be consistent with the other and integrated so that transition on declaration of a disaster is effective.

Risk Acceptance and Tolerance

The general issues of risk have been addressed in chapter 2. Risk specific to incident response and recovery operations are numerous and must be considered from a magnitude and frequency basis as well as from the perspective of the potential for impact.

Risk tolerance is the acceptable degree of variance to acceptable risk that, in the final analysis, must be determined by management. The essential consideration from an information security perspective is to ensure that an incident or disaster does not result in a security compromise. Because the focus will be on recovery or restoration of services, there is a significant possibility that expediency and procedural shortcuts will pose increased risk exposure.

Business Impact Analysis

No matter how good controls may be, the risk of an incident cannot be eliminated. Accordingly, the information security manager should oversee the development of response and recovery plans to ensure they are properly designed and implemented. The basis and priorities for the plan should, as described previously, be founded on impacts determined by conducting a BIA of all essential organizational functions.

Next, response and recovery strategies should be identified and validated and then approved by senior management. Once senior management approves these strategies, the information security manager should oversee the development of the response and recovery plans. During this process, response and recovery teams should be identified and team members mobilized. The plans must provide the teams guidance concerning the steps to be taken to recover business processes.

Recovery Time Objectives

RTO is defined as the amount of time allowed for the recovery of a business function or resource to a predefined operational level after a disaster occurs. Exceeding this time would mean organization survival would be threatened or the losses would exceed acceptable levels. RTOs are determined as a result of management deciding the level of acceptable impact as a result of the unavailability of information resources.

Recovery Point Objectives

RPO is defined as a measurement of the point prior to an outage to which data are to be restored; that is, the last point of known good data. RTO and RPO must be closely linked to facilitate effective incident management and response. This is because a short RTO may be adversely impacted by the RPO if there is a large amount of data that must be re-created prior to achieving acceptable levels of operation.

Service Delivery Objectives

The SDO is the level of acceptable service that must be achieved within the RTO. In many cases, an acceptable level may be substantially less than normal operations, less costly and easier to achieve. The SDO is determined by various factors, including business requirements, SLAs, costs and sustainability over the likely duration of operation in an alternate facility.

Maximum Tolerable Outage

MTO is the total time that operations can be sustained at an alternate site. A number of factors must be considered to arrive at this value. It must be related to the probable types of events that may require operations to move to an alternate site and their probable duration. If the threats such as a major earthquake are likely to result in long-term damage, the MTO may need to be measured in months whereas other types of events might typically be much shorter.

4.10.10 NOTIFICATION REQUIREMENTS

The recovery plan must cover notification responsibilities and requirements. It should also include a directory of key decision-making personnel, IRT members, information systems owners, end users, and others required to initiate and carry out response efforts. This directory should also include multiple communication methods (telephone, cellular phone, texting, email, etc.) in the event of any communication channel failure.

The directory should also include at least the following individuals:

- Representatives of equipment and software vendors
- Contacts within companies that have been designated to provide supplies and equipment or services
- Contacts at recovery facilities, including hot site representatives or predefined network communications rerouting services
- Contacts at offsite media storage facilities and the contacts within the company who are authorized to retrieve media from the offsite facility
- Insurance company agents
- Contact information for regulatory bodies
- Contacts at HR and/or contract personnel services
- Law enforcement contacts

Note that the decision to bring in law enforcement during such an incident rests solely with senior management. It is generally not the role of an information security manager to directly contact external organizations. There are companies that handle automated emergency response communications. However, as with any other outsourced service, careful consideration should be given to the required processes that must be in place to ensure the effectiveness of this solution.

4.10.11 SUPPLIES

The plan must include provisions for all supplies necessary for continuing normal business activities during the recovery effort. This includes detailed, up-to-date hard-copy procedures that can be followed easily by staff and contract personnel who are unfamiliar with the standard and recovery operations. This is to ensure that the plan can be implemented, even if members of the regular staff are unavailable. Also, a supply of special forms, such as check stock, invoice forms and order forms, should be secured at an offsite location.

If the data entry function is dependent on certain hardware devices and/or software programs, these programs and equipment, including specialized electronic data interchange (EDI) equipment and programs, must also be provided at the recovery site.

4.10.12 COMMUNICATION NETWORKS

The plan must contain details of the organization's telecommunication networks needed to restore business operations. Because of the criticality of these networks, the procedures to ensure continuous telecommunication capabilities should be given a high priority. Telecommunication networks are susceptible to the same natural disasters as data centers but are also are vulnerable to disruptive events unique to telecommunications. These include central switching office disasters, cable cuts, communication software glitches and errors, security breaches from hacking (phone hackers are known as "phreakers"), and a host of other human errors. The local exchange carrier is typically not responsible for providing

Knowledge Check: Incident Response Parameters

Match each parameter with its type.

Parameter	Types
1. If the database is corrupted by an incident, the business will be able to record transactions through a spreadsheet, but customer experience will be negatively affected after 12 hours.	A. RTO
2. If the database is corrupted by an incident, the organization will be able to record transactions through an Excel spreadsheet, but other processes will not be able to run until service is restored.	B. RPO
3. If the database is corrupted by an incident, the backup at the close of work on the previous day should be restored.	C. SDO
4. If the database is corrupted by an incident, access to the database will be restored within eight hours.	D. MTO

Answers on page 253.

backup services. Although many carriers normally back up main components within their systems, the organization should make provisions for backing up its own telecommunication facilities.

Telecommunications capabilities to consider include telephone voice circuits, wide area networks (WANs) (connections to distributed data centers), LANs and third-party EDI providers. Options can include satellite and microwave links and, depending on criticality and location, wireless links or even single sideband radiotelephone communications. Critical capacity requirements should be identified for the various thresholds of outage, such as two hours, eight hours or 24 hours, for each telecommunications capability. Uninterruptable power supplies (UPSs) should be sufficient to provide backup for telecommunications equipment as well as for computer equipment.

4.10.13 METHODS FOR PROVIDING CONTINUITY OF NETWORK SERVICES

Methods for providing continuity of network services include:

- **Redundancy**—Achieving redundancy involves a variety of solutions, including:
  - Providing extra capacity with a plan to use the surplus capacity should the normal primary transmission capability not be available. In the case of a LAN, a second cable could be installed through an alternate route for use in the event that the primary cable is damaged.
  - Providing multiple paths between routers
  - Using special dynamic routing protocols such as the Open Shortest Path First (OSPF) and External Gateway Routing Protocol (EGRP)
  - Providing for failover devices to avoid single points of failure in routers, switches, firewalls, etc.
  - Saving configuration files for recovery of network devices, such as routers and switches, in the event that they fail
- **Alternative routing**—Alternative routing means routing information via an alternate medium such as copper cable or fiber optics. This involves use of different networks, circuits or end points, if the normal network is unavailable. Most local carriers are deploying counter-rotating fiber-optic rings. These rings have fiber-optic cables that transmit information in two different directions and in separate cable sheaths for increased protection. Currently, these rings connect through one central switching office. However, future expansion of the rings may incorporate a second central office in the circuit. Some carriers are offering alternate routes to different points of presence or alternate central offices. Other examples include dial-up circuits as an alternative to dedicated circuits, a cellular phone and microwave communications as alternatives to land circuits, and couriers as an alternative to electronic transmissions.
- **Diverse routing**—This is the method of routing traffic through split cable facilities or duplicate cable facilities. This can be accomplished with different and/or duplicate cable sheaths. If different cable sheaths are used, the cable may be in the same conduit and, therefore, subject to the same interruptions as the cable it is backing up. The communication service subscriber can duplicate the facilities by having alternate routes, although the entrance to and from the customer's premises may be in the same conduit. The subscriber can obtain diverse routing and alternate routing from the local carrier, including dual entrance facilities; however, acquiring this type of access is time-consuming and costly. Most carriers provide facilities for alternate and diverse routing, although most services are transmitted over terrestrial media. These cable facilities are usually located in the ground or the basement of buildings that house computer equipment. Ground-based facilities are at risk due to the aging infrastructures of cities. In addition, cable-based facilities usually share space with mechanical and electrical systems that can impose risk due to human error and disastrous events.
- **Long-haul network diversity**—Many vendors of recovery facilities provide diverse long-distance network availability, using high-speed data circuits among the major long-distance carriers. This ensures long-distance access if any single carrier experiences a network failure. Several of the major carriers have now installed automatic rerouting software and redundant lines that provide instantaneous recovery if a break in their lines occurs. The information security manager should confirm that the recovery facility has these vital telecommunications capabilities.
- **Last-mile circuit protection**—Many recovery facilities provide a redundant combination of local carrier high-speed data circuits, microwave and/or coaxial cable access to the local communications loop. This enables the facility to have access during a local carrier communication disaster. Alternate local carrier routing is also used.
- **Voice recovery**—Many service, financial and retail industries dependent on voice communication. Therefore, their recovery plans should provide for redundant cabling and alternative routing for voice communication lines as well as data communication lines.

4.10.14 HIGH-AVAILABILITY CONSIDERATIONS

The loss or disruption of servers managing sensitive and critical business processes could have catastrophic effects on an organization. Plans should include operational failover methods to prevent servers from going offline for an extended period of time. Server recovery should also be included in the DRP. Some of the techniques for providing failover or fault-tolerant capabilities include UPSs and the use of failover systems to prevent power failures of varying levels.

Direct attached storage (DAS) is a data storage and availability solution in which the storage device (e.g., disk drive) is directly attached to a server or client. In order for multiple users to access the DAS, each needs to have direct access to the server that is housing the storage device. If the need for more storage arises, the server or client needs to be taken offline so additional drives can be installed, which affects availability.

Network attached storage (NAS) is a data storage and availability solution that has its own operating system (such as LUnix), storage and user interface connected to a server through an existing Ethernet network. Adding storage to the NAS has no downtime, because the connection is to the network rather than the server; therefore, availability is not necessarily impacted as storage needs increase.

A storage area network (SAN) is a high-speed, special-purpose network that provides mass storage using remote interconnected devices—such as disk arrays, tape libraries or optical jukeboxes—

with associated data servers that function as if they were attached locally. SANs are typically part of the overall network of computing resources for larger enterprises, but may also serve as remote resources for backup and archival storage. SANs typically support disk mirroring, backup and restore functions, data migration between storage devices, and the sharing of data among different servers in one or more networks.

Compatible with DAS, NAS and SAN storage solutions, a redundant array of inexpensive (or independent) disks (RAID) provides performance improvements and fault-tolerant capabilities via hardware or software solutions, breaking up data and writing them to a series of multiple disks to improve performance and/or save large files simultaneously. These systems provide the potential for cost-effective continuous data availability onsite or offsite.

Strategies to ensure data availability and re-creation in the event of an incident are dependent on the RPOs and RTOs, as discussed in 4.10.9 Integrating Incident Response With Business Continuity. An illustration of techniques implemented in relation to RTOs and RPOs is seen in **figure 4.6**.

After an incident involving data loss, the acceptable time before a system must be restored/re-created (i.e., RTO) and degree of completeness, or the acceptable limit of data loss for the re-created system (i.e., RPO), are business requirements that are enabled through the use of high-availability or fault-tolerant solutions. Their use depends on the needs of the business and varies among industries.

If the RTO is instantaneous and the RPO is equally stringent (i.e., the goal is for users to experience no downtime), then a fault-tolerant storage solution is established. Fault-tolerant servers provide for fail-safe redundancy through mirrored images of the primary server. Using this approach also entails distributed processing of a server load—a concept referred to as "load balancing" or "clustering"—where all servers take part in processing. In this arrangement, an intelligent cluster unit provides for load balancing for improved performance. This type of server architecture is transparent to users. The only thing that may be noticeable to a user is performance degradation if a server fails.

The information security manager must be aware of the high costs associated with fault-tolerant solutions and difficulties in achieving this state of data availability. For example, in a virtual environment, multiple instances (copies) of a virtual machine (VM) running in parallel must exist. Any change in the state of the primary VM (such as modification of a file) must be applied on all of the secondary VMs in real time.

If the RTO and RPO are more flexible (i.e., the business will tolerate a limited amount of downtime), then a high-availability storage solution is established. An example of high availability is a set of servers that have a fail-over relationship. When an application is running and the primary server fails, the application is restarted in the failover server. In this arrangement, the most current work in progress may be lost; however, the application could be up and running within seconds. High availability solutions are less costly to implement than those designed for 100 percent fault tolerance. See **figure 4.7**.

4.10.15 INSURANCE

The IRP should include information regarding the organization's insurance plans, including general coverage, cyber insurance or IT-related insurance. Some elements of the organization's coverage, such as a business interruption policy, may provide some level of protection and should be considered as part of the plan. Current information systems processing insurance policies are usually a multiperil policy designed to provide various types of IT coverage. These policies should be constructed modularly so they can be adapted to the insured's particular IT environment.

Figure 4.6—Techniques Implemented in Relation to RTOs and RPOs

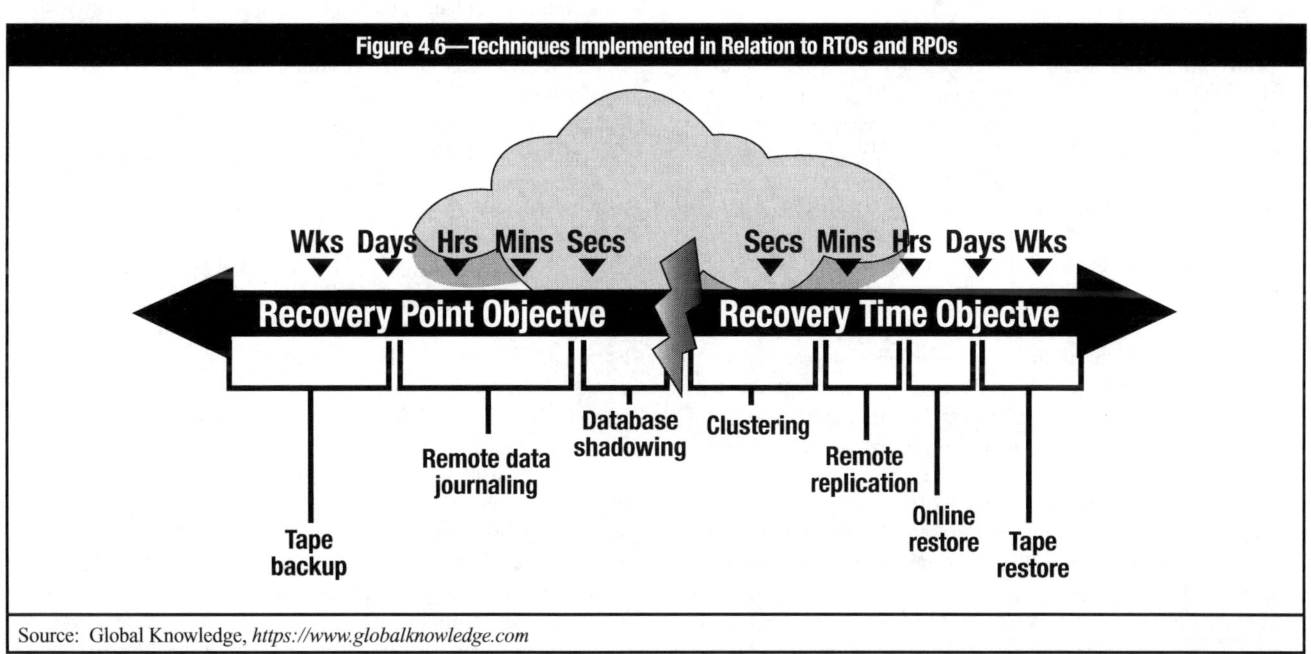

Source: Global Knowledge, *https://www.globalknowledge.com*

Figure 4.7—Relationship of RTOs, RPOs, High Availability and Fault Tolerance

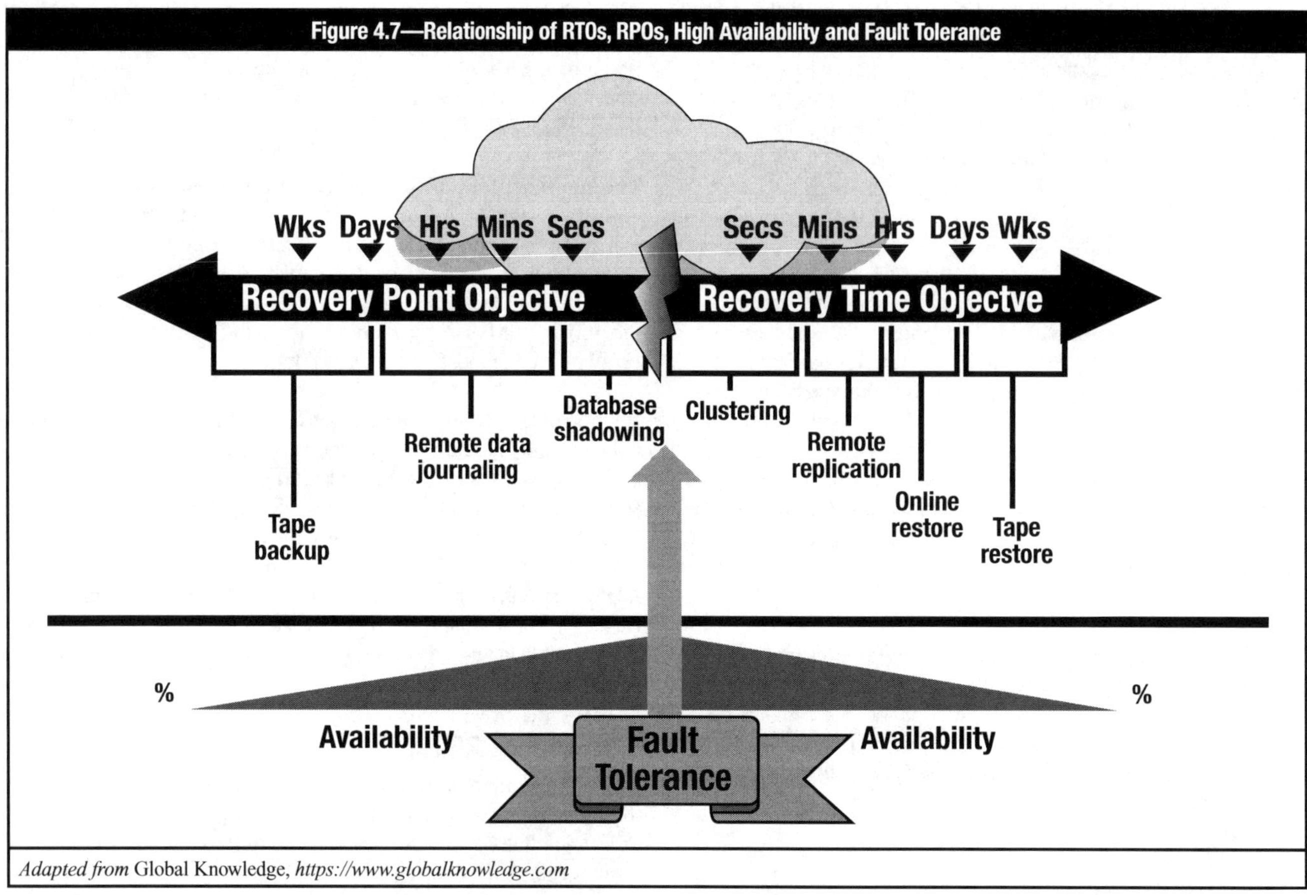

Adapted from Global Knowledge, *https://www.globalknowledge.com*

It should be noted that an organization can typically not insure against failure to comply with legal and regulatory requirements or any other breach of the law. There are usually a number of other exclusions and deductibles, and careful review by the information security manager is required to gain a clear understanding of the actual protection provided.

Some of the specific types of coverage that are available include:

- **IT equipment and facilities**—Provides coverage of physical damage to the IPF and owned equipment. An organization should also insure leased equipment if it is obtained when the lessee is responsible for hazard coverage. The information security manager should review these policies carefully; many policies are worded such that insurers are obligated to replace damaged or destroyed equipment with "like kind and quality," not necessarily the identical brand and model.
- **Media (software) reconstruction**—Covers damage to computer-related media that are the property of the insured and for which the insured may be liable. Insurance is available for on-premises, off-premises or in-transit disasters and covers the actual reproduction cost of the property. Considerations in determining the amount of coverage needed are programming costs to reproduce the media damaged; backup expenses; and physical replacement of media devices such as tapes, cartridges and disks.
- **Cybersecurity**—Relatively recent insurance that provides cover for cyberattacks of various kinds, including malware, DoS and DDoS, and breach losses
- **Professional and commercial liability**—Protection from third-party claims for losses and damages caused by the insured
- **Extra expense**—Designed to cover the extra costs of continuing operations following damage or destruction at the IPF. The amount of insurance needed is based on the availability and cost of backup facilities and operations. Extra expense can also cover the loss of net profits caused by computer media damage. This provides reimbursement for monetary losses resulting from suspension of operations due to the physical loss of equipment or media as in the case when IPFs are on the sixth floor and the first five floors are burned out. In this case, operations would be interrupted even though the IPF remained unaffected.
- **Business interruption**—Covers the loss of profit due to the disruption of the activity of the company caused by any covered IT malfunction or security-related event in which an attacker or malicious code causes loss of availability of computing resources
- **Valuable papers and records**—Covers the actual cash value of papers and records (not defined as media) on the insured's premises against unauthorized disclosure, direct physical loss or damage
- **Errors and omissions**—Provides legal liability protection in the event that the professional practitioner commits an act, error or omission that results in financial loss to a client. This insurance originally was designed for service bureaus, but it is now available from several insurance companies for protecting against actions of systems analysts, software designers, programmers, consultants and other information systems personnel.
- **Fidelity coverage**—Usually takes the form of banker's blanket bonds, excess fidelity insurance and commercial blanket bonds, and covers loss from dishonest or fraudulent acts by employees. This type of coverage is prevalent in financial institutions operating their own IPF.

- **Media transportation**—Provides coverage for potential loss or damage to media in transit to off-premises IPFs. Transit coverage wording in the policy usually specifies that all documents must be filmed or otherwise copied. When the policy does not specifically require that data be filmed prior to being transported and the work is not filmed, management should obtain from the insurance carrier a letter that specifically describes the carrier's position and coverage in the event data are destroyed.

4.10.16 UPDATING RECOVERY PLANS

Because organizations constantly evolve and change, the response and recovery plans also need to change. The information security manager must establish a process in which recovery plans are updated as changes occur within an organization. Assessing the response and recovery plan requirements during the change management process within an organization is an essential part of effective response management.

Plans and strategies for response and recovery should be reviewed and updated according to a schedule to reflect continuing recognition of changing requirements. The following factors as well as others may impact requirements and the need for the plan to be updated:
- A strategy that is appropriate at one point in time may not be adequate as the needs of an organization change.
- New applications may be developed or acquired.
- Changes in business strategy may alter the significance of critical applications or result in additional applications being deemed critical.
- Changes in the software or hardware environment may make current provisions obsolete or inappropriate.
- Changing physical and environmental circumstances may also need to be considered.

The responsibility for maintaining the BCP/DRP often falls to a BCP coordinator and the information security manager may be responsible for maintaining the IRP. However these responsibilities are allocated, specific plan maintenance activities include:
- Developing a schedule for periodic review and maintenance of the plan and advising all personnel of their roles and the deadline for receiving revisions and comments
- Calling for revisions when significant changes have occurred
- Reviewing revisions and comments and updating the plan within a reasonable period (e.g., 30 days) after the review date
- Arranging and coordinating scheduled and unscheduled tests of the plan to evaluate its adequacy
- Participating in scheduled plan tests, which should be performed at least once each year. For scheduled and unscheduled tests, the coordinator should write evaluations and integrate changes to resolve unsuccessful test results into the response plan within a reasonable period (e.g., 30 days).
- Developing a schedule for training personnel in emergency and recovery procedures, as set forth in the plan. Training dates should be scheduled within a reasonable period (e.g., 30 days) after each plan revision and scheduled plan test.
- Maintaining records of plan maintenance activities, such as testing, training and reviews
- Updating, at least quarterly, the notification directory to include all personnel changes, such as changes to phone numbers and responsibilities or status within the company

4.11 TESTING INCIDENT RESPONSE AND BUSINESS CONTINUITY/DISASTER RECOVERY PLANS

Testing all aspects of the IRP is the most important factor in achieving success in an emergency situation. The main objective of testing is to ensure that executing the plans will result in the successful recovery of the infrastructure and critical business processes.

Testing should focus on:
- Identifying gaps
- Verifying assumptions
- Testing timelines
- Determining the effectiveness of strategies
- Evaluating the performance of personnel
- Determining the accuracy and currency of plan information

Testing promotes collaboration and coordination among teams and is a useful training tool. Many organizations require complete testing annually. In addition, testing should be considered on the completion or major revision of each draft plan or complementary plans and following changes in key personnel, technology or the business/regulatory environment.

Testing must be carefully planned and controlled to avoid placing the business at increased risk. To ensure that all plans are regularly tested, the information security manager should maintain a "testing schedule" of dates and tests to be conducted for all critical functions.

Prior to each test, the security manager should ensure that:
- The risk and impact of disruption from testing is minimized.
- The business understands and accepts the risk inherent in testing.
- Fallback arrangements exist to restore operations at any point during the test.

All tests must be fully documented with pretest, test and posttest reports. Test documentation should be retained for audit review and reference. The security manager must ensure that information security is also tested and not compromised during testing.

4.11.1 PERIODIC TESTING OF THE RESPONSE AND RECOVERY PLANS

As discussed in prior sections, the scope and nature of incident response and recovery teams and their capabilities will vary with different organizations as will business continuity and disaster recovery operations. The exact relationship between these functions must be clearly defined and their scope and capabilities understood and integrated. Regardless of the specific scope at any particular organization, it is essential for the information security manager to have a good understanding of the entire business continuity process, including incident management and disaster recovery.

Whatever the structure, the full scope of incident management responsibilities must be tested up to the point of a disaster declaration, including the escalation, and involvement of or handover to the disaster management and recovery operation if this is the responsibility of another group. The discussion in this section includes full disaster recovery, which the information security manager should understand regardless of the specific scope of incident management and response responsibilities.

The information security manager, helped by the recovery team's organization, should implement periodic testing of response and recovery plans. Testing should include:
- Developing test objectives
- Executing the test
- Evaluating the test
- Developing recommendations to improve the effectiveness of testing processes and the response and recovery plans
- Implementing a follow-up process to ensure that the recommendations are implemented

Response and recovery plans that have not been tested leave an organization with an unacceptable likelihood that plans will not work, even though care is taken in developing and documenting these plans. Because testing plans costs time and resources, an organization should carefully plan tests and develop test objectives to be methodical and help ensure that measurable benefits can be achieved.

Once test objectives have been defined, the information security manager should ensure that an independent third party is present to monitor and evaluate the test. Internal or external audit or other assurance personnel can often assume this role. A result of the evaluation step should be a list of recommendations that an organization should complete to improve its response and recovery plans. It is extremely unlikely that no recommendations will result and everything will work as planned. If it does, it is likely that a more challenging test should be planned.

The information security manager should also implement a tracking process to ensure that any recommendations resulting from testing are implemented in a timely fashion. Personnel should be tasked with making any necessary changes.

4.11.2 TESTING FOR INFRASTRUCTURE AND CRITICAL BUSINESS APPLICATIONS

Testing recovery and response plans must include both infrastructure and critical applications, although not necessarily at the same time. With today's organizations relying heavily on IT, the information security manager is tasked with securing these systems not only during normal operations, but also during disaster events.

Based on the risk assessment and business impact information, the information security manager can identify critical applications the organization requires and the infrastructure needed to support them. To ensure that these are recovered in a timely fashion, the information security manager needs to perform appropriate recovery tests.

4.11.3 TYPES OF TESTS

Types of basic tests include:
- **Checklist review**—This is a preliminary step to a real test. Recovery checklists are distributed to all members of a recovery team to review and ensure that the checklist is current.
- **Structured walkthrough**—Team members physically implement the plans on paper and review each step to assess its effectiveness and identify enhancements, constraints and deficiencies.
- **Simulation test**—The recovery team role-plays a prepared disaster scenario without activating processing at the recovery site.
- **Parallel test**—The recovery site is brought to a state of operational readiness, but operations at the primary site continue normally.
- **Full interruption test**—Operations are shut down at the primary site and shifted to the recovery site in accordance with the recovery plan; this is the most rigorous form of testing, but it is also expensive and potentially disruptive.

Testing should start simply and increase gradually, stretching the objectives and success criteria of previous tests so as to build confidence and minimize risk to the business. At a minimum, "full-interruption" tests should be performed annually after individual plans have been tested separately with satisfactory results.

Tests that are progressively more challenging can include:
- Table-top walkthrough of the plans
- Table-top walkthrough with mock disaster scenarios
- Testing the infrastructure and communication components of the recovery plan
- Testing the infrastructure and recovery of the critical applications
- Testing the infrastructure, critical applications and involvement of the end users
- Full restoration and recovery tests with some personnel unfamiliar with the systems
- Surprise tests

There are three main recovery testing categories:
- **Paper tests**—Paper tests are an on-paper walkthrough of the plan involving the major players in the plan's execution who reason out what might happen in a particular type of service disruption. They may walk through the entire plan or just a portion. The paper test usually precedes preparedness tests.
- **Preparedness tests**—Preparedness tests are usually localized versions of a full test, wherein actual resources are expended in the simulation of a system crash. These tests are performed regularly on different aspects of the plan and can be a cost-effective way to gradually obtain evidence about how good the plan is. They also provide a means to improve the plan in increments.
- **Full operational tests**—These tests are one step away from an actual service disruption. An organization should have tested the plan well on paper and locally before endeavoring to completely shut down operations. For purposes of BCP testing, the full operational testing scenario is the disaster.

Most organizations start with paper tests before attempting preparedness and operational tests as these pose increasing difficulty and risk. Regardless of which tests are being performed, during every phase of the test, detailed documentation of observations, problems and resolutions should be maintained. Each team should have a diary with specific steps and information recorded. This documentation serves as important historical information that can facilitate actual recovery during a real disaster. The documentation also aids in performing detailed analysis of the strengths and weaknesses of the plan.

It is common for response and recovery tests to fall short of a full-scale test of all operational portions of the organization. This should not preclude performing full or partial testing because one of the purposes of the business continuity test is to determine how well the plan works or which portions of the plan need improvement. Although surprise tests are potentially advantageous from the standpoint that they are similar to real-life incident response situations, they have some potential downsides. They can be very disruptive to production and operations. The information security manager should carefully consider the ramifications before recommending various testing approaches. There have been instances of serious extended outages resulting from the inability to restore systems as planned.

Potentially disruptive testing should be scheduled during a time that will minimize impact on normal operations. Weekends are generally a good time to conduct tests. It is important for all key recovery team members to be involved in the recovery test process. The test should address all critical components and simulate actual prime-time processing conditions, even if the test is conducted during off hours.

4.11.4 TEST RESULTS

Each test should yield an expected set of results. In other words, there are specific outcomes that should be expected as a result of running a test.

A recovery test should strive to, at a minimum, accomplish the following tasks:

- Verify the completeness and precision of the response and recovery plan.
- Evaluate the performance of the personnel involved in the exercise.
- Appraise the demonstrated level of training and awareness of individuals who are not part of the recovery/response team.
- Evaluate the coordination among the team members and external vendors and suppliers.
- Measure the ability and capacity of the backup site to perform prescribed processing.
- Assess the vital records retrieval capability.
- Evaluate the state and quantity of equipment and supplies that have been relocated to the recovery site.
- Measure the overall performance of operational and information systems processing activities related to maintaining the business entity.

To perform preparedness or operational recovery testing, each of the following test phases should be completed:

- **Pretest**—The pretest consists of the set of actions necessary to set the stage for the actual test. This ranges from placing tables in the proper operations recovery area to transporting and installing backup telephone equipment. These activities are outside the realm of those that would take place in the case of a real emergency, in which there is generally no forewarning of the event and thus no time to take preparatory actions.
- **Test**—Actual operational activities are executed to test the specific objectives of the plan. Data entry; telephone calls; information systems processing; handling orders; and movement of personnel, equipment and suppliers should take place. Evaluators should review staff members as they perform the designated tasks. This is the actual test of preparedness to respond to an emergency.
- **Posttest**—The posttest is the cleanup of group activities. This phase comprises assignments such as returning all resources to their proper place, disconnecting equipment, returning personnel to their normal locations and deleting all company data from third-party systems. The posttest cleanup also includes formally evaluating the plan and implementing indicated improvements.

4.11.5 RECOVERY TEST METRICS

Just as with nearly everything else in information security, metrics should be developed and used in measuring the success of the plan and testing against the stated objectives. Results should be recorded and evaluated quantitatively, as opposed to an evaluation based only on verbal descriptions. The resulting metrics should be used not only to measure the effectiveness of the plan, but more importantly, to improve it. Although specific measurements vary depending on the test and the organization, the following general types of metrics usually apply:

- **Time**—Elapsed time for completion of prescribed tasks, delivery of equipment, assembly of personnel and arrival at a predetermined site. This is essential to refine the response time estimated for every task in the escalation process.
- **Amount**—Amount of work performed at the backup site by clerical personnel and the amount of information systems processing operations
- **Percentage and/or number**—The number of vital records successfully carried to the backup site vs. the required number, and the number of supplies and equipment requested vs. actually received. The number of critical systems successfully recovered can be measured with the number of transactions processed.
- **Accuracy**—Accuracy of the data entry at the recovery site vs. normal accuracy (as a percentage). The accuracy of actual processing cycles can be determined by comparing output results with those for the same period processed under normal conditions.

This testing process enables the information security manager to achieve initial successes and modify the plan based on information gained from the initial tests. It is important to note that performing a preparedness or operational test costs resources and requires coordination among various departments.

4.12 EXECUTING RESPONSE AND RECOVERY PLANS

Given that a major incident usually causes considerable confusion and a host of unexpected conditions, it is essential that the incident management and response plans have been tested under realistic conditions. It is possible that untested plans will not work as expected. It is also safe to assume that the more severe the incident, the greater the potential chaos, confusion and problems facing the incident management and response teams. Incidents can range from a virus attack bringing down IT systems to an earthquake bringing down the building. To provide reasonable assurance that the organization is preserved under foreseeable circumstances, all reasonably possible events must be anticipated and prepared for and the planning must be thorough, realistic and tested.

4.12.1 ENSURING EXECUTION AS REQUIRED

To ensure the response and recovery plans are executed as required, the plans need a facilitator or director to direct the tasks within the plans, oversee their execution, liaise with senior management and make decisions as necessary. The information security manager may or may not be the appropriate person to act as the recovery plan director or coordinator, but must be certain the role is assigned to someone who can perform this critical function.

Developing appropriate response and recovery strategies and alternatives is an essential component in the overall process of executing the response and recovery plans. It will provide reasonable assurance that the organization can recover its key business functions in the event of a disruption and respond appropriately to a security-related incident.

Testing the plans is essential to ensuring that plans can be executed as required. It will also help employees become familiar with the process so when an incident occurs they can respond accordingly.

The information security manager should also appoint an independent observer to record progress and document any exceptions that occur during testing and an actual event. Through a postevent review, the information security manager and key recovery personnel can then review the observations and make adjustments to the plan accordingly.

4.13 POSTINCIDENT ACTIVITIES AND INVESTIGATION

Understanding the purpose and structure of postincident reviews and follow-up procedures enables the information security manager to continuously improve the security program. A consistent methodology should be adopted within the information security organization so when a problem is found, an action plan is developed to reduce/mitigate it. Once the action plan is created, steps should be taken to implement the solution. By repeating these basic principles, the information security program is able to adapt to changes in the organization and the threats it faces. In addition, this reduces the amount of time personnel need to react to security incidents so they are able to spend more time on proactive activities.

The follow-up process in incident response is potentially the most valuable part of the effort. Lessons learned during incident handling can improve a security practice as well as the incident response process itself. Additionally, the information security manager needs to calculate the cost of the incident once all response efforts are done by adding the cost of any loss or damage plus the cost of labor and any special software or hardware needed to handle the incident. The cost provides a useful metric, especially in justifying the existence of the response team to senior management, and may be used as evidence in a court case.

The information security manager should manage postevent reviews to learn from each incident and the resulting response and recovery effort; this information can then be used to improve the organization's response and recovery procedures. The information security manager may perform these reviews with the help of third-party specialists if detailed forensic skills are needed.

4.13.1 IDENTIFYING CAUSES AND CORRECTIVE ACTIONS

Security incidents can be the result of externally initiated attacks, internally initiated attacks or failures in security controls that have been implemented. For a systematic review of security events, the information security manager should appoint an event review team. This team should review any evidence and develop recommendations to enhance the information security program by identifying root causes of a specific event and necessary measures to prevent the same/similar events from recurring. The root cause of many system compromises, for example, is weak or nonexistent vulnerability assessment and patch management efforts.

The analysis should be done to determine answers to questions such as:

- Who is involved?
- What has happened?
- Where did the attack originate?
- When (what time frame)?
- Why did it happen?
- How was the system vulnerable or how did the attack occur?
- What was the reason for the attack (i.e., the perpetrator's motivation)?

4.13.2 DOCUMENTING EVENTS

During and subsequent to any actual or potential security incident, the information security manager should have processes in place to develop a clear record of events. By preserving this information, events can be investigated and provided to a forensics team or authorities if necessary. To ensure this occurs, one or more individuals should be specifically charged with incident documentation and the preservation of evidence. Documentation of any event that has possible security implications can provide clarity as to whether an incident is merely an accident, a mistake or a deliberate attack.

A serious incident is typically chaotic. Good documentation will prove invaluable in postincident investigation and forensics and may also be helpful in incident resolution.

4.13.3 ESTABLISHING PROCEDURES

Having a good legal framework is important to provide options to the organization, including pursuing civil or criminal legal action against the perpetrators. The information security manager should develop data preservation procedures with the advice and assistance of legal counsel, the organization's managers and knowledgeable law enforcement officials to ensure that the procedures provide sufficient guidance to IT and security staff. With the assistance of these specialized resources, the information security manager can develop procedures to handle security events in a manner that preserves evidence, ensures legally sufficient chain of custody and is appropriate to meet business objectives.

There are a few basic actions that the information systems staff must understand. This includes doing nothing that could change, modify or contaminate potential or actual evidence. Trained forensics personnel can inspect computer systems that have been attacked, but if the organization's personnel contaminate the information, the data may not be admissible in a court of law and/or the forensics staff may be unable to use the data in investigating an incident. Computer forensics and gathering and handling information and physical objects relevant to a security incident in a systematic manner so they can be used as evidence in a court of law should usually be performed by specially trained staff, third-party specialists, security IRT or law enforcement officials. The initial response by the system administrator should include:

- Retrieving information needed to confirm an incident
- Identifying the scope and size of the affected environment (e.g., networks, systems, applications)
- Determining the degree of loss, modification or damage (if any)
- Identifying the possible path or means of attack

4.13.4 REQUIREMENTS FOR EVIDENCE

The information security manager should understand that any contamination of evidence following an intrusion could prevent an organization prosecuting a perpetrator and limit its options. In addition, the modification of data can inhibit computer forensic activity necessary to identify the perpetrator and all the changes and effects resulting from an attack. It may also preclude the possibility of determining how the attack occurred and how the security program should be changed and enhanced to reduce the risk of a similar attack in the future.

The usual recommendation for a computer that has been compromised is to disconnect the power to maximize the preservation of evidence on the hard disk. This approach is generally the recommendation of law enforcement based on the risk of the evidence being compromised. This can occur as a result of the system swap files overwriting evidence or an intruder or malware erasing evidence of compromise. There is also the risk of contaminating evidence.

This approach is not universally accepted as the best solution. One argument against disconnecting power is that data in memory are lost and sudden power loss may result in corruption of critical information on the hard disk. Because some malware is only memory-resident, the cause of an incident and the avenue of attack may be difficult to establish.

Since the best approach is subject to controversy, the information security manager will need to establish the most appropriate approach for the organization and train personnel in the appropriate procedures.

Whichever procedure is used to secure a compromised system, trained personnel must use forensic tools to create a bit-by-bit copy (or disk image) of any evidence that may exist on hard drives and other media to ensure legal admissibility. To avoid the potential for alteration or destruction of incident-related data, any testing or data analysis should be conducted using this copy. The original should be given to a designated evidence custodian who must store it in a safe location. The original media must remain unchanged and a record of who has had custody of it—the chain of custody—must be maintained for the evidence to be admissible in court.

When taking a copy of a hard drive, the technician should take a bit-level image of all the data on the drive, using a cable with a write-protect diode to prevent writing anything back onto the source drive. Hash values of both the source and destination drive should be calculated to ensure that the copied drive is an exact image of the original.

4.13.5 LEGAL ASPECTS OF FORENSIC EVIDENCE

As noted above, for evidence to be admissible in legal proceedings, it must have been acquired in a forensically sound manner and its chain of custody maintained. The information security manager in charge of an incident must have established and documented procedures for acquisition of evidence by properly trained personnel.

The required documentation to maintain legally admissible evidence must include:

- Chain of custody forms that include:
  - Name and contact information of custodians
  - When, why and by whom an evidence item was acquired or moved
  - Detailed identification of the evidence (serial numbers, model information, etc.)
  - Where it is stored (physically or logically)
  - When/if it was returned
- Checklists for acquiring technicians (including details of legally acceptable forensic practices)
- Detailed activity log templates for acquiring technicians
- Signed nondisclosure/confidentiality forms for all technicians involved in recovering evidence
- An up-to-date case log that outlines:
  - Dates when requests were received
  - Dates investigations were assigned to investigators
  - Name and contact information of the investigator and requestor
  - Identifying case number
  - Basic notes about the case and its requirements and completed procedures
  - Date when completed
- Investigation report templates that include:
  - Name and contact information of investigators
  - Date of investigation and an identifying case number

- Details of any interviews or communications with management or staff regarding the investigation
- Details of devices or data that were acquired (serial numbers, models, physical or logical locations)
- Details of software or hardware tools used for acquisition or analysis (must be recognized forensically sound tools)
- Details of findings including samples or copies of relevant data and/or references to their storage location
- Final signatures of investigator in charge

Procedures for initiating a forensics investigation need to be agreed to, documented, followed carefully and understood by everyone in the enterprise. The information security manager should work with management and HR (and other stakeholders) to establish a process that ensures that all investigations are fair, unbiased and well documented.

It is important to be aware that legal requirements vary in different jurisdictions. As a result, informed legal advice for appropriate processes that meet judicial standards will be required.

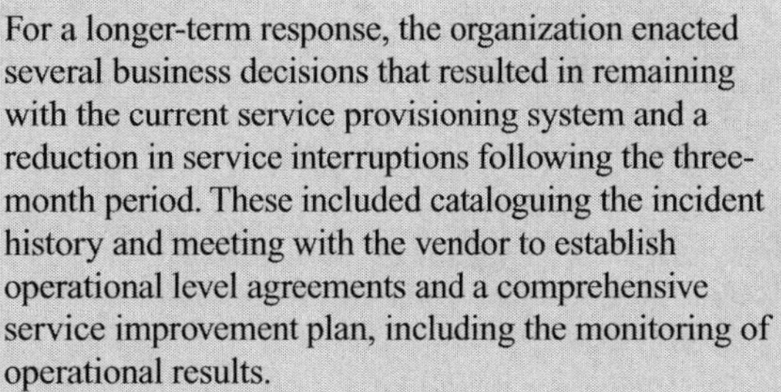

4.14 CASE STUDIES

Case Study 1

A vendor-supplied, business-critical service provisioning system for internal organizational functions (i.e., not customer-facing) has peak usage between the hours of 7:00 a.m. through 7:00 p.m. on weekdays.

On Monday at 6:00 a.m., the system became unresponsive. The network operations center worked with the IMT and opened a conference bridge at 6:30 a.m. At 11:00 a.m., the application owner on the user-organization side was notified of the incident by a user group manager. The application was back online by 1:00 p.m.

Three weeks following the first incident, another application service interruption occurred. On a Wednesday at 2:00 p.m., the system became unresponsive for a portion of the user base (i.e., some areas were unaffected, including the office location of the system operations owner for the user organization). The network operations center opened a conference bridge at 2:40 p.m. At 3:00 p.m., the application owner on the user-organization side was notified of the incident by the vendor. The application was back online for the complete user community by 4:00 p.m.

Going forward, the application experienced a service interruption on average once a month for the next three months. The impact of the service interruptions varied and included partial availability loss for certain functions, as well as a full service provisioning loss.

Following the lapse in notification from the first incident and the inconsistency brought about by notification from the second incident, the organization began to revamp its internal processes. Internal process revisions occurred through the creation of an incident communication procedure, which was distributed to all IT personnel charged with operating the application. The procedures designated communication goals as well as the creation of postincident reviews identifying clear root causes for the incidents no later than a week after resolution.

For a longer-term response, the organization enacted several business decisions that resulted in remaining with the current service provisioning system and a reduction in service interruptions following the three-month period. These included cataloguing the incident history and meeting with the vendor to establish operational level agreements and a comprehensive service improvement plan, including the monitoring of operational results.

1. **Considering the first incident, when should the business owner have been notified about the incident?**
   A. Within incident reporting operational level agreements
   B. No later than 6:15 a.m.
   C. Within the vendor's incident reporting service level agreements
   D. No later than 6:30 a.m.

2. **In a process-mature organization, how should the initial notification for the second incident have been delivered to the business owner?**
   A. Through separate notifications upon incident discovery to stakeholders
   B. By the security information and event management tool
   C. By the application vendor
   D. By the network operations center

3. **When cataloguing incidents in preparation for meeting with the vendor, which of the following is the BEST choice for the initial set of actions to be taken by the information security manager?**
   A. Create a postincident review for each incident involved.
   B. Perform a business impact analysis.
   C. Update the incident response and recovery plans.
   D. Perform a risk assessment.

Answers on page 253.

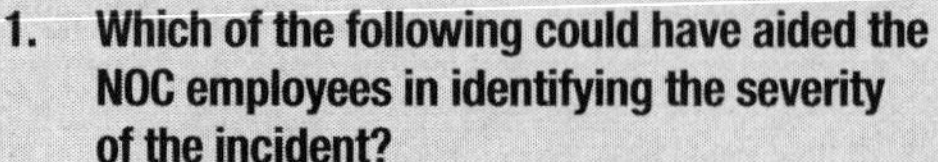

Case Study 2

In a major US financial institution, low-level personnel monitoring the network operation center (NOC) noticed unusual network activity on a Sunday evening when the bank was closed. Uncertain of what they were seeing, they decided to watch the event rather than risk disturbing management on a weekend. No severity criteria, notification requirements or escalation processes had been developed by the organization. By 3 a.m. on Monday, traffic continued to increase at the main facility and then suddenly began to grow dramatically at the mirror site, hundreds of miles away. The NOC staff remained unconcerned and felt they could still handle the increase in traffic.

By 7 a.m. on Monday, the NOC personnel were sufficiently concerned to notify the IT managers that there was a problem and the monitors showed the network was becoming saturated. An hour later when the external CIRT arrived, the network was totally inoperative and the team determined that the network had been compromised by the Slammer worm. The CIRT team manager informed the IT manager that the Slammer worm was memory-resident and restarting the entire network and mirror facility would resolve the issue. The manager stated that he did not have the authority to shut down the system and the CIO would need to approve the action The CIO could not be located, and current emergency phone and pager numbers were kept only in a new emergency paging system that required network access. When asked about the DRP and what it had to say regarding declaration criteria, three different plans were produced that had been prepared by teams in different parts of the organization, unbeknownst to each other. None contained declaration criteria or specified roles, responsibilities or authority. The final resolution ultimately required the CEO, who was traveling overseas and not immediately available, to finally issue instructions the next morning (Tuesday) to shut down the nonfunctioning network. The institution was inoperative for a full day and a half, with over 30,000 employees unable to perform their work. The final direct costs were estimated by the postincident team to exceed US $50 million.

1. **Which of the following could have aided the NOC employees in identifying the severity of the incident?**
   A. Updated security policies
   B. Meaningful security metrics
   C. Additional security training
   D. Increased decision-making autonomy

2. **Which is an example of a metric that would be MOST meaningful to the NOC in this scenario?**
   A. Amount of traffic during a typical downtime
   B. Number of transactions per hour
   C. Duration of increased traffic prior to contacting management
   D. The number of systems showing traffic

3. **Which of the following BEST could have aided the network manager in containing the incident in a more timely manner?**
   A. A single, approved disaster recovery plan
   B. Earlier notification from the NOC
   C. Better organizational governance structure
   D. Additional security training for staff

4. **Who ultimately should be responsible for ensuring that a single DRP is in place for the organization?**

5. **Who should be accountable for the lack of an effective DRP?**

Answers on page 254.

CHAPTER 4 ANSWER KEY

KNOWLEDGE CHECK: GAP ANALYSIS (PAGE 232)

Question 1: No formal incident response plan is in place. The IT team has the skills to identify potential events, but it is not empowered to make timely decisions to respond to events and better contain an incident before it becomes a larger issue.

Question 2: A documented incident response plan that includes an organizationally approved definition for an incident should be determined. This response plan would include levels of severity and an escalation plan that would empower the IT team to make better, timelier decisions to match the response time of similar organizations.

Question 3: There is no agreed-upon definition from the organization as far as what constitutes an incident. Senior leadership may be unsure what "incident" means or levels of severity for incidents. No formal, documented incident response plan is in place. There is no set system for alerting the COO that an incident has been identified. There is no escalation plan in place.

Question 4: Define what constitutes an incident. Develop an incident response plan to include severity/escalation criteria, roles and responsibilities, and metrics for effectiveness, including time to response.

KNOWLEDGE CHECK: INCIDENT RESPONSE PARAMETERS (PAGE 241)

1. D
2. C
3. B
4. A

CASE STUDY 1 (PAGE 251)

Question 1:

A. The timeframe for incident response should be determined by internal user-organization operational level agreements (OLAs).

B. The timeframe for incident response should be determined by internal OLAs. These should be determined based on incident categorizations as applied by the IMT.

C. The user-organization's OLAs take precedence over the vendor's internal SLAs. If the vendor's SLAs are insufficient, then the vendor and the organization should work to develop an improved solution to bring those SLAs to a level accepted by the user-organization. The vendor can provide notification of a service disruption; however, the organization should have internal monitoring systems in place.

D. The timeframe for incident response should be determined by internal OLAs. These should be determined based on incident categorizations as applied by the IMT. 6:30 a.m. represents the time that the conference bridge was opened; however, that may or may not be commensurate with the user-organization's incident reporting OLAs.

Question 2:

A. Depending on the way an organization's security information and event management (SIEM) tool is set up, it is possible for the same incident to be reported multiple times through numerous sources. However, if reporting comes at multiple times to the business owner, then it stops being an "initial" notification. In fact, all notification subsequent to the first become ancillary notifications, and therefore not what the question is asking for (which is the initial notification).

B. The SIEM tool should have notified the application business owner of the incident.

C. The application vendor could have notified the user-organization's business owner of the incident (as stated in the case); however, this should not have been the first (or initial) notification. The initial notification should have come from the SIEM tool. If the alert first came from the vendor, then the incident is either something of an extreme outlier that the SIEM tool did not catch or the organization lacks the process maturity that is presupposed in this question.

D. The user-organization's network operations center could have sent out the initial alert, if that was integrated with the SIEM tool; however, there is not enough information either within the case study or this question to conclude that as a certainty.

Question 3:

A. A postincident review was already created for incidents that have occurred over the previous three months, based on the updated documentation that was created following the first incident.

B. Conducting a business impact analysis is the first step in the incident response management process. In this scenario, the incidents have already occurred and were responded to; however, the combined business impact of the combined incident activity needs to be taken into account prior to discussions with the vendor. The vendor needs to understand the damage that system downtime has caused the business because this, along with the consistency of incidents, will be the basis from which a response strategy can be determined.

C. Updating the incident response and recovery plans can be performed in tandem with preparation for a vendor meeting. However, the response and recovery plans are to help matters going forward, while the vendor meeting is to address the impact of incidents as they occurred and to form a consolidated response to the pervasiveness in the lack of consistent system availability.

D. The incidents have already occurred; therefore, an impact assessment, not a risk assessment, needs to be performed.

CASE STUDY 2 (PAGE 252)

Question 1:

A. Updated security policies would have helped to establish reporting and escalation procedures; however, meaningful security metrics would better help network operation center (NOC) employees recognize that an incident was occurring, which would have helped them to react in a more effective way.

B. For those monitoring the NOC, metrics indicated a problem, but they were not sufficiently meaningful to the employees for them to make any active decisions, much less the correct ones. Either better metrics or greater proficiency of the personnel could have resolved the issue quickly in the initial stages of the incident before it became a problem.

C. Additional security training would help employees know what to do should an incident occur; however, meaningful metrics should be established in order to make training as impactful as possible so incidents can be recognized and handled in a timely manner.

D. Increased decision-making autonomy is helpful when there are meaningful metrics in place to help employees recognize incidents are occurring. Based on that information, they can make informed decisions on how to best contain and respond to incidents.

Question 2:

A. Knowing the normal traffic occurring at that time of night would have made it clear that the situation was not normal and a serious problem existed.

B. Without a reference point, the number of transactions would not have been particularly meaningful.

C. The duration of traffic without a point of reference would not have been meaningful.

D. The number of systems without a point of reference would not be helpful.

Question 3:

A. A single disaster recovery plan could have been useful but without clear authority it would not by itself have helped.

B. Earlier notification would have been helpful but only if authority to act was clear.

C. Better governance would have vested adequate authority in the network manager to take the appropriate action.

D. Training of staff would have been useful in recognizing the event but authority to act was essential.

Question 4: A disaster recovery plan (DRP) for the overall organization is a governance issue; therefore, it is the responsibility of the board of directors to ensure that a DRP is in place.

Question 5: Better governance would have ensured that risk management was in place; however, those involved in governance direct the program. The CISO or CIO—whomever the information security manager reports to—would be accountable for the ineffective DRP. The information security manager and the IRT would be responsible for periodically reviewing and testing the DRP to make sure it was working as expected.

GENERAL INFORMATION

REQUIREMENTS FOR CERTIFICATION

To earn the CISM designation, information security professionals are required to:

1. Successfully pass the CISM exam
2. Adhere to the ISACA Code of Professional Ethics
3. Agree to comply with the CISM continuing professional education policy
4. Submit verified evidence of five (5) years of work experience in the field of information security. Three (3) of the five (5) years of work experience in the role of managing information security. In addition, this work experience must be broad and gained in three of the four job practice domains. A processing fee of must accompany CISM applications for certification.

Substitutions for work performed in the role of an information security manager are not allowed. However, a maximum of two (2) years for general work experience in the field of information security may be substituted as follows:

- Two years of general work experience may be substituted for currently holding one of the following broad security-related certifications or a postgraduate degree:
  - Certified Information Systems Auditor (CISA) in good standing
  - Certified Information Systems Security Professional (CISSP) in good standing
  - Postgraduate degree in information security or a related field (for example, business administration, information systems, information assurance)

OR

- A maximum of one (1) year of work experience may be substituted for one of the following:
  - One full year of information systems management experience
  - Currently holding an information security-related skill-based certification (e.g., SANS Global Information Assurance Certification [GIAC], Microsoft Certified Systems Engineer [MCSE], CompTIA Security+, Disaster Recovery Institute Certified Business Continuity Professional [CBCP])
  - Completion of a bachelor's degree in information security or a related field (e.g., cybersecurity)

Experience must have been gained within the 10-year period preceding the application for certification or within five (5) years from the date of initially passing the exam. Application for certification must be submitted within five (5) years from the passing date of the CISM exam. All experience must be verified independently with employers.

CISM candidates may choose to take the CISM exam prior to meeting the experience requirements.

DESCRIPTION OF THE EXAM

The CISM Certification Working Group oversees the development of the exam and ensures the currency of its content. Questions for the CISM exam are developed through a multitiered process designed to enhance the ultimate quality of the exam. Once the CISM Certification Working Group approves the questions, they go into the item pool from which all CISM exam questions are selected.

The purpose of the exam is to evaluate a candidate's knowledge and experience in information security management. The exam consists of 150 multiple-choice questions, administered during a four-hour session, that cover the CISM job practice domains. The exam covers four information security management domains created from the CISM job practice analysis and reflects the work performed by information security managers. The job practice was developed and validated using prominent industry leaders, subject matter experts and industry practitioners.

REGISTRATION FOR THE CISM EXAM

The CISM exam is administered multiple times annually during predefined testing windows. Please refer to the *ISACA Exam Candidate Information Guide* at *www.isaca.org/examguide* for specific exam registration dates, language offerings and deadlines. Exam registrations can be placed online at *www.isaca.org/examreg.*

CISM PROGRAM ACCREDITATION RENEWED UNDER ISO/IEC 17024:2012

The American National Standards Institute (ANSI) has voted to continue the accreditation for the CISA, CISM, CGEIT and CRISC certifications, under ISO/IEC 17024:2012—*General requirements for bodies operating certification systems of persons.* ANSI, a private, nonprofit organization, accredits other organizations to serve as third-party product, system and personnel certifiers.

ISO/IEC 17024 specifies the requirements to be followed by organizations certifying individuals against specific requirements. ANSI describes ISO/IEC 17024 as "expected to play a prominent role in facilitating global standardization of the certification community, increasing mobility among countries, enhancing public safety, and protecting consumers."

ANSI's accreditation:

- Promotes the unique qualifications and expertise that ISACA's certifications provide
- Protects the integrity of the certifications and provides legal defensibility
- Enhances consumer and public confidence in the certifications and the people who hold them
- Facilitates mobility across borders or industries

Accreditation by ANSI signifies that ISACA's procedures meet ANSI's essential requirements for openness, balance, consensus and due process. With this accreditation, ISACA anticipates that significant opportunities for CISAs, CISMs, CGEITs and CRISCs will continue to open in the USA and around the world.

ANSI Accredited Program
PERSONNEL
CERTIFICATION

PREPARING FOR THE CISM EXAM

The CISM exam evaluates a candidate's practical knowledge of the job practice domains listed in this manual and online at *www.isaca.org/cismjobpractice.* That is, the exam is designed to test a candidate's knowledge, experience and judgment of the proper or preferred application of information security management principles, methods and practices. Since the exam covers a broad spectrum of information security issues, candidates are cautioned not to assume that reading CISM study guides and reference publications will fully prepare them for the exam. CISM candidates are encouraged to refer to their own experiences when studying for the exam and refer to CISM study guides and reference publications for further explanation of concepts or practices with which the candidate is not familiar.

No representation or warranties are made by ISACA in regard to CISM exam study guides, other ISACA publications, references or courses assuring candidates' passage of the exam.

TYPES OF EXAM QUESTIONS

CISM exam questions are developed with the intent of measuring and testing practical knowledge and the application of general concepts and standards. All questions are multiple choice and are designed for one best answer.

Every question has a stem (question) and four options (answer choices). The candidate is asked to choose the correct or best answer from the options. The stem may be in the form of a question or incomplete statement. In some instances, a scenario may also be included. These questions normally include a description of a situation and require the candidate to answer two or more questions based on the information provided. The candidate is cautioned to read each question carefully. An exam question may require the candidate to choose the appropriate answer based on a qualifier, such as **MOST** or **BEST**. In every case, the candidate is required to read the question carefully, eliminate known incorrect answers and then make the best choice possible. To gain a better understanding of the types of question that might appear on the exam and how these questions are developed, refer to the Item Writing Guide available at *www.isaca.org/itemwriting.* Representations of CISM exam questions are available at *www.isaca.org/cismassessment.*

ADMINISTRATION OF THE EXAM

ISACA has contracted with an internationally recognized testing agency that engages in the development and administration of credentialing exams for certification and licensing purposes. It assists ISACA in the construction, administration and scoring of the CISM exam.

SITTING FOR THE EXAM

Candidates are to report to the testing site a minimum of 15 minutes prior to their scheduled testing appointment. Candidates who arrive after their scheduled time will not be allowed to sit for the exam and will forfeit their registration fee. To ensure that candidates arrive in time for the exam, it is recommended that candidates become familiar with the exact location of, and the best travel route to, the exam site prior to their scheduled test appointment.

The following conventions should be observed when completing the exam:

- Do not bring study materials (including notes, paper, books or study guides) or scratch paper or notepads into the exam site.
- Candidates are not allowed to bring any type of communication, surveillance or recording device (e.g., cell phone, tablet, smart watches or glasses, mobile devices, etc.) into the test center. If candidates are viewed with any such device during the exam administration, their exams will be voided and they will be asked to immediately leave the exam.
- Read the provided instructions carefully before attempting to answer questions. Skipping over these directions or reading them too quickly could result in missing important information and possibly losing credit points.
- Remember to answer all questions since there is no penalty for wrong answers. Grading is based solely on the number of questions answered correctly. Do not skip any questions. The exam will be scored based on your answered questions.
- Identify key words or phrases in the question (**MOST, BEST, FIRST …**) before selecting and recording the answer.

BUDGETING TIME

The following are time-management tips for the exam:

- It is recommended that candidates become familiar with the exact location of, and the best travel route to, the exam site prior to the date of the exam.
- Candidates should arrive at the exam testing site a minimum of 15 minutes prior to their scheduled testing appointment. This will give time for candidates to be seated and get acclimated.
- The exam is administered over a four-hour period. This allows for a little over one minute per question. Therefore, it is advisable that candidates pace themselves to complete the entire exam. In order to do so, candidates must complete an average of 38 questions per hour.

RULES AND PROCEDURES

- Upon the discretion of the CISM Certification Working Group, any candidate can be disqualified who is discovered engaging in any kind of misconduct, such as giving or receiving help; using notes, papers or other aids; attempting to take the exam for someone else; or removing test materials or notes from the testing room. The testing agency will provide the CISM Certification Working Group with records regarding such irregularities. The working group will review reported incidents, and all working group decisions are final.
- Additional information on exam rules is available in the *ISACA Exam Candidate Information Guide (www.isaca.org/examguide).*

GRADING THE CISM EXAM AND RECEIVING RESULTS

The exam consists of 150 items. Candidate scores are reported as a scaled score. A scaled score is a conversion of a candidate's raw score on an exam to a common scale. ISACA uses and reports scores on a common scale from 200 to 800. A candidate must receive a score of 450 or higher to pass the exam. A score of 450 represents a minimum consistent standard of knowledge as established by ISACA's CISM Certification Working Group. A candidate receiving a passing score may then apply for certification if all other requirements are met.

The CISM exam contains some questions that are included only for research and analysis purposes. These questions are not separately identified and are not used to calculate the candidate's final score.

Passing the exam does not grant the CISM designation. To become a CISM, each candidate must complete all requirements, including submitting an application and receiving approval for certification.

A candidate receiving a score less than 450 is not successful and can retake the exam by registering and paying the appropriate exam fee for any future exam administration. To assist with future study, the result letter each candidate receives includes a score analysis by content area. There are no limits to the number of times a candidate can take the exam.

Preliminary pass/fail results will be provided at the testing site immediately upon completion of the exam. Official CISM Exam scores will be emailed approximately 10 days after the exam date. This email notification is only sent to the email address listed in the candidate's profile at the time of the initial release of the results. To ensure the confidentiality of scores, exam results are not reported by telephone or fax. To prevent email notification from being sent to a spam folder, the candidate should add *certification@isaca.org* to their address book, white list or safe senders list.

In order to become CISM-certified, candidates must pass the CISM exam and must complete and submit an application for certification within five years of the passing date (and must receive confirmation from ISACA that the application is approved). The application is available on the ISACA web site at *www.isaca.org/cismapp*. Once the application is approved, the applicant will be sent confirmation of the approval. The candidate is not CISM certified, and cannot use the CISM designation, until the candidate's application is approved. A processing fee must accompany CISM applications for certification.

For those candidates not passing the examination, the score report contains a subscore for each job domain. The subscores can be useful in identifying those areas in which the candidate may need further study before retaking the exam. Unsuccessful candidates should note that taking either a simple or weighted average of the subscores does not derive the total scaled score. Candidates receiving a failing score on the exam may request a rescoring of their answer sheet. This procedure ensures that no conditions interfered with computer scoring. Candidates should understand, however, that all scores are subjected to several quality control checks before they are reported; therefore, rescores most likely will not result in a score change. Requests for hand scoring must be made in writing to the certification department within 90 days following the release of the exam results. Requests for a hand score after the deadline date will not be processed. All requests must include a candidate's name, exam identification number and mailing address.

Page intentionally left blank

GLOSSARY

Note: Glossary terms are provided for reference within the *CISM Review Manual*. As terms and definitions may evolve due to the changing technological environment, please see *www.isaca.org/glossary* for the most up-to-date terms and definitions.

A

Acceptable interruption window
The maximum period of time that a system can be unavailable before compromising the achievement of the organization's business objectives

Acceptable use policy
A policy that establishes an agreement between users and the organization and defines for all parties the ranges of use that are approved before gaining access to a network or the Internet

Access controls
The processes, rules and deployment mechanisms that control access to information systems, resources and physical access to premises

Access path
The logical route that an end user takes to access computerized information. Typically it includes a route through the operating system, telecommunications software, selected application software and the access control system.

Access rights
The permission or privileges granted to users, programs or workstations to create, change, delete or view data and files within a system, as defined by rules established by data owners and the information security policy

Accountability
The ability to map a given activity or event back to the responsible party

Address Resolution Protocol (ARP)
Defines the exchanges between network interfaces connected to an Ethernet media segment in order to map an IP address to a link layer address on demand

Administrative control
The rules, procedures and practices dealing with operational effectiveness, efficiency and adherence to regulations and management policies

Advance encryption standard (AES)
The international encryption standard that replaced 3DES

Alert situation
The point in an emergency procedure when the elapsed time passes a threshold and the interruption is not resolved. The organization entering into an alert situation initiates a series of escalation steps.

Algorithm
A finite set of step-by-step instructions for a problem-solving or computation procedure, especially one that can be implemented by a computer

Alternate facilities
Locations and infrastructures from which emergency or backup processes are executed, when the main premises are unavailable or destroyed. This includes other buildings, offices or data processing centers.

Alternate process
Automatic or manual process designed and established to continue critical business processes from point-of-failure to return-to-normal

Annual loss expectancy (ALE)
The total expected loss divided by the number of years in the forecast period yielding the average annual loss

Anomaly detection
Detection on the basis of whether the system activity matches that defined as abnormal

Anonymous File Transfer Protocol (AFTP)
A method of downloading public files using the File Transfer Protocol (FTP). AFTP does not require users to identify themselves before accessing files from a particular server. In general, users enter the word "anonymous" when the host prompts for a username. Anything can be entered for the password, such as the user's email address or simply the word "guest." In many cases, an AFTP site will not prompt a user for a name and password.

Antivirus software
An application software deployed at multiple points in an IT architecture. It is designed to detect and potentially eliminate virus code before damage is done, and repair or quarantine files that have already been infected.

Application controls
The policies, procedures and activities designed to provide reasonable assurance that objectives relevant to a given automated solution (application) are achieved

Application layer
In the Open Systems Interconnection (OSI) communications model, the application layer provides services for an application program to ensure that effective communication with another application program in a network is possible. The application layer is not the application that is doing the communication; it is a service layer that provides these services.

Application programming interface (API)
A set of routines, protocols and tools referred to as "building blocks" used in business application software development. A good API makes it easier to develop a program by providing all the building blocks related to functional characteristics of an operating system that applications need to specify, for example, when interfacing with the operating system (e.g., provided by Microsoft Windows, different versions of UNIX). A programmer utilizes these APIs in developing applications that can operate effectively and efficiently on the platform chosen.

Application service provider (ASP)
Also known as managed service provider (MSP), it deploys, hosts and manages access to a packaged application to multiple parties from a centrally managed facility. The applications are delivered over networks on a subscription basis.

Architecture
Description of the fundamental underlying design of the components of the business system, or of one element of the business system (e.g., technology), the relationships among them, and the manner in which they support the organization's objectives

Asymmetric key
A cipher technique in which different cryptographic keys are used to encrypt and decrypt a message

Attack signature
A specific sequence of events indicative of an unauthorized access attempt. Typically a characteristic byte pattern used in malicious code or an indicator, or set of indicators, that allows the identification of malicious network activities.

Audit trail
A visible trail of evidence enabling one to trace information contained in statements or reports back to the original input source

Authentication
The act of verifying the identity (i.e., user, system)

Authorization
Access privileges granted to a user, program or process, or the act of granting those privileges

Availability
Information that is accessible when required by the business process now and in the future

B

Backup center
An alternate facility to continue IT/IS operations when the primary data processing (DP) center is unavailable

Baseline security
The minimum security controls required for safeguarding an IT system based on its identified needs for confidentiality, integrity and/or availability protection

Benchmarking
A systematic approach to comparing an organization's performance against peers and competitors in an effort to learn the best ways of conducting business. Examples include benchmarking of quality, logistic efficiency and various other metrics.

Bit
The smallest unit of information storage; a contraction of the term "binary digit"; one of two symbols "0" (zero) and "1" (one) that are used to represent binary numbers

Bit copy
Provides an exact image of the original and is a requirement for legally justifiable forensics

Bit-stream image
Bit-stream backups, also referred to as mirror image backups, involve the backup of all areas of a computer hard disk drive or other type of storage media. Such backups exactly replicate all sectors on a given storage device including all files and ambient data storage areas.

Botnet
A large number of compromised computers that are used to create and send spam or viruses or flood a network with messages such as a denial-of-service attack

Brute force attack
Repeatedly trying all possible combinations of passwords or encryption keys until the correct one is found

Business case
Documentation of the rationale for making a business investment, used both to support a business decision on whether to proceed with the investment and as an operational tool to support management of the investment through its full economic life cycle

Business continuity plan (BCP)
A plan used by an organization to respond to disruption of critical business processes. Depends on the contingency plan for restoration of critical systems.

Business dependency assessment
A process of identifying resources critical to the operation of a business process

Business impact
The **net** effect, positive or negative, on the achievement of business objectives

Business impact analysis (BIA)
Evaluating the criticality and sensitivity of information assets. An exercise that determines the impact of losing the support of any resource to an organization, establishes the escalation of that loss over time, identifies the minimum resources needed to recover, and prioritizes the recovery of processes and supporting system. This process also includes addressing: income loss, unexpected expense, legal issues (regulatory compliance or contractual), interdependent processes, and loss of public reputation or public confidence.

Business Model for Information Security (BMIS)
A holistic and business-oriented model that supports enterprise governance and management information security, and provides a common language for information security professionals and business management

C

Capability Maturity Model Integration (CMMI)
Contains the essential elements of effective processes for one or more disciplines. It also describes an evolutionary improvement path from ad hoc, immature processes, to disciplined, mature processes, with improved quality and effectiveness.

Certificate (certification) authority (CA)
A trusted third party that serves authentication infrastructures or enterprises and registers entities and issues them certificates

Certificate revocation list (CRL)
An instrument for checking the continued validity of the certificates for which the certification authority (CA) has responsibility. The CRL details digital certificates that are no longer valid. The time gap between two updates is very critical and is also a risk in digital certificates verification.

Certification practice statement
A detailed set of rules governing the certificate authority's operations. It provides an understanding of the value and trustworthiness of certificates issued by a given certificate authority (CA).

Stated in terms of the controls that an organization observes, the method it uses to validate the authenticity of certificate applicants and the CA's expectations of how its certificates may be used.

Chain of custody
A legal principle regarding the validity and integrity of evidence. It requires accountability for anything that will be used as evidence in a legal proceeding to ensure that it can be accounted for from the time it was collected until the time it is presented in a court of law. This includes documentation as to who had access to the evidence and when, as well as the ability to identify evidence as being the exact item that was recovered or tested. Lack of control over evidence can lead to it being discredited. Chain of custody depends on the ability to verify that evidence could not have been tampered with. This is accomplished by sealing off the evidence, so it cannot be changed, and providing a documentary record of custody to prove that the evidence was, at all times, under strict control and not subject to tampering.

Chain of evidence
A process and record that shows who obtained the evidence, where and when the evidence was obtained, who secured the evidence, and who had control or possession of the evidence. The "sequencing" of the chain of evidence follows this order: collection and identification, analysis, storage, preservation, presentation in court, return to owner.

Challenge/response token
A method of user authentication that is carried out through use of the Challenge Handshake Authentication Protocol (CHAP). When a user tries to log onto the server using CHAP, the server sends the user a "challenge," which is a random value. The user enters a password, which is used as an encryption key to encrypt the "challenge" and return it to the server. The server is aware of the password. It, therefore, encrypts the "challenge" value and compares it with the value received from the user. If the values match, the user is authenticated. The challenge/response activity continues throughout the session and this protects the session from password sniffing attacks. In addition, CHAP is not vulnerable to "man-in-the-middle" attacks because the challenge value is a random value that changes on each access attempt.

Change management
A holistic and proactive approach to managing the transition from a current to a desired organizational state

Checksum
A mathematical value that is assigned to a file and used to "test" the file at a later date to verify that the data contained in the file have not been maliciously changed.

A cryptographic checksum is created by performing a complicated series of mathematical operations (known as a cryptographic algorithm) that translates the data in the file into a fixed string of digits called a hash value, which is then used as the checksum. Without knowing which cryptographic algorithm was used to create the hash value, it is highly unlikely that an unauthorized person would be able to change data without inadvertently changing the corresponding checksum. Cryptographic checksums are used in data transmission and data storage. Cryptographic checksums are also known as message authentication codes, integrity check values, modification detection codes or message integrity codes.

Chief information officer (CIO)
The most senior official of the enterprise who is accountable for IT advocacy, aligning IT and business strategies, and planning, resourcing and managing the delivery of IT services, information and the deployment of associated human resources. In some cases, the CIO role has been expanded to become the chief knowledge officer (CKO) who deals in knowledge, not just information. Also see chief technology officer.

Chief information security officer (CISO)
Responsible for managing information risk, the information security program, and ensuring appropriate confidentiality, integrity and availability of information assets

Chief security officer (CSO)
Typically responsible for physical security in the organization although increasingly the CISO and CSO roles are merged.

Chief technology officer (CTO)
The individual who focuses on technical issues in an organization

Cloud computing
An approach using external services for convenient on-demand IT operations using a shared pool of configurable computing capability. Typical capabilities include infrastructure as a service (IaaS), platform as a service (PaaS) and software as a service (SaaS) (e.g., networks, servers, storage, applications and services) that can be rapidly provisioned and released with minimal management effort or service provider interaction. This cloud model is composed of five essential characteristics (on-demand self-service, ubiquitous network access, location independent resource pooling, rapid elasticity, and measured service). It allows users to access technology-based services from the network cloud without knowledge of, expertise with, or control over, the technology infrastructure that supports them and provides four models for enterprise access (private cloud, community cloud, public cloud and hybrid cloud).

COBIT 5
Formerly known as Control Objectives for Information and related Technology (COBIT); now used only as the acronym in its fifth iteration. A complete, internationally accepted framework for governing and managing enterprise information and technology (IT) that supports enterprise executives and management in their definition and achievement of business goals and related IT goals. COBIT describes five principles and seven enablers that support enterprises in the development, implementation, and continuous improvement and monitoring of good IT-related governance and management practices.

Earlier versions of COBIT focused on control objectives related to IT processes, management and control of IT processes and IT governance aspects. Adoption and use of the COBIT framework are supported by guidance from a growing family of supporting products. (See *www.isaca.org/cobit* for more information.)

COBIT 4.1 and earlier
Formerly known as Control Objectives for Information and related Technology (COBIT). A complete, internationally accepted process framework for IT that supports business and IT executives and management in their definition and achievement of business goals and related IT goals by providing a comprehensive IT governance, management, control and assurance model. COBIT describes IT processes and associated control objectives, management guidelines (activities, accountabilities, responsibilities and performance metrics) and maturity models. COBIT supports enterprise management in the development, implementation, continuous improvement and monitoring of good IT-related practices.

Common vulnerabilities and exposures (CVE)
A system that provides a reference method for publicly known information-security vulnerabilities and exposures. MITRE Corporation maintains the system, with funding from the National Cyber Security Division of the United States Department of Homeland Security.

Compensating control
An internal control that reduces the risk of an existing or potential control weakness resulting in errors and omissions

Computer forensics
The application of the scientific method to digital media to establish factual information for judicial review. This process often involves investigating computer systems to determine whether they are or have been used for illegal or unauthorized activities. As a discipline, it combines elements of law and computer science to collect and analyze data from information systems (e.g., personal computers, networks, wireless communication and digital storage devices) in a way that is admissible as evidence in a court of law.

Confidentiality
The protection of sensitive or private information from unauthorized disclosure

Configuration management
The control of changes to a set of configuration items over a system life cycle

Content filtering
Controlling access to a network by analyzing the contents of the incoming and outgoing packets and either letting them pass or denying them based on a list of rules. Differs from packet filtering in that it is the data in the packet that are analyzed instead of the attributes of the packet itself (e.g., source/target IP address, transmission control protocol [TCP] flags).

Contingency plan
A plan used by an organization or business unit to respond to a specific systems failure or disruption

Continuous monitoring
The process implemented to maintain a current security status for one or more information systems or for the entire suite of information systems on which the operational mission of the enterprise depends.

The process includes: 1) the development of a strategy to regularly evaluate selected IS controls/metrics, 2) recording and evaluating IS-relevant events and the effectiveness of the enterprise in dealing with those events, 3) recording changes to IS controls, or changes that affect IS risks, and 4) publishing the current security status to enable information-sharing decisions involving the enterprise.

Control
The means of managing risk, including policies, procedures, guidelines, practices or organizational structures which can be of an administrative, technical, management or legal nature

Control center
Hosts the recovery meetings where disaster recovery operations are managed

Controls policy
A policy defining control operational and failure modes (e.g., fail secure, fail open, allowed unless specifically denied, denied unless specifically permitted)

Corporate governance
The system by which enterprises are directed and controlled. The board of directors is responsible for the governance of their enterprise. It consists of the leadership and organizational structures and processes that ensure the enterprise sustains and extends strategies and objectives.

COSO
Committee of Sponsoring Organizations of the Treadway Commission. Its report "Internal Control--Integrated Framework" is an internationally accepted standard for corporate governance. See *www.coso.org*.

Cost-benefit analysis
A systematic process for calculating and comparing benefits and costs of a project, control or decision

Countermeasures
Any process that directly reduces a threat or vulnerability

Criticality
A measure of the impact that the failure of a system to function as required will have on the organization

Criticality analysis
An analysis to evaluate resources or business functions to identify their importance to the organization, and the impact if a function cannot be completed or a resource is not available

Cryptographic algorithm
A well-defined computational procedure that takes variable inputs, including a cryptographic key, and produces an output

Cryptographic strength
A measure of the expected number of operations required to defeat a cryptographic mechanism

Cryptography
The art of designing, analyzing and attacking cryptographic schemes

Cyclical redundancy check (CRC)
A method to ensure that data have not been altered after being sent through a communication channel

Damage evaluation
The determination of the extent of damage that is necessary to provide for an estimation of the recovery time frame and the potential loss to the organization

Data classification
The assignment of a level of sensitivity to data (or information) that results in the specification of controls for each level of classification. Levels of sensitivity of data are assigned according to predefined categories as data are created, amended, enhanced, stored or transmitted. The classification level is an indication of the value or importance of the data to the organization.

Data custodian
The individual(s) and/or department(s) responsible for the storage and safeguarding of computerized data

Data Encryption Standard (DES)
An algorithm for encoding binary data. It is a secret key cryptosystem published by the National Bureau of Standards (NBS), the predecessor of the US National Institute of Standards and Technology (NIST). DES and its variants have been replaced by the Advanced Encryption Standard (AES).

Data integrity
The property that data meet with a priority expectation of quality and that the data can be relied on

Data leakage
Siphoning out or leaking information by dumping computer files or stealing computer reports and tapes

Data leak protection (DLP)
A suite of technologies and associated processes that locate, monitor and protect sensitive information from unauthorized disclosure

Data mining
A technique used to analyze existing information, usually with the intention of pursuing new avenues to pursue business

Data normalization
A structured process for organizing data into tables in such a way that it preserves the relationships among the data

Data owner
The individual(s), normally a manager or director, who has responsibility for the integrity, accurate reporting and use of computerized data

Data warehouse
A generic term for a system that stores, retrieves and manages large volumes of data. Data warehouse software often includes sophisticated comparison and hashing techniques for fast searches, as well as advanced filtering.

Decentralization
The process of distributing computer processing to different locations within an organization

Decryption key
A digital piece of information used to recover plaintext from the corresponding ciphertext by decryption

Defense in depth
The practice of layering defenses to provide added protection. Defense in depth increases security by raising the effort needed in an attack. This strategy places multiple barriers between an attacker and an organization's computing and information resources.

Degauss
The application of variable levels of alternating current for the purpose of demagnetizing magnetic recording media. The process involves increasing the alternating current field gradually from zero to some maximum value and back to zero, leaving a very low residue of magnetic induction on the media. Degauss loosely means: to erase.

Demilitarized zone (DMZ)
A screened (firewalled) network segment that acts as a buffer zone between a trusted and untrusted network. A DMZ is typically used to house systems such as web servers that must be accessible from both internal networks and the Internet.

Denial-of-service (DoS) attack
An assault on a service from a single source that floods it with so many requests that it becomes overwhelmed and is either stopped completely or operates at a significantly reduced rate

Digital certificate
A process to authenticate (or certify) a party's digital signature; carried out by trusted third parties

Digital code signing
The process of digitally signing computer code to ensure its integrity

Disaster declaration
The communication to appropriate internal and external parties that the disaster recovery plan is being put into operation

Disaster notification fee
The fee the recovery site vendor charges when the customer notifies them that a disaster has occurred and the recovery site is required. The fee is implemented to discourage false disaster notifications.

Disaster recovery plan (DRP)
A set of human, physical, technical and procedural resources to recover, within a defined time and cost, an activity interrupted by an emergency or disaster

Disaster recovery plan desk checking
Typically a read-through of a disaster recovery plan without any real actions taking place. Generally involves a reading of the plan, discussion of the action items and definition of any gaps that might be identified.

Disaster recovery plan walk-through
Generally a robust test of the recovery plan requiring that some recovery activities take place and are tested. A disaster scenario is often given and the recovery teams talk through the steps they would need to take to recover. As many aspects of the plan should be tested as possible.

Discretionary access control (DAC)
A means of restricting access to objects based on the identity of subjects and/or groups to which they belong. The controls are discretionary in the sense that a subject with a certain access permission is capable of passing that permission (perhaps indirectly) on to any other subject.

Disk mirroring
The practice of duplicating data in separate volumes on two hard disks to make storage more fault tolerant. Mirroring provides data protection in the case of disk failure because data are constantly updated to both disks.

Distributed denial-of-service (DDoS) attack
A denial-of-service (DoS) assault from multiple sources

Domain name system (DNS)
A hierarchical database that is distributed across the Internet that allows names to be resolved into IP addresses (and vice versa) to locate services such as web and email servers

Dual control
A procedure that uses two or more entities (usually persons) operating in concert to protect a system resource so that no single entity acting alone can access that resource

Due care
The level of care expected from a reasonable person of similar competency under similar conditions

Due diligence
The performance of those actions that are generally regarded as prudent, responsible and necessary to conduct a thorough and objective investigation, review and/or analysis

Dynamic Host Configuration Protocol (DHCP)
A protocol used by networked computers (clients) to obtain IP addresses and other parameters such as the default gateway, subnet mask and IP addresses of domain name system (DNS) servers from a DHCP server. The DHCP server ensures that all IP addresses are unique (e.g., no IP address is assigned to a second client while the first client's assignment is valid [its lease has not expired]). Thus, IP address pool management is done by the server and not by a human network administrator.

E

Electronic data interchange (EDI)
The electronic transmission of transactions (information) between two enterprises. EDI promotes a more efficient paperless environment. EDI transmissions can replace the use of standard documents, including invoices or purchase orders.

Electronic funds transfer (EFT)
The exchange of money via telecommunications. EFT refers to any financial transaction that originates at a terminal and transfers a sum of money from one account to another.

Encryption
The process of taking an unencrypted message (plaintext), applying a mathematical function to it (encryption algorithm with a key) and producing an encrypted message (ciphertext)

Enterprise governance
A set of responsibilities and practices exercised by the board and executive management with the goal of providing strategic direction, ensuring that objectives are achieved, ascertaining that risks are managed appropriately and verifying that the enterprise's resources are used responsibly

Exposure
The potential loss to an area due to the occurrence of an adverse event

External storage
The location that contains the backup copies to be used in case recovery or restoration is required in the event of a disaster

F

Fail-over
The transfer of service from an incapacitated primary component to its backup component

Fail safe
Describes the design properties of a computer system that allow it to resist active attempts to attack or bypass it

Fall-through logic
An optimized code based on a branch prediction that predicts which way a program will branch when an application is presented

Firewall
A system or combination of systems that enforces a boundary between two or more networks typically forming a barrier between a secure and an open environment such as the Internet

Flooding
An attack that attempts to cause a failure in a system by providing more input than the system can process properly

Forensic copy
An accurate bit-for-bit reproduction of the information contained on an electronic device or associated media, whose validity and integrity has been verified using an accepted algorithm

Forensic examination
The process of collecting, assessing, classifying and documenting digital evidence to assist in the identification of an offender and the method of compromise

G

Guideline
A description of a particular way of accomplishing something that is less prescriptive than a procedure

H

Harden
To configure a computer or other network device to resist attacks

Hash function
An algorithm that maps or translates one set of bits into another (generally smaller) so that a message yields the same result every time the algorithm is executed using the same message as input. It is computationally infeasible for a message to be derived or reconstituted from the result produced by the algorithm or to find two different messages that produce the same hash result using the same algorithm.

Help desk
A service offered via telephone/Internet by an organization to its clients or employees that provides information, assistance and troubleshooting advice regarding software, hardware or networks. A help desk is staffed by people who can either resolve the problem on their own or escalate the problem to specialized personnel. A help desk is often equipped with dedicated customer relationship management (CRM) software that logs the problems and tracks them until they are solved.

Honeypot
A specially configured server, also known as a decoy server, designed to attract and monitor intruders in a manner such that their actions do not affect production systems

Hot site
A fully operational offsite data processing facility equipped with hardware and system software to be used in the event of a disaster

Hypertext Transfer Protocol (HTTP)
A communication protocol used to connect to servers on the World Wide Web. Its primary function is to establish a connection with a web server and transmit hypertext markup language (HTML), extensible markup language (XML) or other pages to the client browsers.

I

Identification
The process of verifying the identity of a user, process or device, usually as a prerequisite for granting access to resources in an information system

Impact analysis
A study to prioritize the criticality of information resources for the organization based on costs (or consequences) of adverse events. In an impact analysis, threats to assets are identified and potential business losses determined for different time periods. This assessment is used to justify the extent of safeguards that are required and recovery time frames. This analysis is the basis for establishing the recovery strategy.

Incident
Any event that is not part of the standard operation of a service and that causes, or may cause, an interruption to, or a reduction in, the quality of that service

Incident handling
An action plan for dealing with intrusions, cybertheft, denial-of-service attack, fire, floods, and other security-related events. It is comprised of a six-step process: Preparation, Identification, Containment, Eradication, Recovery, and Lessons Learned.

Incident response
The response of an enterprise to a disaster or other significant event that may significantly affect the enterprise, its people or its ability to function productively. An incident response may include evacuation of a facility, initiating a disaster recovery plan (DRP), performing damage assessment and any other measures necessary to bring an enterprise to a more stable status.

Information security
Ensures that only authorized users (confidentiality) have access to accurate and complete information (integrity) when required (availability)

Information security governance
The set of responsibilities and practices exercised by the board and executive management with the goal of providing strategic direction, ensuring that objectives are achieved, ascertaining that risks are managed appropriately and verifying that the enterprise's resources are used responsibly

Information security program
The overall combination of technical, operational and procedural measures, and management structures implemented to provide for the confidentiality, integrity and availability of information based on business requirements and risk analysis

Integrity
The accuracy, completeness and validity of information

Internal controls
The policies, procedures, practices and organizational structures designed to provide reasonable assurance that business objectives will be achieved and undesired events will be prevented or detected and corrected

Internet protocol
Specifies the format of packets and the addressing scheme

Internet service provider (ISP)
A third party that provides individuals and organizations access to the Internet and a variety of other Internet-related services

Interruption window
The time the company can wait from the point of failure to the restoration of the minimum and critical services or applications. After this time, the progressive losses caused by the interruption are excessive for the organization.

Intrusion detection
The process of monitoring the events occurring in a computer system or network to detect signs of unauthorized access or attack

Intrusion detection system (IDS)
Inspects network and host security activity to identify suspicious patterns that may indicate a network or system attack

Intrusion prevention system (IPS)
Inspects network and host security activity to identify suspicious patterns that may indicate a network or system attack and then blocks it at the firewall to prevent damage to information resources

IP Security (IPSec)
A set of protocols developed by the Internet Engineering Task Force (IETF) to support the secure exchange of packets

ISO/IEC 15504
ISO/IEC 15504 *Information technology—Process assessment.* ISO/IEC 15504 provides a framework for the assessment of processes. The framework can be used by organizations involved in planning, managing, monitoring, controlling and improving the acquisition, supply, development, operation, evolution and support of products and services.

ISO/IEC 17799
Originally released as part of the British Standard for *Information Security* in 1999 and then as the *Code of Practice for Information Security Management* in October 2000, it was elevated by the International Organization for Standardization (ISO) to an international code of practice for information security management. This standard defines information's confidentiality, integrity and availability controls in a comprehensive information security management system. The latest version is ISO/IEC 17799:2005.

ISO/IEC 27001
An international standard, released in 2005 and revised in 2013, that defines a set of requirements for an information security management system. Prior its adoption by the ISO, this standard was known as BS 17799 Part 2, which was originally published in 1999.

ISO/IEC 27002
A code of practice that contains a structured list of suggested information security controls for organizations implementing an information security management system. Prior to its adoption by ISO/IEC, this standard existed as BS 77799.

ISO/IEC 31000
ISO 31000:2009 *Risk management—Principles and guidelines.* Provides principles and generic guidelines on risk management. It is industry- and sector-agnostic and can be used by any public, private or community enterprise, association, group or individual.

IT governance
The responsibility of executives and the board of directors; consists of the leadership, organizational structures and processes that ensure that the enterprise's IT sustains and extends the organization's strategies and objectives

IT steering committee
An executive management-level committee that assists the executive in the delivery of the IT strategy, oversees day-to-day management of IT service delivery and IT projects and focuses on implementation aspects

IT strategic plan
A long-term plan (i.e., three- to five-year horizon) in which business and IT management cooperatively describe how IT resources will contribute to the enterprise's strategic objectives (goals)

IT strategy committee
A committee at the level of the board of directors to ensure that the board is involved in major IT matters and decisions. The committee is primarily accountable for managing the portfolios of IT-enabled investments, IT services and other IT resources. The committee is the owner of the portfolio.

K

Key goal indicator (KGI)
A measure that tells management, after the fact, whether an IT process has achieved its business requirements; usually expressed in terms of information criteria

Key performance indicator (KPI)
A measure that determines how well the process is performing in enabling the goal to be reached. A KPI is a lead indicator of whether a goal will likely be reached, and a good indicator of capability, practices and skills. It measures an activity goal, which is an action that the process owner must take to achieve effective process performance.

Key risk indicator (KRI)
A subset of risk indicators that are highly relevant and possess a high probability of predicting or indicating important risk

L

Least privilege
The principle of allowing users or applications the least amount of permissions necessary to perform their intended function

M

Mail relay server
An electronic mail (email) server that relays messages so that neither the sender nor the recipient is a local user

Malicious code
Software (e.g., Trojan horse) that appears to perform a useful or desirable function, but actually gains unauthorized access to system resources or tricks a user into executing other malicious logic

Malware
Software designed to infiltrate, damage or obtain information from a computer system without the owner's consent

Malware is commonly taken to include computer viruses, worms, Trojan horses, spyware and adware. Spyware is generally used for marketing purposes and, as such, is not malicious, although it is generally unwanted. Spyware can, however, be used to gather information for identity theft or other clearly illicit purposes.

Mandatory access control (MAC)
A means of restricting access to data based on varying degrees of security requirements for information contained in the objects and the corresponding security clearance of users or programs acting on their behalf

Man-in-the-middle attack (MITM)
An attack strategy in which the attacker intercepts the communication stream between two parts of the victim system and then replaces the traffic between the two components with the intruder's own system, eventually assuming control of the communication

Masqueraders
Attackers that penetrate systems by using the identity of legitimate users and their login credentials

Maximum tolerable outage (MTO)
Maximum time the organization can support processing in alternate mode

Media access control (MAC)
Applied to the hardware at the factory and cannot be modified, MAC is a unique, 48-bit, hard-coded address of a physical layer device, such as an Ethernet local area network (LAN) or a wireless network card.

Message authentication code
An American National Standards Institute (ANSI) standard checksum that is computed using the Data Encryption Standard (DES)

Message digest
A cryptographic checksum, typically generated for a file that can be used to detect changes to the file; Secure Hash Algorithm-1 (SHA-1) is an example of a message digest algorithm.

Mirrored site
An alternate site that contains the same information as the original. Mirror sites are set up for backup and disaster recovery as well as to balance the traffic load for numerous download requests. Such download mirrors are often placed in different locations throughout the Internet.

Mobile site
The use of a mobile/temporary facility to serve as a business resumption location. They can usually be delivered to any site and can house information technology and staff.

Monitoring policy
Rules outlining or delineating the way in which information about the use of computers, networks, applications and information is captured and interpreted.

Multipurpose Internet mail extension (MIME)
A specification for formatting non-ASCII messages so that they can be sent over the Internet. Many email clients now support MIME, which enables them to send and receive graphics, audio and video files via the Internet mail system. In addition, MIME supports messages in character sets other than ASCII.

N

Net present value (NPV)
Calculated by using an after-tax discount rate of an investment and a series of expected incremental cash outflows (the initial investment and operational costs) and cash inflows (cost savings or revenues) that occur at regular periods during the life cycle of the investment. To arrive at a fair NPV calculation, cash inflows accrued by the business up to about five years after project deployment also should be taken into account.

Network address translation (NAT)
Basic NATs are used when there is a requirement to interconnect two IP networks with incompatible addressing. However, it is common to hide an entire IP address space, usually consisting of private IP addresses, behind a single IP address (or in some cases a small group of IP addresses) in another (usually public) address space. To avoid ambiguity in the handling of returned packets, a one-to-many NAT must alter higher level information such as Transmission Control Protocol (TCP)/User Datagram Protocol (UDP) ports in outgoing communications and must maintain a translation table so that return packets can be correctly translated back.

Network-based intrusion detection (NID)
Provides broader coverage than host-based approaches but functions in the same manner detecting attacks using either an anomaly-based or signature-based approach or both

Nonintrusive monitoring
The use of transported probes or traces to assemble information, track traffic and identify vulnerabilities

Nonrepudiation
The assurance that a party cannot later deny originating data; that is, it is the provision of proof of the integrity and origin of the data and can be verified by a third party. A digital signature can provide nonrepudiation.

Offline files
Computer file storage media not physically connected to the computer; typically tapes or tape cartridges used for backup purposes

Open Shortest Path First (OSPF)
A routing protocol developed for IP networks. It is based on the shortest path first or link state algorithm.

Open Source Security Testing Methodology
An open and freely available methodology and manual for security testing

Outcome measure
Represents the consequences of actions previously taken; often referred to as a lag indicator. An outcome measure frequently focuses on results at the end of a time period and characterizes historical performance. It is also referred to as a key goal indicator (KGI) and is used to indicate whether goals have been met. Can be measured only after the fact and, therefore, is called a lag indicator.

P

Packet
Data unit that is routed from source to destination in a packet-switched network. A packet contains both routing information and data. Transmission Control Protocol/Internet Protocol (TCP/IP) is such a packet-switched network.

Packet filtering
Controlling access to a network by analyzing the attributes of the incoming and outgoing packets, and either letting them pass or denying them based on a list of rules

Packet sniffer
Software that observes and records network traffic

Packet switched network
Individual packets follow their own paths through the network from one endpoint to another and reassemble at the destination.

Partitions
Major divisions of the total physical hard disk space

Passive response
A response option in intrusion detection in which the system simply reports and records the problem detected, relying on the user to take subsequent action

Password cracker
A tool that tests the strength of user passwords searching for passwords that are easy to guess. It repeatedly tries words from specially crafted dictionaries and often also generates thousands (and in some cases, even millions) of permutations of characters, numbers and symbols.

Penetration testing
A live test of the effectiveness of security defenses through mimicking the actions of real-life attackers

Personally Identifiable Information (PII)
Information that can be used alone or with other sources to uniquely identify, contact or locate a single individual

Pharming
This is a more sophisticated form of a man-in-the-middle (MITM) attack. A user's session is redirected to a masquerading web site. This can be achieved by corrupting a domain name system (DNS) server on the Internet and pointing a URL to the masquerading web site's IP address.

Phishing
This is a type of electronic mail (email) attack that attempts to convince a user that the originator is genuine, but with the intention of obtaining information for use in social engineering. Phishing attacks may take the form of masquerading as a lottery organization advising the recipient or the user's bank of a large win; in either case, the intent is to obtain account and personal identification number (PIN) details. Alternative attacks may seek to obtain apparently innocuous business information, which may be used in another form of active attack.

Platform as a Service (PaaS)
Offers the capability to deploy onto the cloud infrastructure customer-created or -acquired applications that are created using programming languages and tools supported by the provider

Policy
Overall intention and direction as formally expressed by management

Port
A hardware interface between a CPU and a peripheral device. Can also refer to a software (virtual) convention that allows remote services to connect to a host operating system in a structured manner.

Privacy
Freedom from unauthorized intrusion or disclosure of information of an individual

Private key
A mathematical key (kept secret by the holder) used to create digital signatures and, depending on the algorithm, to decrypt messages or files encrypted (for confidentiality) with the corresponding public key

Procedure
A document containing a detailed description of the steps necessary to perform specific operations in conformance with applicable standards. Procedures are defined as part of processes.

Proxy server
A server that acts on behalf of a user. Typically proxies accept a connection from a user, make a decision as to whether or not the user or client IP address is permitted to use the proxy, perhaps perform additional authentication, and then complete a connection to a remote destination on behalf of the user.

Public key
In an asymmetric cryptographic scheme, the key that may be widely published to enable the operation of the scheme

R

Reciprocal agreement
Emergency processing agreements among two or more organizations with similar equipment or applications. Typically, participants promise to provide processing time to each other when an emergency arises.

Recovery action
Execution of a response or task according to a written procedure

Recovery point objective (RPO)
Determined based on the acceptable data loss in case of a disruption of operations. It indicates the earliest point in time to which it is acceptable to recover data. It effectively quantifies the permissible amount of data loss in case of interruption.

Recovery time objective (RTO)
The amount of time allowed for the recovery of a business function or resource after a disaster occurs

Redundant Array of Inexpensive Disks (RAID)
Provides performance improvements and fault-tolerant capabilities, via hardware or software solutions, by writing to a series of multiple disks to improve performance and/or save large files simultaneously

Redundant site
A recovery strategy involving the duplication of key information technology components, including data or other key business processes, whereby fast recovery can take place

Request for proposal (RFP)
A document distributed to software vendors requesting them to submit a proposal to develop or provide a software product

Residual risk
The remaining risk after management has implemented risk response

Resilience
The ability of a system or network to resist failure or to recover quickly from any disruption, usually with minimal recognizable effect

Return on investment (ROI)
A measure of operating performance and efficiency, computed in its simplest form by dividing net income by the total investment over the period being considered

Return on security investment (ROSI)
An estimate of return on security investment based on how much will be saved by reduced losses divided by the investment

Risk
The combination of the probability of an event and its consequence. (ISO/IEC 73). Risk has traditionally been expressed as Threats × Vulnerabilities = Risk.

Risk analysis
The initial steps of risk management: analyzing the value of assets to the business, identifying threats to those assets and evaluating how vulnerable each asset is to those threats. It often involves an evaluation of the probable frequency of a particular event, as well as the probable impact of that event.

Risk appetite
The amount of risk, on a broad level, that an entity is willing to accept in pursuit of its mission

Risk assessment
A process used to identify and evaluate risk and potential effects. Risk assessment includes assessing the critical functions necessary for an organization to continue business operations, defining the controls in place to reduce organization exposure and evaluating the cost for such controls. Risk analysis often involves an evaluation of the probabilities of a particular event.

Risk avoidance
The process for systematically avoiding risk, constituting one approach to managing risk

Risk mitigation
The management and reduction of risk through the use of countermeasures and controls

Risk tolerance
The acceptable level of variation that management is willing to allow for any particular risk while pursuing its objective

Risk transfer
The process of assigning risk to another organization, usually through the purchase of an insurance policy or outsourcing the service

Robustness
The ability of systems to withstand attack, operate reliably across a wide range of operational conditions and to fail gracefully outside of the operational range

Role-based access control
Assigns users to job functions or titles. Each job function or title defines a specific authorization level.

Root cause analysis
A process of diagnosis to establish origins of events, which can be used for learning from consequences, typically of errors and problems

Rootkit
A software suite designed to aid an intruder in gaining unauthorized administrative access to a computer system

S

Secret key
A cryptographic key that is used with a secret key (symmetric) cryptographic algorithm, that is uniquely associated with one or more entities and is not made public. The same key is used to both encrypt and decrypt data. The use of the term "secret" in this context does not imply a classification level, but rather implies the need to protect the key from disclosure.

Secure hash algorithm (SHA)
A hash algorithm with the property that is computationally infeasible 1) to find a message that corresponds to a given message digest, or 2) to find two different messages that produce the same message digest

Secure shell (SSH)
Network protocol that uses cryptography to secure communication, remote command line login and remote command execution between two networked computers

Security information and event management (SIEM)
SIEM solutions are a combination of the formerly disparate product categories of SIM (security information management) and SEM (security event management). SIEM technology provides real-time analysis of security alerts generated by network hardware and applications. SIEM solutions come as software, appliances or managed services, and are also used to log security data and generate reports for compliance purposes. Capabilities include:
- **Data aggregation:** SIEM/LM (log management) solutions aggregate data from many sources, including network, security, servers, databases and applications, providing the ability to consolidate monitored data to help avoid missing crucial events.
- **Correlation:** Looks for common attributes, and links events together into meaningful bundles. This technology provides the ability to perform a variety of correlation techniques to integrate different sources, in order to turn data into useful information.
- **Alerting:** The automated analysis of correlated events and production of alerts, to notify recipients of immediate issues.
- **Dashboards:** SIEM/LM tools take event data and turn them into informational charts to assist in seeing patterns, or identifying activity that is not forming a standard pattern

- **Compliance:** SIEM applications can be employed to automate the gathering of compliance data, producing reports that adapt to existing security, governance and auditing processes.
- **Retention:** SIEM/SIM solutions employ long-term storage of historical data to facilitate correlation of data over time, and to provide the retention necessary for compliance requirements.

Security metrics
A standard of measurement used in management of security-related activities

Segregation/separation of duties (SoD)
A basic internal control that prevents or detects errors and irregularities by assigning to separate individuals the responsibility for initiating and recording transactions and for the custody of assets. Segregation/separation of duties is commonly used in large IT organizations so that no single person is in a position to introduce fraudulent or malicious code without detection.

Sensitivity
A measure of the impact that improper disclosure of information may have on an organization

Service delivery objective (SDO)
Directly related to business needs, SDO is the level of services to be reached during the alternate process mode until the normal situation is restored.

Service level agreement (SLA)
An agreement, preferably documented, between a service provider and the customer(s)/user(s) that defines minimum performance targets for a service and how they will be measured

Session key
A single-use symmetric key used for a defined period of communication between two computers, such as for the duration of a single communication session or transaction set

Shell programming
A script written for the shell, or command line interpreter, of an operating system; it is often considered a simple domain-specific programming language. Typical operations performed by shell scripts include file manipulation, program execution and printing text. Usually, shell script refers to scripts written for a UNIX shell, while COMMAND.COM (DOS) and cmd.exe (Windows) command line scripts are usually called batch files. Others, such as AppleScript, add scripting capability to computing environments lacking a command line interface. Other examples of programming languages primarily intended for shell scripting include digital command language (DCL) and job control language (JCL).

Sniffing
The process by which data traversing a network are captured or monitored

Social engineering
An attack based on deceiving users or administrators at the target site into revealing confidential or sensitive information

Split knowledge/split key
A security technique in which two or more entities separately hold data items that individually convey no knowledge of the information that results from combining the items; a condition under which two or more entities separately have key components that individually convey no knowledge of the plaintext key that will be produced when the key components are combined in the cryptographic module

Spoofing
Faking the sending address of a transmission in order to gain illegal entry into a secure system

Software as a service (SaaS)
Offers the capability to use the provider's applications running on cloud infrastructure. The applications are accessible from various client devices through a thin client interface such as a web browser (e.g., web-based email).

Standard
A mandatory requirement, code of practice or specification approved by a recognized external standards organization, such as International Organization for Standardization (ISO)

Symmetric key encryption
System in which a different key (or set of keys) is used by each pair of trading partners to ensure that no one else can read their messages. The same key is used for encryption and decryption.

System owner
Person or organization having responsibility for the development, procurement, integration, modification, operation and maintenance, and/or final disposition of an information system

T

Threat
Anything (e.g., object, substance, human) that is capable of acting against an asset in a manner that can result in harm. A potential cause of an unwanted incident. (ISO/IEC 13335)

Threat agent
Methods and things used to exploit a vulnerability. Examples include determination, capability, motive and resources.

Threat analysis
An evaluation of the type, scope and nature of events or actions that can result in adverse consequences; identification of the threats that exist against information assets. The threat analysis usually also defines the level of threat and the likelihood of it materializing.

Threat assessment
The identification of types of threats to which an organization might be exposed

Threat event
Any event where a threat element/actor acts against an asset in a manner that has the potential to directly result in harm

Threat model
Used to describe a given threat and the harm it could to do a system if it has a vulnerability

Threat vector
The method a threat uses to exploit the target

Token
A device that is used to authenticate a user, typically in addition to a user name and password. A token is usually a device that displays a pseudo random number that changes every few minutes.

Total cost of ownership (TCO)
Includes the original cost of the computer plus the cost of: software, hardware and software upgrades, maintenance, technical support, training, and certain activities performed by users

Transmission Control Protocol (TCP)
A connection-based Internet protocol that supports reliable data transfer connections

> **Scope Notes:** Packet data are verified using checksums and retransmitted if they are missing or corrupted. The application plays no part in validating the transfer.

Trusted system
A system that employs sufficient hardware and software assurance measures to allow its use for processing simultaneously a range of sensitive or classified information

Tunneling
Commonly used to bridge between incompatible hosts/routers or to provide encryption; a method by which one network protocol encapsulates another protocol within itself

Two-factor authentication
The use of two independent mechanisms for authentication, (e.g., requiring a smart card and a password); typically the combination of something you know, are or have

U

Uniform resource locator (URL)
The global address of documents and other resources on the World Wide Web. The first part of the address indicates what protocol to use; the second part specifies the IP address or the domain name where the resource is located (e.g., *http://www.isaca.org*).

V

Virtual private network (VPN)
A secure private network that uses the public telecommunications infrastructure to transmit data. In contrast to a much more expensive system of owned or leased lines that can only be used by one company, VPNs are used by enterprises for both extranets and wide areas of intranets. Using encryption and authentication, a VPN encrypts all data that pass between two Internet points, maintaining privacy and security.

Virus signature files
The file of virus patterns that are compared with existing files to determine if they are infected with a virus or worm

Voice-over IP (VoIP)
Also called IP Telephony, Internet Telephony and Broadband Phone, a technology that makes it possible to have a voice conversation over the Internet or over any dedicated Internet Protocol (IP) network instead of over dedicated voice transmission lines

Vulnerability
A weakness in the design, implementation, operation or internal controls in a process that could be exploited to violate system security

Vulnerability analysis
A process of identifying and classifying vulnerabilities

W

Warm site
Similar to a hot site, but not fully equipped with all of the necessary hardware needed for recovery

Web hosting
The business of providing the equipment and services required to host and maintain files for one or more web sites and provide fast Internet connections to those sites. Most hosting is "shared," which means that web sites of multiple companies are on the same server to share/reduce costs.

Web server
Using the client-server model and the World Wide Web's Hypertext Transfer Protocol (HTTP), Web server is a software program that serves web pages to users.

Wide area network (WAN)
A computer network connecting different remote locations that may range from short distances, such as a floor or building, to long transmissions that encompass a large region or several countries

Worm
A programmed network attack in which a self-replicating program does not attach itself to programs, but rather spreads independently of users' action

Wi-Fi protected access 2 (WPA2)
The replacement security method for WPA for wireless networks that provides stronger data protection and network access control. It provides enterprise and consumer Wi-Fi users with a high level of assurance that only authorized users can access their wireless networks. Based on the ratified IEEE 802.11i standard, WPA2 provides government-grade security by implementing the National Institute of Standards and Technology (NIST) FIPS 140-2 compliant advanced encryption standard (AES) encryption algorithm and 802.1X-based authentication.

ACRONYMS

The following is a list of common acronyms used throughout the *CISM Review Manual 15th Edition*. These may be defined in the text for clarity.

AESRM	Alliance for Enterprise Security Risk Management
AIW	Acceptable interruption window
ALE	Annual loss expectancy
API	Application programming interface
AS/NZS	Australian Standard/New Zealand Standard
ASCII	American Standard Code for Information Interchange
ASIC	Application-specific integrated circuit
ASP	Application service provider
ATM	Asynchronous Transfer Mode
AV	Asset value
BCI	Business Continuity Institute
BCM	Business continuity management
BCP	Business continuity planning
BGP	Border Gateway Protocol
BI	Business intelligence
BIA	Business impact analysis
BIMS	Biometric information management and security
BIOS	Basic input/output system
BITS	Banking Information Technology Standards
BLP	Bell-LaPadula
BLP	Bypass label process
BS	British Standard
CA	Certificate authority
CASPR	Commonly accepted security practices and recommendations
CD	Compact disk
CD-ROM	Compact disk-read only memory
CEO	Chief executive officer
CERT	Computer emergency response team
CFO	Chief financial officer
CIM	Computer-integrated manufacturing
CIO	Chief information officer
CIRT	Computer incident response team
CISO	Chief information security officer
CMM	Capability Maturity Model
CMU/SEI	Carnegie Mellon University Software Engineering Institute
COO	Chief operating officer
COOP	Continuity of operations plan
CORBA	Common Object Request Broker Architecture
COSO	Committee of Sponsoring Organizations of the Treadway Commission
CPO	Chief privacy officer
CPU	Central processing unit
CRM	Customer relationship management
CSA	Control self-assessment
CSF	Critical success factor
CSIRT	Computer security incident response team
CSO	Chief security officer
CSRC	Computer Security Resources Center (USA)
CRO	Chief risk officer
CTO	Chief technology officer
CVE	Common vulnerabilities and exposures
DAC	Discretionary access controls
DBMS	Database management system
DCE	Distributed control environment
DCE	Data communications equipment
DCE	Distributed computing environment
DCL	Digital command language
DDoS	Distributed denial of service
DES	Data Encryption Standard
DHCP	Dynamic Host Configuration Protocol
DMZ	Demilitarized zone
DNS	Domain name system
DNSSEC	Domain Name Service Secure
DoS	Denial of service
DOSD	Data-oriented system development
DR	Disaster recovery
DRII	Disaster Recovery Institute International
DRP	Disaster recovery planning
EDI	Electronic data interchange
EER	Equal error rate
EFT	Electronic funds transfer
EF	Exposure factor
EGRP	External Gateway Routing Protocol
EIGRP	Enhanced Interior Gateway Routing Protocol
EU	European Union
FAIR	Factor analysis of information risk
FAR	False-acceptance rate
FCPA	Foreign Corrupt Practices Act
FIPS	Federal Information Processing Standards (USA)
FISMA (USA)	Federal Information Security Modernization Act
FSA	Financial Security Authority (USA)
GLBA	Gramm-Leach-Bliley Act (USA)
GMI	Governance Metrics International
HD-DVD	High definition/high-density-digital video disc
HIDS	Host-based intrusion detection system
HIPAA	Health Insurance Portability and Accountability Act (USA)
HIPO	Hierarchy Input-Process-Output
HR	Human resources
HTML	Hypertext Markup Language
HTTP	Hypertext Transfer Protocol
I/O	Input/output
ICMP	Internet Control Message Protocol
ICT	Information and communication technologies
ID	Identification
IDEFIX	Integration Definition for Information Modeling
IDS	Intrusion detection system
IEC	International Electrotechnical Commission
IETF	Internet engineering task force
IFAC	International Federation of Accountants
IIA	Institute of Internal Auditors
IMT	Incident management team
IP	Internet Protocol
IPF	Information processing facility
IPL	Initial program load
IPMA	International Project Management Association
IPRs	Intellectual property rights
IPS	Intrusion prevention system
IPSec	Internet Protocol Security
IRP	Incident response plan
IRT	Incident response team

IS	Information systems
ISF	Information Security Forum
ISO	International Organization for Standardization
ISP	Internet service provider
ISSA	Information Systems Security Association
ISSEA	International Systems Security Engineering Association
IT	Information technology
ITGI	IT Governance Institute
ITIL	Information Technology Infrastructure Library
JCL	Job control language
KGI	Key goal indicator
KLOC	Kilo lines of code
KPI	Key performance indicator
KRI	Key risk indicator
L2TP	Layer 2 Tunneling Protocol
LAN	Local area network
LCP	Link Control Protocol
M&A	Mergers and acquisitions
MAC	Mandatory access control
MAO	Maximum allowable outage
MIME	Multipurpose Internet mail extension
MIS	Management information system
MitM	Man-in-the-middle
MTD	Maximum tolerable downtime
MTO	Maximum tolerable outage
NAT	Network address translation
NCP	Network Control Protocol
NDA	Nondisclosure agreement
NIC	Network interface card
NIDS	Network intrusion detection system
NIST	National Institute of Standards and Technology (USA)
NPV	Net present value
OCSP	Online Certificate Status Protocol
OCTAVE	Operationally Critical Threat, Asset and Vulnerability Evaluation
OECD	Organisation for Economic Co-operation and Development
OEP	Occupant emergency plan
OS	Operating system
OSI	Open systems interconnection
OSPF	Open Shortest Path First
PaaS	Platform as a Service
PAN	Personal area network
PCI DSS	Payment Card Industry Data Security Standard
PDCA	Plan-do-check-act
PKI	Public key infrastructure
PMBOK	Project Management Body of Knowledge
POS	Point-of-sale
PPPoE	Point-to-point Protocol over Ethernet
PRA	Probabilistic risk assessment
PSTN	Public switched telephone network
PVC	Permanent virtual circuit
QA	Quality assurance
RACI	Responsible, accountable, consulted, informed
RAID	Redundant array of inexpensive disks
ROI	Return on investment
ROSI	Return on security investment
RPO	Recovery point objective
RRT	Risk Reward Theorem/Tradeoff
RSA	Rivest, Shamir and Adleman (RSA stands for the initials of the developers last names)
RTO	Recovery time objective
S/HTTP	Secure Hypertext Transfer Protocol
SaaS	Software as a Service
SABSA	Sherwood Applied Business Security Architecture
SCADA	Supervisory Control and Data Acquisition
SDLC	System development life cycle
SDO	Service delivery objective
SEC	Securities and Exchange Commission (USA)
SIEM	Security information and event management
SIM	Security information management
SLA	Service level agreement
SMART	Specific, measurable, achievable, relevant, time-bound
SMF	System management facility
SOP	Standard operating procedure
SPI	Security Parameter Index
SPICE	Software process improvement and capability determination
SPOC	Single point of contact
SPOOL	Simultaneous peripheral operations online
SQL	Structured Query Language
SSH	Secure Shell
SSL	Secure sockets layer
SSO	Single sign-on
TCO	Total cost of ownership
TCP	Transmission Control Protocol
TLS	Transport layer security
UDP	User Datagram Protocol
URL	Uniform resource locator
USB	Universal Serial Bus
VAR	Value at risk
VoIP	Voice-over IP
VPN	Virtual private network
XBRL	Extensible Business Reporting Language
XML	Extensible Markup Language
XSS	Cross-site scripting

Page intentionally left blank

INDEX

A

B

CISM
Certified Information Security Manager
An ISACA Certification

C

D

E

CISM
Certified Information Security Manager
An ISACA Certification

F

G

H

I

J

K

L

P

Q

R

CISM
Certified Information Security Manager
An ISACA Certification

S

CISM
Certified Information Security Manager
An ISACA Certification

T

U

V

W